Permutation group algorithms comprise one of the workhorses of symbolic algebra systems computing with groups and play an indispensable role in the proof of many deep results, including the construction and study of sporadic finite simple groups. This book describes the theory behind permutation group algorithms, up to the most recent developments based on the classification of finite simple groups. Rigorous complexity estimates, implementation hints, and advanced exercises are included throughout.

The central theme is the description of nearly linear-time algorithms, which are extremely fast in terms of both asymptotic analysis and practical running time. A significant part of the permutation group library of the computational group algebra system GAP is based on nearly linear-time algorithms.

The book fills a significant gap in the symbolic computation literature. It is recommended for everyone interested in using computers in group theory and is suitable for advanced graduate courses.

Ákos Seress is a Professor of Mathematics at The Ohio State University.

CAMBRIDGE TRACTS IN MATHEMATICS

General Editors

B. BOLLOBÁS, W. FULTON, A. KATOK, F. KIRWAN, P. SARNAK

152 Permutation Group Algorithms

Ákos Seress
The Ohio State University

Permutation Group Algorithms

CAMBRIDGE
UNIVERSITY PRESS

PUBLISHED BY THE PRESS SYNDICATE OF THE UNIVERSITY OF CAMBRIDGE
The Pitt Building, Trumpington Street, Cambridge, United Kingdom

CAMBRIDGE UNIVERSITY PRESS
The Edinburgh Building, Cambridge CB2 2RU, UK
40 West 20th Street, New York, NY 10011-4211, USA
477 Williamstown Road, Port Melbourne, VIC 3207, Australia
Ruiz de Alarcón 13, 28014 Madrid, Spain
Dock House, The Waterfront, Cape Town 8001, South Africa

http://www.cambridge.org

First published 2003

Printed in the United States of America

Typeface Times 10/13 pt. *System* LaTeX 2_ε [TB]

A catalog record for this book is available from the British Library.

Library of Congress Cataloging in Publication Data available

ISBN 0 521 66103 X hardback

Contents

1
Introduction

Computational group theory (CGT) is a subfield of symbolic algebra; it deals with the design, analysis, and implementation of algorithms for manipulating groups. It is an interdisciplinary area between mathematics and computer science. The major areas of CGT are the algorithms for finitely presented groups, polycyclic and finite solvable groups, permutation groups, matrix groups, and representation theory.

The topic of this book is the third of these areas. Permutation groups are the oldest type of representations of groups; in fact, the work of Galois on permutation groups, which is generally considered as the start of group theory as a separate branch of mathematics, preceded the abstract definition of groups by about a half a century. Algorithmic questions permeated permutation group theory from its inception. Galois group computations, and the related problem of determining all transitive permutation groups of a given degree, are still active areas of research (see [Hulpke, 1996]). Mathieu's constructions of his simple groups also involved serious computations.

Nowadays, permutation group algorithms are among the best developed parts of CGT, and we can handle groups of degree in the hundreds of thousands. The basic ideas for handling permutation groups appeared in [Sims, 1970, 1971a]; even today, Sims's methods are at the heart of most of the algorithms.

At first glance, the efficiency of permutation group algorithms may be surprising. The input consists of a list of generators. On one hand, this representation is very efficient, since a few permutations in S_n can describe an object of size up to $n!$. On the other hand, the succinctness of such a representation $G = \langle S \rangle$ necessitates nontrivial algorithms to answer even such basic questions as finding the order of G or testing membership of a given permutation in G.

Initially, it is not even clear how to prove in polynomial time in the input length that a certain permutation g is in G, because writing g as a product of the given generators S for G may require an exponentially long word. Sims's seminal idea was to introduce the notions of base and strong generating set. This data structure enables us to decide membership in G constructively, by writing any given element of G as a short product of the strong generators. The technique for constructing a strong generating set can also be applied to other tasks such as computing normal closures of subgroups and handling homomorphisms of groups. Therefore, a significant part of this book is devoted to the description of variants and applications of Sims's method.

A second generation of algorithms uses divide-and-conquer techniques by utilizing the orbit structure and imprimitivity block structure of the input group, thereby reducing the problems to primitive groups. Although every abstract group has a faithful transitive permutation representation, the structure of primitive groups is quite restricted. This extra information, partly obtained as a consequence of the classification of finite simple groups, can be exploited in the design of algorithms.

We shall also describe some of the latest algorithms, which use an even finer divide-and-conquer technique. A tower of normal subgroups is constructed such that the factor groups between two consecutive normal subgroups are the products of isomorphic simple groups. Abelian factors are handled by linear algebra, whereas the simple groups occurring in nonabelian factors are identified with standard copies of these groups, and the problems are solved in the standard copies. This identification process works in the more general black-box group setting, when we do not use the fact that the input group is represented by permutations: The algorithms only exploit the facts that we can multiply and invert group elements and decide whether two group elements are equal. This generality enables us to use the same algorithms for matrix group inputs. Computations with matrix groups is currently the most active area of CGT.

Dealing with permutation groups is the area of CGT where the complexity analysis of algorithms is the most developed. The initial reason for interest in complexity analysis was the connection of permutation group algorithms with the celebrated graph isomorphism problem. The decisive result in establishing the connection is the polynomial-time algorithm in [Luks, 1982] for testing isomorphism of graphs with bounded valence, where the isomorphism problem is reduced to finding setwise stabilizers of subsets in the permutation domain of groups with composition factors of bounded size. This paper not only established a link between complexity theory and CGT but provided new methodology for permutation group algorithms.

Up until the end of the 1980s, permutation group algorithms were developed in two different contexts. In one of these, the primary goal was efficient implementation, to handle the groups occurring in applications. In the other context, the main goal was the rigorous asymptotic analysis of algorithms.

Algorithms for numerous tasks were developed separately in the two contexts, and the two previous books on permutation group algorithms reflect this division: [Butler, 1991] deals mostly with the practical approach, whereas [Hoffmann, 1982] concentrates on the asymptotic analysis. In the past decade, a remarkable convergence of the approaches occurred, and algorithms with fast asymptotic running times that are suitable for implementation were developed. The main purpose of this book is to describe this new development. We consider the interaction of theory and implementation to be of great importance to each side: Symbolic algebra can benefit considerably by the influx of ideas of algorithmic complexity theory and rigorous asymptotic analysis; conversely, the implementations help demonstrate the power of the asymptotic paradigm, which is at the foundation of the theory of computing.

The major theme of this book is the description of nearly linear-time algorithms. These are the algorithms representing the convergence of theoretical and practical considerations. Their running time is $O(n|S|\log^c |G|)$ for input groups $G = \langle S \rangle \leq S_n$; in particular, in the important subcase of small-base groups, when $\log |G|$ is bounded from above by a polylogarithmic function of n, the running time is a nearly linear, $O(N \log^c N)$, function of the input length $N = n|S|$. The category of small-base groups includes all permutation representations of finite simple groups except the alternating ones and all primitive groups that do not have alternating composition factors in their socle. Most practical computations are performed with small-base input groups.

Quite different methods give the asymptotically fastest solutions for computational problems in large-base groups, where $\log |G|$ is bounded only by $\log n!$. Most of these algorithms have not yet been implemented. We shall also describe backtrack methods, which are the practical algorithms for problems with no known polynomial-time solutions. For small-base input groups, backtrack methods may be practical in groups of degree in the tens of thousands.

Our main goal is to present the mathematics behind permutation group algorithms, and implementation details will be mostly omitted. We shall give details only in the cases where the implemented version differs significantly from the one described by the theoretical result or when the reason for the fast asymptotic running time is a nontrivial data structure. Most of the algorithms described in this book have been implemented in the GAP system [GAP, 2000], which, along with its source code, is freely available. GAP code is written in a high-level, Pascal-like language, and it can be read as easily as the customary pseudocode in other books and articles on computational group theory. The addresses of ftp servers for GAP can be obtained from the World Wide Web page

`http://www-gap.dcs.st-and.ac.uk/~gap.`

The other large computer algebra system particularly suitable for computations with groups is MAGMA (see [Bosma et al., 1997]). The World Wide Web page

`http://www.maths.usyd.edu.au:8000/u/magma`

describes how to access MAGMA on a subscription basis.

Acknowledgments

The writing of this book began in 1993, on the suggestion of Joachim Neubüser, who envisioned a series of books covering the major areas of computational group theory. The fact that the writing is finished in less than a decade is in no small part the consequence of my wife Sherry's continuous encouragement. I am thankful to both of them and to the editors of Cambridge University Press for their patience. During this period, I was partially supported by the National Science Foundation.

Alexander Hulpke, William Kantor, Joachim Neubüser, Cheryl Praeger, Charles Sims, and Leonard Soicher read parts of the manuscript, and their comments improved the presentation significantly. I am especially indebted to William Kantor and Joachim Neubüser for their help.

1.1. A List of Algorithms

In this book, most algorithms are described in the proofs of theorems or just in the narrative, without any display or pseudocode. Whenever it is possible, algorithms given in the narrative are preceded by a centered paragraph header. The following list serves as a reference guide; it is organized roughly along the lines of the lists in Sections 3.1 and 3.3. The input is a permutation group $G \leq \mathrm{Sym}(\Omega)$.

- Orbit of some $\alpha \in \Omega$: Section 2.1.1; in particular, Theorem 2.1.1
- Blocks of imprimitivity
  - (i) A minimal nontrivial block: Section 5.5.1 (algorithm MinimalBlock)
  - (ii) The minimal block containing a given subset of Ω: Section 5.5.2
- Shallow Schreier tree construction
  - (i) Deterministic: Lemma 4.4.2, Remark 4.4.3, Lemma 4.4.8
  - (ii) Las Vegas: Theorem 4.4.6, Remark 4.4.7
- Strong generating set construction
  - (i) Deterministic: Section 4.2 (Schreier–Sims algorithm), Theorem 5.2.3 (with known base), Section 7.1 (for solvable groups), Theorem 10.1.3 (stored in a labeled branching)
  - (ii) Monte Carlo: Section 4.5, Theorems 5.2.5 and 5.2.6, Lemma 5.4.1, Section 10.3

  (iii) Heuristic: Section 4.3 (random Schreier–Sims algorithm)
  (iv) GAP implementation: Section 4.5.1, Remark 5.2.7
- Strong generating set verification
  (i) Deterministic: Section 8.1 (Schreier–Todd–Coxeter–Sims algorithm), Section 8.2 (Verify routine)
  (ii) Monte Carlo: Lemma 4.5.6
  (iii) Las Vegas: Theorem 8.3.1
- Membership test (sifting): Section 4.1
- Reduction of the size of generating sets
  (i) Strong generators, deterministic: Lemma 4.4.8, Exercise 4.7
  (ii) Arbitrary generators, Monte Carlo: Lemma 2.3.4, Theorem 2.3.6
- Random element generation
  (i) With an SGS: Section 2.2, first paragraph
  (ii) Without an SGS: Section 2.2 (random walk on a Cayley graph, product replacement algorithm)
  (iii) In alternating and symmetric groups: Exercises 2.1 and 2.2
- Isomorphism with other representations
  (i) With a black-box group: Section 5.3
  (ii) Solvable groups, with a power-commutator presentation: Section 7.2
  (iii) A_n and S_n, with natural action: Theorem 10.2.4
  (iv) $\mathrm{PSL}_d(q)$, with the action on projective points: Section 5.3
- Operations with base images and with words in generators: Lemmas 5.2.1, 5.2.2, and 5.3.1
- Base change (transposing and conjugating base points, deterministic and Las Vegas algorithms): Section 5.4, Exercise 5.5
- Presentations: Section 7.2 (for solvable groups), Section 8.1, Exercise 5.2, Theorem 8.4.1
- Pointwise stabilizer of a subset of Ω: Section 5.1.1
- Handling of homomorphisms (kernel, image, preimage)
  (i) Transitive constituent and block homomorphisms: Section 5.1.3
  (ii) General case: Section 5.1.2
- Closure for G-action, normal closure
  (i) With membership test in substructures, deterministic: Sections 2.1.2 and 5.1.4, Lemma 6.1.1
  (ii) Without membership test, Monte Carlo: Theorems 2.3.9 and 2.4.5
- Commutator subgroup computation, derived series, lower central series
  (i) With membership test in substructures, deterministic: Sections 2.1.2 and 5.1.4
  (ii) Without membership test, Monte Carlo: Theorems 2.3.12 and 2.4.8
- Upper central series in nilpotent groups: Section 7.4.2
- Solvability test: Sections 2.1.2, 5.1.4, and 7.1

- Nilpotency test: Sections 2.1.2, 5.1.4, and 7.4.1
- Subnormality test: Section 2.1.2
- Commutativity test (Monte Carlo): Lemma 2.3.14
- Regularity test: Exercises 5.12–5.15
- Center: Section 6.1.3
- Permutation representation with a normal subgroup N in the kernel: Lemmas 6.2.2 and 6.2.4, Theorem 6.3.1 (if N is abelian)
- Composition series
  - (i) Reduction to the primitive group case: Section 6.2.1
  - (ii) Finding normal subgroups in various types of primitive groups (Monte Carlo): Sections 6.2.3 and 6.2.4
  - (iii) Verification of composition series, if SGS is known: Section 6.2.6
  - (iv) GAP implementation: Section 6.2.5
  - (v) Composition series without the classification of finite simple groups: Section 6.2.6
- Chief series: Section 6.2.7
- Sylow subgroups and Hall subgroups in solvable groups: Section 7.3.1, Exercise 7.5 (Theorem 7.3.3 for conjugating Sylow subgroups)
- Core of a subnormal subgroup: Section 6.1.5
- p-core and solvable radical: Section 6.3.1
- Backtrack, general description: Section 9.1 (traditional), Section 9.2 (partition backtrack)
- Setwise stabilizer of a subset of Ω: Section 9.1.2, Example 2
- Centralizer
  - (i) In the full symmetric group: Section 6.1.2
  - (ii) Of a normal subgroup: Section 6.1.4
  - (iii) General case: Section 9.1.2, Example 1
- Intersection of groups: Corollary 6.1.3 (if one of the groups normalizes the other), Section 9.1.2, Example 3 (general case)
- Conjugating element: Section 9.1.2, Example 4
- Conjugacy classes: Section 7.3.2 (in solvable groups), Section 9.4 (general case)
- Normalizer: Section 9.3

1.2. Notation and Terminology

We assume that the reader is familiar with basic notions concerning groups covered in introductory graduate courses and with elementary probability theory. A background area with which we do *not* suppose reader familiarity is the detailed properties of finite simple groups, and the occasional references to these properties can be ignored without impeding understanding of the subsequent material. However, readers interested in further research in permutation group algorithms are strongly advised to acquire knowledge of groups of Lie type. One of the current largest obstacles, both in the permutation group and matrix group setting, is our inability to exploit algorithmically properties of exceptional groups of Lie type.

The required (minimal) background material about permutation groups can be found for example in the first chapter of the recent books [Dixon and Mortimer, 1996] and [Cameron, 1999]. Here we only summarize our notation and terminology. In this book, *all groups are finite*.

All statements (i.e., theorems, lemmas, propositions, corollaries, and remarks) are numbered in a common system. For example, Theorem X.Y.Z denotes the Zth statement in Chapter X, Section Y, if this statement happens to be a theorem. Definitions are just part of the text and are not displayed with a number. Any unknown items (hopefully) can be found in the index. In the index, boldface type is used for the page number where an item or notation is defined. There are exercises at the end of some chapters, numbered as Exercise X.Y in Chapter X. A third numbering system is used for the displayed formulas, in the form (X.Y) in Chapter X.

1.2.1. Groups

If G is a group and $S \subseteq G$ then we denote by $\langle S \rangle$ the subgroup generated by S. We write $H \leq G$ to indicate that H is a subgroup of G and $H \lneqq G$ if $H \leq G$ and $H \neq G$. If H is isomorphic to a subgroup of G then we write $H \lesssim G$. The symbol $|G : H|$ denotes the number $|G|/|H|$, and $H \lhd G$ denotes that H is normal in G. A subgroup $H \leq G$ is *subnormal* in G, in notation $H \lhd\lhd G$, if there exists a chain of subgroups $H = H_0 \lhd H_1 \lhd \cdots \lhd H_k = G$. If $N \lhd G$ and $H \leq G$ such that $N \cap H = 1$ and $G = NH$ then we call H a *complement* of N in G.

The group of automorphisms, outer automorphisms, and inner automorphisms of G are denoted by $\mathrm{Aut}(G)$, $\mathrm{Out}(G)$, and $\mathrm{Inn}(G)$, respectively. We say that G *acts on a group* H if a homomorphism $\varphi : G \to \mathrm{Aut}(H)$ is given. If φ is clear from the context, for $g \in G$ and $h \in H$ we sometimes denote $\varphi(g)(h)$, the image of h under the automorphism $\varphi(g)$, by h^g. If G acts on H and $U \subseteq H$ then $U^G := \{U^g \mid g \in G\}$ is the *orbit* of U under the G-action, and $\langle U^G \rangle$ is the *G-closure* of U. In the special case $H = G$, the group $\langle U^G \rangle$ is called the *normal closure* of U. For $U \leq H$, $C_G(U) := \{g \in G \mid (\forall u \in U)(u^g = u)\}$ is the *centralizer* of U in G and $N_G(U) := \{g \in G \mid U^g = U\}$ is the *normalizer* of U in G. In particular, $Z(G) := C_G(G)$ is the *center* of G.

The *commutator* of $a, b \in G$ is $[a, b] := a^{-1}b^{-1}ab$ and the *conjugate* of a by b is $a^b := b^{-1}ab$. For $H, K \leq G$, the *commutator* of H and K is defined as $[H, K] := \langle [h, k] \mid h \in H, k \in K \rangle$. In particular, $[G, G]$, also called the *derived subgroup* of G, is denoted by G'. A group G is *perfect* if $G = G'$. The *derived series* of G is the sequence $D_0 \geq D_1 \geq \cdots$ of subgroups of G, defined recursively by the rules $D_0 := G$ and $D_{i+1} := D_i'$ for $i \geq 0$. The *lower central series* $L_0 \geq L_1 \geq \cdots$ of G is defined as $L_0 := G$ and $L_{i+1} := [L_i, G]$ for $i \geq 0$. The *upper central series* $Z_0 \leq Z_1 \leq \cdots$ of G is defined as $Z_0 := 1$ and Z_{i+1} is the preimage of $Z(G/Z_i)$ in G for $i \geq 0$. A group G is called *solvable* if $D_m = 1$ for some m, and it is called *nilpotent* if $L_m = 1$ or $Z_m = G$ for some m.

The *direct product* of groups $A_1, \ldots, A_m$ is denoted by $A_1 \times \cdots \times A_m$ or by $\prod_{i=1}^m A_i$. For $i = 1, 2, \ldots, m$, the *projection function* $\pi_i \colon A_1 \times \cdots \times A_m \to A_i$ is defined by the rule $\pi_i \colon (a_1, \ldots, a_m) \mapsto a_i$. A group $H \leq A_1 \times \cdots \times A_m$ is a *subdirect product* of the A_i if all functions π_i restricted to H are surjective, i.e., $\{\pi_i(h) \mid h \in H\} = A_i$.

For $H \leq G$, a *transversal* G mod H is a set of representatives from the right cosets of H in G. For a fixed transversal T and $g \in G$, we denote by $\bar{g}$ the coset representative in T such that $g \in H\bar{g}$. Unless stated explicitly otherwise, cosets always mean *right* cosets.

If Σ is any collection of simple groups, $O_\Sigma(G)$ denotes the largest normal subgroup of G such that each composition factor of $O_\Sigma(G)$ is isomorphic to a member of Σ, and $O^\Sigma(G)$ denotes the smallest normal subgroup of G such that each composition factor of $G/O^\Sigma(G)$ is isomorphic to a member of Σ. In particular, if Σ consists of a single group of prime order p then $O_\Sigma(G)$ is denoted by $O_p(G)$; this is the largest normal p-subgroup, the *p-core*, of G. When Σ consists of all cyclic simple groups, $O_\Sigma(G)$ is denoted by $O_\infty(G)$; this is the largest solvable normal subgroup, the *solvable radical* of G. Similarly, $O^\infty(G)$ denotes the smallest normal subgroup of G with solvable factor group and it is called the *solvable residual* of G. For $H \leq G$, $\mathrm{Core}_G(H) := \bigcap\{H^g \mid g \in G\}$ is the largest normal subgroup of G contained in H; it is the kernel of the permutation representation of G on the (right) cosets of H. The *socle* of G is the subgroup of G generated by all minimal normal subgroups of G and is denoted by $\mathrm{Soc}(G)$.

The cyclic group of order n is denoted by C_n. The group of invertible $d \times d$ matrices over the q-element field $\mathrm{GF}(q)$ is denoted by $\mathrm{GL}_d(q)$. Similar notation is used for the other classical matrix groups of Lie type and for their projective factor groups: $\mathrm{SL}_d(q)$, $\mathrm{PSL}_d(q)$, and so on. The unitary groups $\mathrm{GU}_d(q)$, $\mathrm{SU}_d(q)$, and $\mathrm{PSU}_d(q)$ are defined over $\mathrm{GF}(q^2)$. For exceptional groups of Lie type, we use the Lie-theoretic notation ${}^2B_2(q)$, ${}^2G_2(q)$, and so on. As mentioned earlier, no detailed knowledge of the groups of Lie type is required in this book.

1.2.2. Permutation Groups

We shall use the cycle notation for permutations, and the identity permutation is denoted by $()$. The group of all permutations of an n-element set Ω is denoted $\mathrm{Sym}(\Omega)$, or S_n if the specific set is inessential. Subgroups of S_n are the *permutation groups* of *degree* n. We use lowercase Greek letters to denote elements of Ω; lower- and uppercase italics denote elements and subgroups of S_n, respectively. For $\alpha \in \Omega$ and $g \in \mathrm{Sym}(\Omega)$, we write α^g for the image of α under the permutation g. The alternating group on Ω is denoted by $\mathrm{Alt}(\Omega)$ (or A_n). The *support* of $g \in \mathrm{Sym}(\Omega)$, denoted by $\mathrm{supp}(g)$, consists of those elements of Ω that are actually displaced by g: $\mathrm{supp}(g) = \{\omega \in \Omega \mid \omega^g \neq \omega\}$. The set of fixed points of g is defined as $\mathrm{fix}(g) := \Omega \backslash \mathrm{supp}(g)$. The *degree* of g is $\deg(g) = |\mathrm{supp}(g)|$.

We say that a group G *acts on* Δ if a homomorphism $\varphi : G \to \mathrm{Sym}(\Delta)$ is given (by specifying the image of a generator set of G). This action is *faithful* if its kernel $\ker(\varphi)$ is the identity. The image $\varphi(G) \leq \mathrm{Sym}(\Delta)$ is also denoted by G^Δ. In the special case when $G \leq \mathrm{Sym}(\Omega)$, $\Delta \subseteq \Omega$ is fixed by G, and φ is the restriction of permutations to Δ, we also denote G^Δ by $G|_\Delta$. The *orbit* of $\omega \in \Omega$ under $G \leq \mathrm{Sym}(\Omega)$ is the set of images $\omega^G := \{\omega^g \mid g \in G\}$. For $\Delta \subseteq \Omega$ and $g \in \mathrm{Sym}(\Omega)$, $\Delta^g := \{\delta^g \mid \delta \in \Delta\}$. A group $G \leq \mathrm{Sym}(\Omega)$ is *transitive* on Ω if it has only one orbit, and G is t*-transitive* if the action of G induced on the set of ordered t-tuples of distinct elements of Ω is transitive ($t \leq n$). The maximum such t is the *degree of transitivity* of G.

If $G \leq \mathrm{Sym}(\Omega)$ is transitive and $\Delta \subseteq \Omega$, then Δ is called a *block of imprimitivity* for G if for all $g \in G$ either $\Delta^g = \Delta$ or $\Delta^g \cap \Delta = \emptyset$. The group G is called *primitive* if all blocks have 0, 1, or $|\Omega|$ elements. If Δ is a block then the set of images of Δ is a partition of Ω, which is called a *block system*, and an action of G is induced on the block system. A block is called *minimal* if it has more than one element and its proper subsets of size at least two are not blocks. A block is called *maximal* if the only block properly containing it is Ω. A block system is *maximal* if it consists of minimal blocks, whereas a block system is *minimal* if it consists of maximal blocks. The action of G on a minimal block system is primitive.

For $\Delta \subseteq \Omega$ and $G \leq \mathrm{Sym}(\Omega)$, $G_{(\Delta)}$ denotes the *pointwise stabilizer* of Δ, namely, $G_{(\Delta)} = \{g \in G \mid (\forall \delta \in \Delta)(\delta^g = \delta)\}$. If Δ has only one or two elements, we often drop the set braces and parentheses from the notation; in particular, G_δ denotes the stabilizer of $\delta \in \Omega$. The *setwise stabilizer* of Δ is denoted by G_Δ (i.e., $G_\Delta = \{g \in G \mid \Delta^g = \Delta\}$). If $\Delta = (\delta_1, \ldots, \delta_m)$ is a *sequence* of elements of Ω then G_Δ denotes the pointwise stabilizer of that sequence (i.e., $G_\Delta = G_{(\{\delta_1,\ldots,\delta_m\})}$).

A group $G \le \mathrm{Sym}(\Omega)$ is *semiregular* if $G_\delta = 1$ for all $\delta \in \Omega$, whereas G is *regular* if it is transitive and semiregular. A *Frobenius group* is a transitive group $G \le \mathrm{Sym}(\Omega)$ that is not regular but for which $G_{\alpha\beta} = 1$ for all distinct $\alpha, \beta \in \Omega$.

If $g \in \mathrm{Sym}(\Omega)$ then a bijection $\varphi : \Omega \to \Delta$ naturally defines a permutation $\bar{\varphi}(g) \in \mathrm{Sym}(\Delta)$ by the rule $\varphi(\omega)^{\bar{\varphi}(g)} := \varphi(\omega^g)$ for all $\omega \in \Omega$. We say that $G \le \mathrm{Sym}(\Omega)$ and $H \le \mathrm{Sym}(\Delta)$ are *permutation isomorphic*, $H \sim G$, if there is a bijection $\varphi\colon \Omega \to \Delta$ such that $\bar{\varphi}(G) := \{\bar{\varphi}(g) \mid g \in G\} = H$.

Let G be an arbitrary group and let $H \le S_k$ be a transitive permutation group. The *wreath product* $G \wr H$ consists of the sequences $(g_1, \ldots, g_k; h)$ where $g_i \in G$ for $i = 1, \ldots, k$ and $h \in H$. The product of $(g_1, \ldots, g_k; h)$ and $(\bar{g}_1, \ldots, \bar{g}_k; \bar{h})$ is defined as $(g_1\bar{g}_{1^h} \ldots, g_k\bar{g}_{k^h}; h\bar{h})$.

1.2.3. Algorithmic Concepts

Groups in algorithms will always be input and output by specifying a list of generators.

Given $G = \langle S \rangle$, a *straight-line program of length m* reaching some $g \in G$ is a sequence of expressions $(w_1, \ldots, w_m)$ such that, for each i, w_i is a symbol for some element of S, or $w_i = (w_j, -1)$ for some $j < i$, or $w_i = (w_j, w_k)$ for some $j, k < i$, such that if the expressions are evaluated in G the obvious way then the value of w_m is g. Namely, the evaluated value of a symbol for a generator is the generator itself; the evaluated value of $w_i = (w_j, -1)$ is the inverse of the evaluated value of w_j; and the evaluated value of $w_i = (w_j, w_k)$ is the product of the evaluated values of w_j and w_k. Hence a straight-line program is an encoding of a sequence of group elements $(g_1, \ldots, g_m)$ such that $g_m = g$ and for each i one of the following holds: $g_i \in S$, or $g_i = g_j^{-1}$ for some $j < i$, or $g_i = g_j g_k$ for some $j, k < i$. However, the more abstract definition as a sequence of expressions not only requires less memory but also enables us to construct a straight-line program in one representation of G and evaluate it in another, which is an important feature of some algorithms.

The symbols $\mathbb{Z}$, $\mathbb{N}$, and $\mathbb{R}$ denote the set of integers, nonnegative integers, and real numbers, respectively. Let

$$\mathcal{F} := \{f\colon \mathbb{N} \to \mathbb{R} \mid (\exists n_0 \in \mathbb{N})(\forall n > n_0)(f(n) > 0)\}$$

(i.e., functions that take positive values with finitely many exceptions). For

$f \in \mathcal{F}$, we define

$$O(f) := \{t \in \mathcal{F} \mid (\exists c > 0)(\exists n_0 \in \mathbb{N})(\forall n > n_0)(t(n) < cf(n))\},$$
$$\Omega(f) := \{t \in \mathcal{F} \mid (\exists c > 0)(\exists n_0 \in \mathbb{N})(\forall n > n_0)(t(n) > cf(n))\},$$
$$o(f) := \{t \in \mathcal{F} \mid (\forall c > 0)(\exists n_0 \in \mathbb{N})(\forall n > n_0)(t(n) < cf(n))\},$$

and

$$\Theta(f) := O(f) \cap \Omega(f).$$

Stated less formally, $t \in O(f)$ means that for large enough n, $t(n)/f(n)$ is bounded from above by an absolute constant c; $t \in \Omega(f)$ means that for large enough n, $t(n)/f(n)$ is bounded from below by an absolute positive constant c; and $t \in o(f)$ means that the limit of $t(n)/f(n)$ is 0.

For $t \in O(f)$, we also say that t is $O(f)$, and we use similar statements for Ω and Θ as well. We believe that this notation is more correct than the traditional $t = O(f)$ (see [Brassard and Bratley, 1988, Chap. 2] for the different variants of these notations).

We also introduce a "soft version" of the big-O notation. We write $t \in O^{\sim}(f)$ if $t(n) \leq Cf(n) \log^c n$ for large enough n (where c, C are positive constants). Logarithms are always of base 2.

1.2.4. Graphs

Let V be a set and $\mathcal{E}$ a subset of the two-element subsets of V. The pair $(V, \mathcal{E})$ is called a *graph* $\mathcal{X}(V, \mathcal{E})$. The elements of V and $\mathcal{E}$ are the *vertices* and the *edges* of $\mathcal{X}$, respectively. For $v \in V$, the number of edges containing v is the *degree* or *valency* of v, which we shall denote by $\deg(v)$. A graph $\mathcal{X}$ is called *regular* if all vertices have the same valency. We also say that an edge $\{u, v\} \in \mathcal{E}$ connects u and v. The set $N(v) := \{u \in V \mid \{u, v\} \in \mathcal{E}\}$ is the *neighborhood* of v in $\mathcal{X}$. A graph $\mathcal{X}$ is called *bipartite* if V can be partitioned into two sets A, B so that all edges of $\mathcal{X}$ connect some vertex in A with some vertex in B. The *automorphism group* $\mathrm{Aut}(\mathcal{X})$ consists of those permutations of V that leave $\mathcal{E}$ invariant.

These notions can be generalized by requiring only that the elements of $\mathcal{E}$ are subsets of V, but of arbitrary size. Then $\mathcal{X}$ is called a *hypergraph*. A hypergraph $\mathcal{X}$ is *uniform* if all elements of $\mathcal{E}$ are of the same size.

Another generalization is when $\mathcal{E}$ consists of ordered pairs of V. Then $\mathcal{X}$ is called a *directed graph*. If we want to emphasize that a graph is directed then we shall use arrows above $\mathcal{E}$ and above the ordered pairs in $\mathcal{E}$. The *out-degree* of a vertex u is the number of edges $\overrightarrow{(u, v)}$ in $\mathcal{E}$. For a directed graph $\mathcal{X}(V, \vec{\mathcal{E}})$, the *underlying graph* of $\mathcal{X}$ is the graph $\mathcal{U}(V, E)$ with edge set $E = \{\{u, v\} \mid \overrightarrow{(u, v)} \in \vec{\mathcal{E}}\}$.

Let $\mathcal{X}(V, \mathcal{E})$ be a graph. A *walk* in $\mathcal{X}$ is a sequence of vertices $(v_0, v_1, \ldots, v_k)$ such that $\{v_i, v_{i+1}\} \in \mathcal{E}$ for all $i \in [0, k-1]$. A *path* is a walk with $v_i \neq v_j$ for all i, j with $0 \leq i < j \leq k$; a *cycle* is a walk with $v_0 = v_k$ and $v_i \neq v_j$ for all i, j with $0 \leq i < j \leq k-1$. We can define a binary relation $\mathcal{R}$ on V by letting $(u, v) \in \mathcal{R}$ if and only if there is a walk $(u = v_0, v_1, \ldots, v_k = v)$. Then $\mathcal{R}$ is an equivalence relation; the equivalence classes are called the *components* of $\mathcal{X}$. The graph $\mathcal{X}$ is *connected* if it has only one component.

If $\mathcal{X}$ is a directed graph then walks, cycles, and paths are defined similarly, but the binary relation $\mathcal{R}$ is not necessarily an equivalence relation. We may define another binary relation $\hat{\mathcal{R}}$ on V: $(u, v) \in \hat{\mathcal{R}}$ if and only if $(u, v) \in \mathcal{R}$ and $(v, u) \in \mathcal{R}$. Then $\hat{\mathcal{R}}$ is an equivalence relation, and its equivalence classes are called the *strongly connected components* of $\mathcal{X}$. The directed graph $\mathcal{X}$ is *strongly connected* if it has only one strongly connected component.

Cycle-free graphs are called *forests* and connected, cycle-free graphs are called *trees*. A *rooted tree* is a tree with a distinguished vertex. If $\mathcal{X}(V, \mathcal{E})$ is a rooted tree with root r then the *parent* of $v \in V\backslash\{r\}$ is the first vertex after v on the unique path from v to r. The *children* of $v \in V$ are those vertices whose parent is v. A vertex without children is called a *leaf*.

Let G be a group and $S \subseteq G$. The *Cayley graph* $\Gamma(G, S)$ is defined to have vertex set G; for $g, h \in G$, $\{g, h\}$ is an edge if and only if $gs = h$ for some $s \in S \cup S^{-1}$. The Cayley graph $\Gamma(G, S)$ is connected if and only if $G = \langle S \rangle$. Cayley graphs are regular and vertex-transitive (i.e., $\mathrm{Aut}(\Gamma(G, S))$ is a transitive subgroup of $\mathrm{Sym}(G)$).

Sequences will be denoted by enclosing their elements in parentheses. For a sequence L, $L[i]$ denotes the ith element of L.

The most abused notation is (a, b) for integers a and b. Depending on the context, it may mean a sequence with elements a and b, the set of real numbers between a and b, the set of integers between a and b (although, in this case, we shall prefer to use the closed interval $[a+1, b-1]$), or the permutation exchanging a and b.

1.3. Classification of Randomized Algorithms

As we shall see in numerous examples in this book, randomization can speed up many algorithms handling permutation groups. In other parts of computational group theory, randomization is even more important: For example, when dealing with matrix groups, the scope of efficient deterministic algorithms is very limited, both in the practical and theoretical sense of efficiency. Randomization also seems to be an indispensable tool in algorithms for black-box groups.

In this section, we describe the different types of randomized algorithms. Our discussion follows [Babai, 1997].

Computational tasks can be described by a relation $R(x, y)$ between an input string x and output string y. The relation $R(x, y)$ holds if y is a correct output for the task described by x. In group algorithms, the input usually contains a set of generators for a group. The output may consist of group elements, generating a desired subgroup of the input, but other types of output are also conceivable: The output could be a number (for example, the order of the input group) or a statement that group elements with the desired property do not exist (for example, when asking for a group element conjugating one given element to another one).

Note that there may be different correct outputs for the same input. We call a computational task *functional* if it has exactly one correct output for all inputs. For example, finding generators for a Sylow 2-subgroup is not a functional computational task, whereas finding the order of a group is. A special category of (functional) computational tasks is the class of *decision problems*. Here, the answer is a single bit, representing "yes" or "no." For example, determining solvability of the input group is a decision problem.

A (correct) *deterministic algorithm* computes an output $f(x)$ for all inputs x, so that $R(x, f(x))$ holds. A *randomized algorithm* uses a string r of random bits ("coin flippings") and returns the output $f(x, r)$. The output may not be correct for every sequence r.

We call a randomized algorithm *Monte Carlo* if for all inputs x,

$$\text{Prob}(R(x, f(x, r)) \text{ holds}) \geq 1 - \varepsilon,$$

where the value for the error term $\varepsilon < 1/2$ is specified in the description of the Monte Carlo algorithm. In most practical situations, the reliability of the algorithms can be improved by repeated applications, or by running the algorithms longer. Although we cannot formulate a theorem in the general setting considered in this section, at least the following holds for all Monte Carlo algorithms for permutation groups described in this book: The probability of an incorrect answer can be bounded from above by an arbitrary $\varepsilon > 0$, prescribed by the user. The associated cost is the running time of the algorithm multiplied by a factor $O(\log(1/\varepsilon))$.

A situation we can analyze here is the case of decision problems. Suppose that we have a Monte Carlo algorithm for a decision problem, with error probability $\varepsilon < 1/2$. Running the algorithm t times, and taking a majority vote of the results, we can increase the reliability to at least $1 - \delta^t$, with $\delta := 2\sqrt{\varepsilon(1-\varepsilon)} < 1$: The probability of error is at most

$$\sum_{k=t/2}^{t} \binom{t}{k} \varepsilon^k (1-\varepsilon)^{t-k} < (1-\varepsilon)^t \sum_{k=t/2}^{t} \binom{t}{k} \left(\frac{\varepsilon}{1-\varepsilon}\right)^{t/2} < \delta^t.$$

Babai also discusses *one-sided Monte Carlo algorithms* (1MC algorithms) for decision problems. These are algorithms where at least one of the outputs is guaranteed to be correct: If the correct answer is "yes," a 1MC algorithm may err; if the correct answer is "no," the algorithm must always output "no." Hence, an output "yes" is always correct, whereas an output "no" may not be. The *co-1MC algorithms* are defined by exchanging the words "yes" and "no" in the definition of 1MC algorithms.

In the context of group theoretical algorithms, the notion of 1MC algorithms can be extended to most computational tasks, since the error is usually one-sided: For example, when computing generators U for the normal closure of some $H \leq G$ (cf. Section 2.3.3), we never place an element of G on the generator list U that is not in $\langle H^G \rangle$. The error we may commit is that $\langle U \rangle$ is a proper subgroup of $\langle H^G \rangle$. Another example is the Monte Carlo order computation in permutation groups (cf. Section 4.5): The result we obtain is always a lower bound for $|G|$, because we do not place permutations not belonging to G into the strong generating set of G.

An important subclass of Monte Carlo algorithms is the class of *Las Vegas algorithms*. This term was introduced in [Babai, 1979] to denote Monte Carlo algorithms that never give an incorrect answer. The output is either correct (with the prescribed probability at least $1 - \varepsilon$) or the algorithm reports failure. Here, ε may be any given constant less than 1, since the probability of an (always correct) output can be increased to at least $1 - \varepsilon^t$ by running the algorithm t times.

Las Vegas algorithms are preferable over general Monte Carlo algorithms for many reasons. One of them is the certainty of the answer. Another one is that recognizing the correct answer often allows early termination. Finally, if we can guarantee that the output is always correct then we may use heuristics to speed up the algorithm, even in cases when we cannot estimate the probability of error for our heuristics.

In practice, if a Las Vegas algorithm reports failure then we rerun it until the correct output is produced. The definition of Las Vegas algorithms can be reformulated so that they always return a correct answer, with the running time estimate replaced by an estimate of the *expected* running time.

A Monte Carlo algorithm, combined with a deterministic checking of the result, becomes a Las Vegas algorithm, because we can recognize that the Monte

Carlo algorithm returned an incorrect output, and report failure. Another way to obtain Las Vegas algorithms is by using both 1MC and co-1MC algorithms for a decision problem. Running both of these algorithms, with probability at least $1 - \varepsilon$ one of them will return a guaranteed correct output. If both algorithms return an answer that may not be correct, then we report failure. An important direction of current research is the upgrading of Monte Carlo algorithms for permutation groups to Las Vegas type. We shall describe a result in this direction in Section 8.3.

2

Black-Box Groups

Groups can be given in different ways: The group elements can be, for example, permutations, invertible matrices, or words in generators satisfying some relations. Algorithms usually try to exploit the specific features of the given representation. As an example, suppose that we have to compute the mth power of some $g \in G$. If G is a permutation group then for large values of m the fastest method to compute g^m is the following. We compute the mth power of each cycle of g separately, first reducing m modulo the cycle lengths. If C is a cycle of length k and j is the remainder of m divided by k, then for all $\alpha \in C$ we have $\alpha^{g^m} = \alpha^{g^j}$. The arithmetic operations required in the computation of j are much cheaper than group operations. Unfortunately, no analogue of this method is available in other representations of groups.

Another way to compute g^m is by repeated squaring. First, we compute g^{2^i} for $1 \le i \le \log m$. Then, if the binary expansion of m is $m = \sum_{i \in I} 2^i$ then $g^m = \prod_{i \in I} g^{2^i}$. This method requires $O(\log m)$ group operations and it is independent of the actual representation of G: All we use is that – somehow – we can compute the product of two given group elements.

In this chapter we deal with algorithms of the latter type. A *black-box group* G is a group whose elements are encoded as strings of length at most N over an alphabet Q, for some positive integer N and finite set Q. We do not require that group elements have a unique representation as a string, and not all strings need to correspond to group elements.

Group operations are performed by an oracle (the black box). Given strings representing $g, h \in G$, we can

(i) compute a string representing gh;
(ii) compute a string representing g^{-1}; and
(iii) decide whether $g = 1$.

It is possible that the oracle accepts strings representing elements of an overgroup $\bar{G}$ of G, and it performs the group operations in $\bar{G}$. If this is the case then we assume that we can decide whether a string represents an element of $\bar{G}$, but we do *not* assume that we can decide whether a string represents an element of G or $\bar{G}\backslash G$.

Combining (i)–(iii), we can compare group elements. In practical situations, usually it is possible to decide whether $g = h$ directly, without computing gh^{-1} and then comparing it with 1. A *black-box group algorithm* is an algorithm that does not use specific features of the group representation or particulars of how the group operations are performed; it can use only the operations described in the list above. Note that we have an upper bound for the order of G, namely, $|G| \leq \sum_{i=1}^{N} |Q|^i \leq N|Q|^N$.

The definition we gave here is a slight generalization of the original one in [Babai and Szemerédi, 1984]. There, only the alphabet $Q = \{0, 1\}$ was allowed, and the group elements had to be represented by strings of uniform length. It is easy to reduce our more general definition to the original one: The elements of Q can be encoded by $1 + \lceil \log |Q| \rceil$-long 0–1 sequences corresponding to the binary representations of the numbers $1, 2, \ldots, |Q|$, and short strings can be padded by zeros to achieve uniform length. Although we did not require in the definition that group elements have a unique representation as a string over the alphabet Q, in a number of practical situations this additional property is satisfied. Unique representation allows ordering of group elements (for example, by the lexicographic ordering of strings) that may speed up algorithms dealing with lists of group elements (see, for example, Section 2.1.1).

Two examples of black-box groups are matrix groups (with a finite field as Q) and permutation groups (with $Q = \{1, 2, \ldots, n\}$). In these examples, the set of all invertible matrices and the set of all permutations, respectively, are natural overgroups. The oracles for group operations are the usual matrix and permutation multiplications and inverses. The strings representing group elements are of uniform length, and each group element has a unique representation as a string.

Another example is the polycyclic representation of solvable groups (cf. Section 7.2). In this case, the alphabet Q is a set of generators, and multiplication is performed by a collection method. Every string up to length N represents a group element, and group elements may be represented by more than one string. However, each group element has a string representation in a normal form, the so-called collected word. In fact, the oracle performs a group multiplication by concatenating the input strings and computing the collected word corresponding to this concatenation. By always using the collected word to represent group elements, we can consider the black-box group as having unique representation. For details, we again refer to Section 7.2.

In Section 5.3, we shall introduce another way to consider permutation groups as black-box groups. Similarly to the case of polycyclic representations, the alphabet is a set of generators, and every string represents a group element. Again, we shall have a normal form for representing group elements, the so-called standard word, and we can restrict ourselves to working with standard words. For small-base permutation groups, a class of groups important for practical computations, this new representation allows much faster group operations than are possible with permutation multiplication. The ability to perform group operations quickly offsets the disadvantage that we lose the information stored implicitly in the cycle structure of permutations; all we can do is to perform the black-box operations listed above.

An important example of black-box groups with no unique representation of elements is when G is a factor group H/M. Here H can be any black-box group, and the oracle for group operations in G is the oracle of H. However, to satisfy condition (iii), we must be able to decide whether a string representing an element of H is in M. For example, if $M = Z(H)$ then we can test $g \in M$ by checking whether g commutes with the generators of H. Another example where we can test membership in M is when $M = O_\infty(H)$ is the solvable radical of H or M is the Fitting subgroup of H. In Sections 2.3 and 2.4, we shall describe Monte Carlo algorithms for computing normal closures and commutator subgroups. Using these methods, we can test whether $g \in M$ by computing $\langle g^H \rangle$ and the derived or lower central series of $\langle g^H \rangle$, respectively. These last examples are oracles with the possibility of incorrect output, and the membership computation in M is quite time consuming and so cannot be applied frequently in practice. However, these examples play an important role in the theoretical complexity investigations of matrix group algorithms.

Finitely presented groups, however, are *not* black-box groups. Although we can perform group multiplications by simply concatenating the words representing g and h, there is no uniform bound for the word length. Even more importantly, the innocent looking condition (iii) corresponds to the word problem, which is, by the celebrated results of Novikov and Boone [Rotman, 1995, Chap. 12], in general undecidable.

2.1. Closure Algorithms

2.1.1. Orbit Computations

A frequently occurring situation is that a black-box group G acts on a permutation domain Ω and we have to compute the orbit $\alpha^G := \{\alpha^g \mid g \in G\}$ of some $\alpha \in \Omega$. Naturally, the primary example is when the black-box group G is actually a permutation group $G \leq \mathrm{Sym}(\Omega)$. Two other examples are when $G = \Omega$ and we have to compute the conjugacy class of some element of G, or when Ω consists of all subgroups of G (specified by sets of generators) and we need the conjugacy class of a subgroup. These latter examples explain why we consider orbit computations here rather than among the permutation group algorithms. When working with permutation groups, we always assume that the generators of G are input as permutations of Ω, while here we may have permutation domains Ω that are too large to be listed explicitly and yet the orbit α^G is small enough to be handled. Therefore, all we assume here is that given $\beta \in \Omega$ and $g \in G$, we can compute the image β^g and that we can compare elements of Ω. Note that the latter task may be nontrivial: In the case when we compute the conjugacy class of a subgroup, comparing elements of Ω amounts to the problem of deciding whether two lists of generators define the same subgroup of G.

Suppose that $G = \langle S \rangle$. The orbit α^G is the smallest subset of Ω containing α and closed under the action of the generators of G. Hence, α^G can be obtained by a standard algorithm for computing connected components in graphs. We may define a directed graph $D(\Omega, \vec{E})$ with vertex set Ω and edge set $\vec{E} = \{\overrightarrow{(\beta, \gamma)} \mid \beta, \gamma \in \Omega \wedge (\exists g \in S)(\beta^g = \gamma)\}$ (i.e., the ordered pair (β, γ) is an edge if and only if one of the generators of G carries β to γ). Note that if $\beta^g = \gamma$ then $\gamma^{g^k} = \beta$ for a suitable power g^k of g (because in this book we deal only with finite groups), so the connected components of D are strongly connected.

The Basic Orbit Algorithm

The orbit α^G is the vertex set of a *breadth-first-search tree* rooted at α in the directed graph D. Let $L_0 := \{\alpha\}$ and, recursively, define

$$L_i := \{\gamma \in \Omega \mid (\exists \beta \in L_{i-1})(\overrightarrow{(\beta, \gamma)} \in \vec{E})\} \Big\backslash \bigcup_{j<i} L_j.$$

In human language, L_i consists of those vertices that are endpoints of directed edges starting in L_{i-1} but do not already occur in the previously defined sets L_j. The L_i are called the *levels* of the breadth-first-search tree. The algorithm stops when $L_m = \emptyset$ for some m. Then $\mathcal{O} := \bigcup_{j<m} L_j$ is the connected component of $D(\Omega, \vec{E})$ containing α. Note that since $\mathcal{O}^s = \mathcal{O}$ for all generators $s \in S$, we also have $\mathcal{O}^g = \mathcal{O}$ for all $g \in G$ since any $g \in G$ can be written as a product of elements of S. Hence indeed $\mathcal{O} = \alpha^G$.

As an example, let $\Omega = [1, 5] = \{1, 2, 3, 4, 5\}$, $G = \langle S \rangle = \langle s_1, s_2 \rangle \leq \mathrm{Sym}(\Omega)$ with $s_1 = (1, 2)(4, 5)$ and $s_2 = (1, 4, 5)$, and let $\alpha = 4$. The orbit algorithm starts with setting $L_0 := \{4\}$. Then we compute the set $L_0^S = \{4^{s_1}, 4^{s_2}\} = \{5\}$ and $L_1 := \{5\} \backslash L_0 = \{5\}$. Next, we compute $L_1^S = \{5^{s_1}, 5^{s_2}\} = \{1, 4\}$ and $L_2 := \{1, 4\} \backslash (L_0 \cup L_1) = \{1\}$. In the next step, $L_2^S = \{1^{s_1}, 1^{s_2}\} = \{2, 4\}$ and $L_3 :=$

$\{2,4\}\backslash(L_0 \cup L_1 \cup L_2) = \{2\}$ are computed. Finally, $L_3^S = \{2^{s_1}, 2^{s_2}\} = \{1,2\}$ and $L_4 := \{1,2\}\backslash(L_0 \cup L_1 \cup L_2 \cup L_3) = \emptyset$ are computed. Since $L_4 = \emptyset$, the algorithm terminates and outputs the orbit $4^G = L_0 \cup L_1 \cup L_2 \cup L_3 = \{1,2,4,5\}$.

In implementations, we collect the elements of the sets L_i in a list U. We define $U[1] := \alpha$ and, successively for each element $\beta \in U$, compute β^g for all $g \in S$. If β^g is not already in U then we add β^g to U. The algorithm terminates when all elements of U are processed.

The timing of the algorithm depends on $|\alpha^G|$. We have to compute the image of each $\beta \in \alpha^G$ under each $g \in S$, and so the total number of image computations is $|\alpha^G||S|$. The cost of an image computation depends heavily on the actual representation of G. If the black-box group G is a permutation group then computing β^g for some $\beta \in \Omega$ and $g \in G$ is the basic operation whose time requirement is the unit in running time estimates; in the other two examples mentioned earlier, an image computation amounts to performing some group operations to compute a conjugate of a group element or the conjugates of a list of generators.

Also, we have to decide whether an image β^g just computed already occurs in the list U. As with image computations, the cost of this operation varies according to the actual representation of the black-box group and Ω. We consider three possible situations. In the first one, a superset Δ of manageable size is known for α^G and the position of elements of α^G in Δ can be determined easily. This occurs, for example, when G is a permutation group acting on Ω: We may simply choose $\Delta := \Omega$. In the second situation, no such superset Δ is available. The third situation is a special case of the second: There is no superset Δ at our disposal, but a linear ordering of Ω is defined and we can compare any two elements of Ω in this ordering. This latter situation occurs, for example, when $\Omega = G$ and we have to compute the conjugacy class of some $\alpha \in G$, and the elements of the black-box group G have unique representations as strings. In this case, we may simply consider the lexicographic ordering of strings as a linear ordering of Ω.

Theorem 2.1.1. *Suppose that a group $G = \langle S \rangle$ acts on a set Ω. Then the orbit of some $\alpha \in \Omega$ can be computed using $O(|\alpha^G||S|)$ image computations. Moreover,*

(i) *if a superset $\Delta \supseteq \alpha^G$ is known in advance and it is possible to allocate $|\Delta|$ units of memory then the image computations dominate the running time;*
(ii) *if no such set Δ is available, the additional cost is the comparison of $O(|\alpha^G|^2|S|)$ pairs in Ω;*

(iii) *if comparison of elements also recognizes which comes first in a linear ordering on* Ω*, then the additional cost drops to the comparison of* $O(|\alpha^G| \log^2 |\alpha^G||S|)$ *pairs.*

Proof. (i) We define an auxiliary list L of size $|\Delta|$. Each time an element is added to U, we also mark the corresponding position in L. Thus, when we have to decide whether some newly computed image $\gamma = \beta^g$ is already in U, it is enough to look up whether the position of γ in L is already marked.

(ii) If no such superset Δ is available, we may compare a newly computed image β^g to the already defined elements of U. This results in an algorithm with $O(|\alpha^G|^2|S|)$ comparisons of elements of Ω.

(iii) Suppose that there is a linear ordering on Ω. If, besides the list U, we maintain a list V containing a linear ordering of U then it is possible to decide whether some $\gamma \in \Omega$ already occurs in U by a binary search of V, which is fast; however, if $\gamma \notin U$ then the insertion of γ into the list V is potentially very expensive, because we have to shift by one position all elements of the list that are greater than γ. Therefore, instead of a long list V containing a linear ordering of U, we maintain shorter lists $V_1, V_2, \ldots$ that contain linear orderings of certain segments of U. Namely, if the binary expansion of $|U|$ is $|U| = 2^{i_1} + 2^{i_2} + \cdots + 2^{i_k}$ with $i_1 > i_2 > \cdots > i_k$, then V_j contains the linear ordering of the 2^{i_j} elements of U in positions $U[l+1], U[l+2], \ldots, U[l+2^{i_j}]$, for $l = 2^{i_1} + \cdots + 2^{i_{j-1}}$. Deciding whether some $\gamma \in \Omega$ already occurs in U can be done by binary searches of all V_j, requiring $O(i_1 + \cdots + i_k) = O(\log^2 |U|)$ comparisons.

If some $\gamma \in \Omega$ is added to U, the linear orderings V_j can be updated in the following way. We create a new set $V_{k+1} := \{\gamma\}$ and, starting with $j := k$ and proceeding recursively through decreasing values of j, we merge the linear orders V_j and V_{j+1} if $|V_j| = |V_{j+1}|$. We stop when we encounter the first j such that $|V_j| > |V_{j+1}|$. For example, if $|U| = 55$ then adding an element to U triggers three merges, since $55 = 32 + 16 + 4 + 2 + 1$ and $55 + 1 = 32 + 16 + 4 + 2 + (1 + 1) = 32 + 16 + 4 + (2 + 2) = 32 + 16 + (4 + 4) = 32 + 16 + 8$.

We made a total of $O(|\alpha^G| \log^2 |\alpha^G||S|)$ comparisons while deciding whether we should add some elements of Ω to U. In addition, we must make comparisons when we merge some sets V_j and V_{j+1}. We claim that during the entire orbit computation, the total number of comparisons at updating the V_j is $O(|\alpha^G| \log |\alpha^G|)$. This claim follows from the observation that for each fixed $\beta \in U$, if the set V_j containing β is merged then the new set containing β is twice the size of the original. Hence, β is processed $\log |\alpha^G|$ times. Because $|\alpha^G| \log |\alpha^G| \in O(|\alpha^G| \log^2 |\alpha^G||S|)$, we are done. □

Remark 2.1.2. Another possibility for searching the list U is to use *hash functions*. A hash function is a function $h : \Omega \to K$ for some set K; in particular, h computes a value associated to each element of U, and accordingly the elements of U can be stored in $|K|$ sets. When we have to decide whether a new $\omega \in \Omega$ is already in U, it is enough to compare ω with previous elements that have the same hash value $h(\omega)$.

The efficiency of the method depends on finding a hash function that is easily computable, and relatively few elements of U will get the same hash value. Among elements with the same hash value, one of the methods described in Theorem 2.1.1 may be applied, but hopefully on an input of much smaller size.

Finding good hash functions is the subject of extensive research (see [Knuth, 1973, Chap. 6]). Here we only mention that if the orbit U is expected to be a "random" subset of Ω then choosing a few random bits works well in practice. For example, if G is a matrix group of dimension d over GF(2), $\Omega = \mathrm{GF}(2)^d$, and we need the orbit of some vector $\alpha \in \Omega$ then we may try the hash function that retrieves, for example, the middle ten coordinates of vectors.

Remark 2.1.3. Using more sophisticated data structures (e.g., 2–3 trees; cf. [Sedgewick, 1988, Chap. 15]), the number of comparisons in Theorem 2.1.1(iii) can be decreased to $O(|\alpha^G| \log |\alpha^G||S|)$.

Often the information that some $\gamma \in \Omega$ occurs in α^G is not enough; we also need an element $h \in G$ such that $\alpha^h = \gamma$. Such elements can be obtained during the orbit computation. For this purpose, besides U, we maintain a list W of group elements such that for all $i \leq |U|$, $\alpha^{W[i]} = U[i]$. When a new image β^g is added to U for some $\beta = U[i]$ and $g \in S$, we define $W[|U| + 1] := W[i]g$.

The additional cost is that of $|\alpha^G|$ group multiplications and, often more restrictively, the storage of $|\alpha^G|$ group elements. Hence, in practice, if β^g is added to U then we store in $W[|U| + 1]$ only (a pointer to) $g \in S$. This method saves storage, and we do not even have to perform group multiplications during the orbit computation. We pay the price when an element $h \in G$ with $\alpha^h = \gamma$ is actually needed. Using W, we have to trace back the preimages of γ until we reach α and compute the product of generators encountered at the tracing. Note that in the version mentioned in Theorem 2.1.1(i), the array of length $|\Delta|$ can be used to store pointers to generators, instead of our having to define an additional array W.

The orbit algorithms can also be used to find the permutation action of the generators of G on α^G. For all $\gamma \in \alpha^G$ and $s \in S$, we have computed γ^s and identified it with an element of α^G. Keeping track of these identifications, we obtain the desired permutation actions.

Finally, we remark that for any $A \subseteq \Omega$, the orbit algorithm can be used to compute $A^G := \{\alpha^g \mid \alpha \in A, g \in G\}$. The only modification needed is to initialize $U := A$.

2.1.2. Closure of Algebraic Structures

An important variant of orbit computations is the situation when the set Ω on which the group G acts has an algebraic structure. Given $A \subseteq \Omega$, we have to determine $\langle A^G \rangle$, the smallest subset of Ω containing A that is closed for the G-action *and* for the operation(s) of the algebraic structure on Ω. The most important example occurring in this book is when $\Omega = G$ and we have to compute the normal closure of some $\langle A \rangle \leq G$ or, slightly more generally, when G acts on a group Ω and we need the G-closure of some subgroup $\langle A \rangle \leq \Omega$.

As customary when the output of an algorithm is closed for some algebraic operations, we seek only *generators* for $\langle A^G \rangle$. This enables us to handle objects that are too large to have all their elements listed.

In this section, we present a version that supposes that we are able to *test membership* in the already constructed substructures of $\langle A^G \rangle$. Versions without this assumption (using randomized algorithms) will be given in Sections 2.3 and 2.4.

The Closure Algorithm

Given $G = \langle S \rangle$ and $A \subseteq \Omega$, generators for the G-closure $\langle A^G \rangle$ can be obtained by a slight modification of the orbit algorithm. We collect the generators for $\langle A^G \rangle$ in a list U. We define $U := A$ and, successively for each element $h \in U$, compute h^g for all $g \in S$. If h^g is not in the algebraic structure $\langle U \rangle$ then we add h^g to U. The algorithm terminates when all elements of U are processed. At that moment, $\langle U \rangle$ contains A and is closed for the action of G on Ω. Also, each element of U is of the form x^y for some $x \in A$ and $y \in G$; therefore $\langle U \rangle = \langle A^G \rangle$.

Note that each time an element was added to U, the algebraic structure $\langle U \rangle$ increased. Hence, if Ω has at least a group structure (which will be the case in all applications) then Lagrange's theorem implies that $|\langle U \rangle|$ increased by an integer factor and so the number of generators added to the generator list U for $\langle A^G \rangle$ is at most $\log |\langle A^G \rangle|$. The time-critical part of the algorithm is usually the computation that allows us to test membership in $\langle U \rangle$.

Algorithms Based on Normal Closure

As a consequence, we obtain algorithms for tasks that are based on normal closure computations. In particular, we can compute the derived series and lower central series in a group and test solvability and nilpotence. Also, it can be determined whether a given $H \leq G$ is subnormal in G. Recall that H is subnormal in G if and only if the series of subgroups defined by $G_0 := G$, $G_{i+1} := \langle H^{G_i} \rangle$ for $i \geq 0$ reaches H. Note, however, the difference in difficulty for these tasks when no membership test is available. Because normal closure can be computed by randomized algorithms, in each case we can construct generators for the required series of subgroups. However, at testing solvability and nilpotence we have to decide only whether a group in the subgroup chain is trivial, whereas at testing subnormality we have to check whether some subgroup we constructed is contained in H. The latter task does not seem possible without membership testing.

2.2. Random Elements of Black-Box Groups

Randomized algorithms for groups frequently call for random elements. In permutation groups, the situation is quite satisfactory *after* the basic data structure bases and strong generators (cf. Chapter 4) are constructed: We can easily construct uniformly distributed, independent random elements of the group, as a product of randomly chosen coset representatives from the transversals along the point stabilizer chain defined by the base and strong generating set. Choosing a random element from a transversal amounts to the choice of a uniformly distributed random integer in an interval, because we have to specify a position in the list of transversal elements. This is possible if independent, uniformly distributed random bits are available: To choose an element from $[1, k]$, we evaluate a sequence of $\lceil \log k \rceil$ random bits as a number in base 2. In this book, we take the availability of random bits for granted, although generating random numbers is a large and important area of computer science (see [Knuth, 1969, Chap. 3]).

The situation is more complicated if we need random elements in a permutation group without a strong generating set. The primary example for this phenomenon is the construction of a strong generating set itself, but this need may arise in other situations as well: We may need random elements of certain subgroups and we do not want to spend time to construct strong generators for these subgroups. In very special situations (cf. Exercises 2.1 and 2.2) we may be able to construct uniformly distributed random elements, but in general, we settle for less: We use a heuristic or try to obtain *nearly uniformly distributed* random elements.

We say that an algorithm outputs an ε-uniformly distributed element x in a group G if $(1-\varepsilon)/|G| < \text{Prob}(x = g) < (1+\varepsilon)/|G|$ for all $g \in G$. Nearly uniformly distributed means ε-uniformly distributed for some $\varepsilon < 1/2$.

The construction of nearly uniformly distributed elements is even more important in matrix groups, where no practical alternative of strong generating sets is known. It turns out that all methods proposed so far for obtaining nearly uniformly distributed elements in matrix groups work in the black-box group setting; they only multiply and invert already available group elements.

We shall describe three methods for constructing random elements in black-box groups. For this description, we need to define the notion of Markov chains. A *Markov chain* M is a sequence $(x_0, x_1, \ldots)$ of random elements from a set $V = \{v_1, \ldots, v_n\}$. The elements of V are called the *states* of the Markov chain. The Markov chain is specified by a vector $\mathbf{a} = (a_1, \ldots, a_n)$ with nonnegative entries and $\sum_j a_j = 1$ and by an $n \times n$ matrix $P = (p_{ij})$ with nonnegative entries and unit row sums. The vector $\mathbf{a}$ describes the distribution of the initial element x_0, namely, $\mathrm{Prob}(x_0 = v_i) = a_i$. The matrix P contains the *transition probabilities*; the entry p_{ij} is the conditional probability that $x_{m+1} = v_j$, provided that $x_m = v_i$.

The matrix P defines a directed graph $\mathcal{X}$ on the vertex set V : $\overrightarrow{(v_i, v_j)}$ is an edge if and only if $p_{ij} > 0$. The Markov chain M can be interpreted as a random walk on $\mathcal{X}$, with starting point x_0. If the random walk arrived at v_i after m steps then the probability that it continues with the edge $\overrightarrow{(v_i, v_j)}$ is p_{ij}.

We defined only Markov chains with a finite number of states and stationary transition probabilities (i.e., the matrix P did not depend on m). The higher transition probabilities $p_{ij}^{(k)}$ are defined as the conditional probabilities $p_{ij}^{(k)} := \mathrm{Prob}(x_{m+k} = v_j \mid x_m = v_i)$. It is easy to see that these are independent of m (Exercise 2.3). We say that the state v_j can be reached from v_i if $p_{ij}^{(k)} > 0$ for some $k \geq 0$. The Markov chain is called *irreducible* if any state can be reached from any state.

We say that the state v_i has *period* $t > 1$ if $p_{ii}^{(k)} = 0$ for all k that are not multiples of t, and t is the greatest number with this property. If no period $t > 1$ exists then v_i is called *aperiodic*. In an irreducible Markov chain, all states are aperiodic or all of them have the same period (Exercise 2.4).

Let $\mathbf{u} = (u_1, \ldots, u_n)$ be a vector with positive coordinates and satisfying $\sum_j u_j = 1$. We say that $\mathbf{u}$ is the *stationary distribution* for M if $\mathbf{u}P = \mathbf{u}$. The major result we need concerning Markov chains is the following well-known theorem (see, e.g., [Feller, 1968]).

Theorem 2.2.1. *Let M be a finite state, irreducible, and aperiodic Markov chain with n states. Then there exists a stationary distribution vector $(u_1, \ldots, u_n)$ for M and a constant $\delta \in (0, 1)$ such that for all states v_i and $m \geq 0$, we have* $|\mathrm{Prob}(x_m = v_i) - u_i| \leq \delta^m$.

Random Elements as Products of Generators

One method for constructing random elements in a black-box group $G = \langle S \rangle$ is to take products $x = s_1 s_2 \cdots s_k$, where each s_i is a randomly chosen element of $S \cup S^{-1}$, and where k is "sufficiently large." Such an x corresponds to a random walk of length k on the Cayley graph $\Gamma(G, S)$, starting from the identity element of G. At each vertex v_i of the Cayley graph, the random walk chooses the next vertex uniformly among the neighbors of v_i. The transition probability matrix P of the Markov chain M associated with this random walk has entries

$$p_{g,h} := \begin{cases} 1/d, & \{g, h\} \in \mathcal{E}(\Gamma(G, S)) \\ 0, & \text{otherwise}, \end{cases} \tag{2.1}$$

where d is the common valency of the vertices of $\Gamma(G, S)$. Because $\Gamma(G, S)$ is connected, this Markov chain is irreducible. We have $p^{(2)}_{g,g} > 0$ for all $g \in G$, because we can walk to any neighbor of g and then back to g in two steps; in fact, $p^{(2)}_{g,g} = 1/d$. Hence M is aperiodic, or each vertex has period 2. In fact, period 2 is possible if $\Gamma(G, S)$ is bipartite; so, to obtain an aperiodic Markov chain M^*, we consider *lazy random walks* on the Cayley graph. At each step, the walk stays at its end vertex v with probability $1/2$, or it continues to a neighbor u of v with probability $1/2d$. The transition probability matrix P^* has entries

$$p^*_{g,h} := \begin{cases} 1/2d, & \{g, h\} \in \mathcal{E}(\Gamma(G, S)) \\ 1/2, & g = h \\ 0, & \text{otherwise}. \end{cases}$$

Since $p^*_{g,g} > 0$ for all $g \in G$, the Markov chain M^* is aperiodic. Staying at a vertex during the random walk means inserting the identity element of G in the product $g = s_1 s_2 \cdots s_k$, so the computation of random elements using M and M^* is actually the same. The difference is the length of products we consider: It is the predetermined number k in M, but it is a random variable in M^*.

Since M^* is irreducible and aperiodic, Theorem 2.2.1 applies. The stationary distribution vector of M^* has entries $1/|G|$ in each coordinate (cf. Exercise 2.5), so x converges to the uniform distribution on G. Since $|\text{Prob}(x = g) - 1/|G|| < \delta^k$ for some $0 < \delta < 1$, stopping after k steps for $k > (1 + \log|G|)/\log(1/\delta)$ means that we obtain a nearly uniformly distributed element of G. The problem with this estimate is that although we have an upper bound for $|G|$ (we can always use $|G| < N|Q|^N$ if the black-box group is encoded by strings of length at most N over the alphabet Q), usually we do not have any idea about δ. The Perron–Frobenius theorem implies that 1 is an eigenvalue of the transition probability matrix P^*, and all other eigenvalues have absolute value less than 1. The value of δ is roughly the second largest among the absolute values of eigenvalues, but obtaining good estimates for this absolute value in an arbitrary

Cayley graph is quite difficult. Also, it is possible that the difference between 1 and the second largest absolute value of an eigenvalue, the *eigenvalue gap* of P^*, is $O(1/|G|)$, which means slow convergence to the uniform distribution. Examples for slow convergence are abelian groups with large cyclic subgroups or, more generally, groups with large abelian factor groups. Ideally, we would like an eigenvalue gap polynomial in $1/\log|G|$, as this means that we can construct a nearly uniformly distributed group element as a product of length bounded by a polynomial of the input length.

If an algorithm needs a sequence of random elements, a practical solution is to construct a random walk of length k_0, for a large enough k_0 with which the user feels comfortable; after that, we continue the random walk and output the sequence of vertices that are added to the walk.

The Product Replacement Algorithm

Another method for generating random elements is the *product replacement algorithm* from [Celler et al., 1995]. Let $(x_1, \ldots, x_m)$ be a sequence of elements of G that generate G. We pick an ordered pair $(i, j), i \neq j$ from the uniform distribution of pairs with $1 \leq i, j \leq m$, and we replace x_i either by the product $x_i x_j$ or by $x_i x_j^{-1}$. The resulting new sequence still generates G, because we can recover x_i as the product or ratio of two elements of the new sequence. Product replacement corresponds to a Markov chain M where the set V of states is the set of m-long generating sequences of G. From each state, we can proceed in $2m(m-1)$ ways to a new state, but it is possible that choosing different pairs (i, j) or a different exponent in the product $x_i x_j^{\pm 1}$ carries us to the same new state. Hence the positive transition probabilities may not have a common value, as in (2.1).

Our goal is to show that, under some mild assumptions, the Markov chain M associated with the product replacement algorithm is irreducible and aperiodic (cf. Theorem 2.2.3). In the proof, we need the following simple lemma.

Lemma 2.2.2. *Let $X = (x_1, \ldots, x_m)$ and $Y = (y_1, \ldots, y_m)$ be two states of M that differ only in the kth coordinate (i.e., $x_i = y_i$ for all $i \neq k$). Suppose further that $X \backslash \{x_k\}$ also generates G. Then Y can be reached from X.*

Proof. Let $z := x_k^{-1} y_k$, and write z as a product $z = x_{i_1} \cdots x_{i_l}$ of elements from $X \backslash \{x_k\}$. Starting with X and replacing the element in position k by the product of elements in position k and i_j for $j = 1, \ldots, l$ takes X to Y. □

Theorem 2.2.3. *Let m_1 be the minimal size of a generating set of G and let m_2 be the maximal size of a nonredundant generating set. If $m \geq m_1 + m_2$*

then the Markov chain M associated with the product replacement algorithm is irreducible and aperiodic.

Proof. Let $X = (x_1, \ldots, x_{m_1}, 1, \ldots, 1) \in V$ be fixed, where $\{x_1, \ldots, x_{m_1}\}$ is a generating set of minimal size of G. First, we prove that the period of the state X is 1. If G is not cyclic of prime order then $m_2 \geq 2$ and so the transition probability $p_{X,X}$ is positive, since we can replace the last coordinate of X by the product of the last two coordinates. If G is cyclic of prime order then $X = (x, 1)$ for some $x \in G\backslash\{1\}$. The sequence $((x, 1), (x, x), (x, 1))$ shows that $p^{(2)}_{X,X} > 0$ and the sequence $((x, 1), (x, x), (x^2, x), (x^2, x^{-1}), (x, x^{-1}), (x, 1))$ shows that $p^{(5)}_{X,X} > 0$; therefore, the period of X is 1 in this case as well. Because the period of X is 1, if we prove that any state $Y \in V$ can be reached from X and X can be reached from Y then irreducibility and aperiodicity follow immediately.

First we show that if X^* is a permutation of the sequence X then X and X^* can be reached from each other. It is enough to prove this if X^* is obtained by a single transposition from X, since any permutation is a product of transpositions. If we want to exchange some x_i, x_j with $i \leq m_1 < j$ then we apply Lemma 2.2.2 to change the jth coordinate of X to x_i, and then, by the same lemma, change the ith coordinate to 1. If we want to exchange some x_i, x_j with $i, j \leq m_1$ then first we exchange x_i with x_k for some $k > m_1$, then we exchange positions i and j (we can do this because now we have the identity element in position i), and finally we exchange positions j and k.

Let now $Y = (y_1, \ldots, y_m)$ be an arbitrary state, and let $(y_{i_1}, \ldots, y_{i_l})$ be a subsequence of Y containing a nonredundant generating set. Then $l \leq m_2$. We can reach Y from X by first reaching a permutation X^* of X where the nontrivial coordinates of X^* comprise a subset of the positions $[1, m]\backslash\{i_1, \ldots, i_l\}$, then applying Lemma 2.2.2 to change positions $i_1, \ldots, i_l$ of X^* to $y_{i_1}, \ldots, y_{i_l}$, and then applying Lemma 2.2.2 again to change the entries in the positions $[1, m]\backslash \{i_1, \ldots, i_l\}$ to the values y_j. The reversal of these steps reaches X from Y. □

Each element of V can be obtained in $2m(m-1)$ ways as a result of a product replacement. Therefore, the transition probability matrix has constant column sums and so the stationary distribution of M is uniform on the m-element generating sets of G. As in the case of random walks on Cayley graphs, the problem is that we cannot estimate the parameter δ, which shows the speed of the convergence to the stationary distribution. Note that because the number of states in this Markov chain is much larger than $|G|$, an eigenvalue gap proportional to $|V|$ would be disastrous. An additional difficulty is that we need random *elements* of G, not random generating sets. The practical solution is to output a random element of the generating set created by the product replacement algorithm. If a sequence of random elements is needed then we perform k_0 steps of product replacement to initialize a random generating set, and at each further product replacement we output the element that is inserted into the new generating set. Hence, after initialization, random elements are created with the cost of one group operation. Note that the elements of G are not necessarily distributed uniformly in generating sets (the case of cyclic groups of nonprime order is simple enough to analyze).

Despite the difficulties mentioned in the previous paragraph, the product replacement algorithm performs amazingly well in practice. In matrix groups of dimension in the low hundreds, the GAP implementation uses about $k_0 = 100$ product replacements to initialize. After that, the distribution of the sequence of random elements output by the algorithm seems to be close to the distribution in the entire group with respect to the properties most frequently needed in matrix group algorithms. [Celler et al., 1995] contains statistical analysis of the output in some linear groups and sporadic groups. Serious efforts have been made to estimate the rate of convergence in product replacement (cf. the survey [Pak, 2001] and its references).

The third method for generating random elements is from [Babai, 1991]. We cite the following theorem without proof.

Theorem 2.2.4. *Let $c, C > 0$ be given constants, and let $\varepsilon = K^{-c}$ where K is a given upper bound for the order of a group G. There is a Monte Carlo algorithm, that, given any set S of generators of G, sets up a data structure for the construction of ε-uniformly distributed elements at a cost of $O(\log^5 K + |S| \log\log K)$ group operations. The probability that the algorithm fails is at most K^{-C}.*

If the algorithm succeeds, it permits the construction of ε-uniformly distributed random elements of G at a cost of $O(\log K)$ group operations per random element.

The algorithm mentioned in Theorem 2.2.4 has not been implemented, because $O(\log^5 K)$ group operations are prohibitively expensive in practice. This is especially true in the case of matrix groups, which is the potentially most important application of black-box group techniques. The significance of Theorem 2.2.4 is in theoretical investigations: It allows, in polynomial time in the input length, the construction of random elements that are *provably* nearly uniformly distributed.

In certain situations, both in theoretical investigations and in practice, we can avoid the use of random elements by applying *random subproducts*, which can emulate some properties of truly random elements. This is the topic of our next section.

2.3. Random Subproducts

2.3.1. Definition and Basic Properties

A *random subproduct* of a sequence of group elements $(g_1, g_2, \dots, g_k)$ is a product of the form $g_1^{\varepsilon_1} g_2^{\varepsilon_2} \cdots g_k^{\varepsilon_k}$, where the ε_i are chosen independently from the uniform distribution over $\{0, 1\}$. Slightly more generally, a random subproduct of *length* l is defined by selecting a random subset of size l of the g_i and forming a random subproduct of them. In some applications, we use the ordering inherited from the sequence $(g_1, g_2, \dots, g_k)$; in other ones, we also construct a random ordering of the chosen l-element subset.

Random subproducts were introduced in [Babai et al., 1988] to speed up certain orbit computations. (The journal version of this paper is [Babai et al., 1997b].)

Lemma 2.3.1. *Suppose that* $G = \langle S \rangle$ *acts on a set* Ω *and let* $\Delta \subseteq \Omega$. *Let* g *be a random subproduct of the elements of* S *(in any fixed ordering of* S*). If* Δ *is not closed for the action of* G *then* $\mathrm{Prob}(\Delta^g \neq \Delta) \geq 1/2$.

Proof. Let $S = \{g_1, \dots, g_k\}$, $g = g_1^{\varepsilon_1} g_2^{\varepsilon_2} \cdots g_k^{\varepsilon_k}$, and $p_i = g_1^{\varepsilon_1} g_2^{\varepsilon_2} \cdots g_i^{\varepsilon_i}$ for $0 \leq i \leq k$. If Δ is not closed for the G-action then $\Delta^{g_i} \neq \Delta$ for some $g_i \in S$. Let us consider the largest index i with this property. If p_{i-1} fixes Δ then p_i does not fix Δ with probability $1/2$, since if $\varepsilon_i = 1$ then $\Delta^{p_i} = \Delta^{p_{i-1} g_i} = \Delta^{g_i} \neq \Delta$. The other case is that p_{i-1} does not fix Δ. Then with probability at least $1/2$ neither does p_i, since if $\varepsilon_i = 0$ then $\Delta^{p_i} = \Delta^{p_{i-1}} \neq \Delta$. Furthermore, if $\Delta^{p_i} \neq \Delta$ then $\Delta^g \neq \Delta$, since all g_j, $j > i$, fix Δ. Formally,

$$\begin{aligned}\mathrm{Prob}(\Delta^g \neq \Delta) &= \mathrm{Prob}\big(\Delta^{g_1^{\varepsilon_1} g_2^{\varepsilon_2} \cdots g_i^{\varepsilon_i}} \neq \Delta\big)\\ &\geq \mathrm{Prob}\big(\varepsilon_i = 1 | \Delta^{g_1^{\varepsilon_1} \cdots g_{i-1}^{\varepsilon_{i-1}}} = \Delta\big)\mathrm{Prob}\big(\Delta^{g_1^{\varepsilon_1} \cdots g_{i-1}^{\varepsilon_{i-1}}} = \Delta\big)\\ &\quad + \mathrm{Prob}\big(\varepsilon_i = 0 | \Delta^{g_1^{\varepsilon_1} \cdots g_{i-1}^{\varepsilon_{i-1}}} \neq \Delta\big)\mathrm{Prob}\big(\Delta^{g_1^{\varepsilon_1} \cdots g_{i-1}^{\varepsilon_{i-1}}} \neq \Delta\big) = 1/2.\end{aligned}$$

(As usual, $\mathrm{Prob}(A|B)$ denotes conditional probability.) □

The systematic development of the random subproduct method started in 1991, in the conference proceedings version of [Babai et al., 1995]. Most applications are based on the following corollary of Lemma 2.3.1.

Lemma 2.3.2. *Suppose that* $G = \langle S \rangle$ *and* $K \lneq G$*. Let* g *be a random subproduct of the elements of* S*. Then* $\mathrm{Prob}(g \notin K) \geq 1/2$.

Proof. G acts transitively (by right multiplication) in its regular representation on $\Omega := G$. The subset $\Delta := K$ of Ω is not closed for this G-action; hence, by Lemma 2.3.1, $\mathrm{Prob}(K^g \neq K) \geq 1/2$. (Note that, in this context, $K^g = \{kg \mid k \in K\}$, *not* the conjugate of K by g.) To finish the proof, observe that $K^g \neq K$ implies that $g \notin K$. □

Numerous algorithms that use uniformly distributed random elements exploit the fact that, if K is a proper subgroup of G and g is a random element of G, then $\mathrm{Prob}(g \notin K) = 1 - 1/|G : K| \geq 1/2$. One of the applications of random subproducts is to provide an efficient alternative in these algorithms via Lemma 2.3.2.

The applications of Lemma 2.3.2 use the following technical lemma from [Babai et al., 1995]. The proof we give is by A. Zempléni.

Lemma 2.3.3. *Let* $X_1, X_2, \ldots$ *be a sequence of* $0-1$ *valued random variables such that* $\mathrm{Prob}(X_i = 1) \geq p$ *for any values of the previous* X_j *(but* X_i *may depend on these* X_j*). Then, for all integers* t *and* $0 < \varepsilon < 1$,

$$\mathrm{Prob}\left(\sum_{i=1}^{t} X_i \leq (1-\varepsilon)pt\right) \leq e^{-\varepsilon^2 pt/2}.$$

Proof. The proof is based on *Chernoff's bound* (cf. [Chernoff, 1952]). This states that if $Y_1, Y_2, \ldots$ are independent Bernoulli trials (coin flippings) with $\mathrm{Prob}(Y_i = 1) = p$ then

$$\mathrm{Prob}\left(\sum_{i=1}^{t} Y_i \leq (1-\varepsilon)pt\right) \leq e^{-\varepsilon^2 pt/2}. \tag{2.2}$$

Hence, to prove the lemma, it is enough to show that for all integers k, t,

$$\mathrm{Prob}\left(\sum_{i=1}^{t} X_i \geq k\right) \geq \mathrm{Prob}\left(\sum_{i=1}^{t} Y_i \geq k\right). \tag{2.3}$$

We prove (2.3) by induction on t. The initial case is obvious. Supposing (2.3) for t, we have

$$\begin{aligned}
\text{Prob}\left(\sum_{i=1}^{t+1} X_i \geq k\right) &= \text{Prob}\left(\sum_{i=1}^{t} X_i \geq k\right) \\
&\quad + \text{Prob}\left(X_{t+1} = 1 \,\middle|\, \sum_{i=1}^{t} X_i = k-1\right) \text{Prob}\left(\sum_{i=1}^{t} X_i = k-1\right) \\
&\geq \text{Prob}\left(\sum_{i=1}^{t} X_i \geq k\right) + p\,\text{Prob}\left(\sum_{i=1}^{t} X_i = k-1\right) \\
&= p\,\text{Prob}\left(\sum_{i=1}^{t} X_i \geq k-1\right) + (1-p)\,\text{Prob}\left(\sum_{i=1}^{t} X_i \geq k\right) \\
&\geq p\,\text{Prob}\left(\sum_{i=1}^{t} Y_i \geq k-1\right) + (1-p)\,\text{Prob}\left(\sum_{i=1}^{t} Y_i \geq k\right). \qquad (2.4)
\end{aligned}$$

We used the inductive hypothesis in the last inequality. Doing the steps of (2.4) backward using the Y_i, and noting that the independence of the Y_i means that we have equality at each step, we obtain

$$p\,\text{Prob}\left(\sum_{i=1}^{t} Y_i \geq k-1\right) + (1-p)\,\text{Prob}\left(\sum_{i=1}^{t} Y_i \geq k\right) = \text{Prob}\left(\sum_{i=1}^{t+1} Y_i \geq k\right),$$

which finishes the proof of (2.3). □

Primarily, we shall apply Lemma 2.3.3 in the following situation. Suppose that we know an upper bound l_H for the length of subgroup chains in a group H and we have a sequence of elements $(x_1, \ldots, x_s)$ from H with the following property: For all $i \leq s$, if $\langle x_1, \ldots, x_{i-1}\rangle \neq H$ then $\text{Prob}(x_i \notin \langle x_1, \ldots, x_{i-1}\rangle) \geq p$. Then, if we define $X_i = 0$ if and only if $\langle x_1, \ldots, x_{i-1}\rangle \neq H$ and $x_i \in \langle x_1, \ldots, x_{i-1}\rangle$, and $X_i = 1$ otherwise, then the condition $\sum X_i \geq l_H$ means that *the x_is generate H*. Indeed, if $\sum X_i \geq l_H$ and $\langle x_1, \ldots, x_s\rangle \neq H$ then the subgroup chain $1 \leq \langle x_1\rangle \leq \langle x_1, x_2\rangle \leq \cdots \leq \langle x_1, \ldots, x_s\rangle \leq H$ increases strictly more than l_H times, which is a contradiction.

We shall refer to such applications of Lemma 2.3.3 as applications of *basic type*. In some cases, we need more complicated variants of the method; here we wanted to present only the basic idea, without including too many hypotheses.

2.3.2. Reducing the Number of Generators

As a first application of random subproducts, we describe algorithms to handle the following situation. Given a black-box group $G = \langle S \rangle$, suppose that an upper bound l_G is known for the length of subgroup chains in G. (If the elements of G are encoded as strings of length at most N over an alphabet Q then we have $|G| \leq N|Q|^N$, and so $l_G \leq \log(N|Q|^N)$.) If $|S| > l_G$ then the generating set S is redundant; if $|S|$ is much larger than l_G then we want to find a smaller generating set.

If membership testing is possible in subgroups of G then the reduction of the generating set can be done by a deterministic algorithm. Namely, if $S = \{g_1, \ldots, g_k\}$ then, recursively for $i = 1, 2, \ldots,$ determine whether $g_{i+1} \in \langle g_1, \ldots, g_i \rangle$. If so, then g_{i+1} is discarded. The remaining elements of S form a generating set of size at most l_G.

If the case when no membership testing is available, there are Monte Carlo algorithms that, with high probability, construct a small generating set.

Lemma 2.3.4. *Suppose that a generating set S and an upper bound l_G for the length of subgroup chains for a group G are given, and let $\delta > 0$ be arbitrary. Then there is a Monte Carlo algorithm that, with probability at least $1 - \delta$, constructs a generating set T of size $O(l_G)$, using $O(|S| l_G)$ group operations.*

Proof. Let the set T consist of cl_G random subproducts made from S, where the constant c depends on the desired error probability. We claim that

$$\text{Prob}(\langle T \rangle = G) \geq 1 - e^{-\left(1-\frac{2}{c}\right)^2 cl_G/4}. \tag{2.5}$$

Let $r_1, r_2, \ldots, r_{cl_G}$ denote the elements of T. If $\langle r_1, \ldots, r_{i-1} \rangle \neq G$ then, by Lemma 2.3.2, $r_i \notin \langle r_1, \ldots, r_{i-1} \rangle$ with probability at least $1/2$. Hence an application of Lemma 2.3.3 of basic type with the parameters $\varepsilon = 1 - 2/c$, $p = 1/2$, and $t = cl_G$ proves (2.5). □

Remark 2.3.5. As is clear from (2.5), to achieve error probability less than δ we may choose the constant $c := \max\{4, 16\ln(\delta^{-1})/l_G\}$. Another interpretation is that we construct a new generating set of size $4l_G$, with the reliability of the algorithm improving exponentially as a function of l_G. In a similar way, we can compute explicit values for the constants in all algorithms presented in Sections 2.3.3 and 2.3.4 as well. In Remark 2.3.13, we shall comment further on the error probability in our algorithms.

The following theorem from [Babai et al., 1995] also constructs a generating set of size $O(l_G)$, while reducing the number of necessary group multiplications significantly. The idea is that at the beginning of the construction of a new generating set, already short random subproducts have a good chance to augment the already constructed subgroup.

Theorem 2.3.6. *Suppose that a generating set S and an upper bound l_G for the length of subgroup chains for a group G are given, and let $\delta > 0$ be arbitrary. Then there is a Monte Carlo algorithm that, with probability at least $1 - \delta$, constructs a generating set T of size $O(l_G)$, using $O(|S| \log l_G)$ group operations.*

Proof. We can suppose that l_G is a power of 2 and also that l_G divides $|S|$ because we may put the first $(|S| \bmod l_G)$ elements of S into T and work with the subgroup of G generated by the remaining elements of S. Also, we can suppose $|S| \geq 10l_G$; otherwise we can define $T := S$.

The elements of T are constructed in two phases. In the first phase, we place cl_G random subproducts of length $|S|/l_G$ made from S into T, where $c \geq 10$ is a constant depending on the desired error probability. We claim that, after this phase,

$$\text{Prob}(|\{g \in S \mid g \notin \langle T \rangle\}| < l_G) \geq 1 - e^{-cl_G/15}. \tag{2.6}$$

To see (2.6), we use an extension of the idea behind the basic-type applications of Lemma 2.3.3. Let $r_1, r_2, \ldots$ be the sequence of random subproducts of length $|S|/l_G$ placed in T and associate a 0–1 valued random variable X_i to each r_i. Set $X_i = 1$ if and only if either $r_i \notin \langle r_1, \ldots, r_{i-1} \rangle$ or $|\{g \in S \mid g \notin \langle r_1, \ldots, r_{i-1} \rangle\}| < l_G$. Clearly, if $\sum_{i \leq cl_G} X_i > l_G$ then fewer than l_G elements of S are not in $\langle T \rangle$, since otherwise the initial segments of the r_i define a subgroup chain of length greater than l_G in G. Hence, it is enough to give an upper estimate for $\text{Prob}(\sum_{i \leq cl_G} X_i \leq l_G)$.

To this end, note that $\text{Prob}(X_i = 1) > 3/10$. This is true because if $|\{g \in S \mid g \notin \langle r_1, \ldots, r_{i-1} \rangle\}| < l_G$ then $\text{Prob}(X_i = 1) = 1$. Otherwise, with probability at least

$$1 - \binom{|S| - l_G}{|S|/l_G} \Big/ \binom{|S|}{|S|/l_G} \geq 1 - \left(\frac{|S| - |S|/l_G}{|S|} \right)^{l_G} > 1 - 1/e > 3/5,$$

the set of size $|S|/l_G$ used at the definition of r_i contains an element of $S \backslash \langle r_1, \ldots, r_{i-1} \rangle$. Hence, by Lemma 2.3.2, $\text{Prob}(r_i \notin \langle r_1, \ldots, r_{i-1} \rangle) > 3/10$. Finally, applying Lemma 2.3.3 with the parameters $\varepsilon = 1 - 10/(3c)$, $p = 3/10$, and $t = cl_G$, we obtain

$$\mathrm{Prob}\left(\sum_{i=1}^{cl_G} X_i \le l_G\right) \le e^{-\frac{1}{2}\left(1-\frac{1}{c(3/10)}\right)^2 \frac{3}{10} cl_G} \le e^{-cl_G/15},$$

which implies (2.6).

In the second phase, there are $\log l_G$ rounds. In the ith round, our goal is to reduce the number $|\{g \in S \mid g \notin \langle T \rangle\}|$ to less than $l_G/2^i$. To this end, we fix a constant $c' \ge 100$ and in the ith round we add $c' l_G/2^i$ random subproducts of length $2^{i-1}|S|/l_G$, made from S, to the set T.

Suppose that, after the $(i-1)$st round, $|\{g \in S \mid g \notin \langle T \rangle\}| < l_G/2^{i-1}$. We claim that, after the ith round,

$$\mathrm{Prob}(|\{g \in S \mid g \notin \langle T \rangle\}| < l_G/2^i) > 1 - e^{-c' l_G/(2^i 64)}. \tag{2.7}$$

To prove (2.7), denote by T_{i-1} the set T after round $i-1$ and by $q_1, q_2, \ldots$ the random subproducts of length $2^{i-1}|S|/l_G$ added to T in round i. Let $S_{ij} := \{g \in S \mid g \notin \langle T_{i-1}, q_1, \ldots, q_{j-1} \rangle\}$. Applying our usual trick, associate a 0–1 valued random variable Z_j to each q_j. Now $Z_j = 1$ if and only if either q_j (considered as a product of elements of S) contains *exactly one* element of S_{ij} or $|S_{ij}| < l_G/2^i$. If q_j contains exactly one element, say h, of S_{ij} then $h \in \langle T_{i-1}, q_1, \ldots, q_{j-1}, q_j \rangle$. Therefore, if $\sum_{j \le c' l_G/2^i} Z_i > l_G/2^i$ then fewer than $l_G/2^i$ elements of S are not in $\langle T_i \rangle$, because the number of such elements cannot decrease more than $l_G/2^i$ and still remain at least $l_G/2^i$. Hence, it is enough to give an upper estimate for $\mathrm{Prob}(\sum_{j \le c' l_G/2^i} Z_j \le l_G/2^i)$.

We claim that $\mathrm{Prob}(Z_j = 1) \ge 1/20$. To see this, let $x = |S_{ij}|$. If $x < l_G/2^i$ then $Z_j = 1$. Otherwise, the subset of S of size $2^{i-1}|S|/l_G$ used at the definition of q_j contains exactly one element h from S_{ij} with probability

$$\begin{aligned}
&x\binom{|S|-x}{2^{i-1}|S|/l_G - 1} \Big/ \binom{|S|}{2^{i-1}|S|/l_G} \\
&= x\frac{2^{i-1}|S|}{l_G}\frac{1}{|S|-x+1}\prod_{k=0}^{x-2}\frac{|S|-2^{i-1}|S|/l_G - k}{|S|-k} \\
&> (x2^{i-1}/l_G)(|S|/(|S|-x+1))\left(\frac{|S|-2^{i-1}|S|/l_G - x + 2}{|S|-x+2}\right)^{x-1} \\
&> \frac{1}{2}\left(1-\frac{2^{i-1}|S|/l_G}{|S|-x+2}\right)^{x-1} > \frac{1}{2}\left(1-\frac{2^{i-1}|S|/l_G}{9|S|/10}\right)^{\frac{l_G}{2^{i-1}}-1} > \frac{1}{10}
\end{aligned}$$

and then h has $1/2$ chance to get exponent 1 in q_j. If h gets exponent 1 then $h \in \langle T_{i-1}, q_1, \ldots, q_j \rangle$. Finally, applying Lemma 2.3.3 with the parameters

$\varepsilon = 1 - 20/c'$, $p = 1/20$, and $t = c'l_G/2^i$, we obtain

$$\text{Prob}\left(\sum_{j \le c'l_G/2^i} Z_i \le l_G/2^i \right) \le e^{-(1/2)(1-20/c')^2 c' l_G/(2^i 20)} < e^{-c'l_G/(2^i 64)},$$

which finishes the proof of (2.7).

Combining the two phases, the algorithm constructs a generating set of size $(c+c')l_G$ with error probability at most $e^{-cl_G/15} + \sum_i e^{-c'l_G/(2^i 64)} < e^{-cl_G/15} + e^{-c'/100}$. The number of group operations is $O(l_G(|S|/l_G)) = O(|S|)$ in the first phase and

$$O\left(\sum_{i=1}^{\log l_G} \frac{l_G}{2^i} \frac{2^{i-1}|S|}{l_G} \right) = O(|S| \log l_G)$$

in the second. □

Lemma 2.3.7. *Suppose that a black-box group $G = \langle S \rangle$ acts transitively on a set Ω, and let $\delta > 0$ be arbitrary. With probability at least $1 - \delta$, $O(\log |\Omega|)$ random subproducts made from S generate a subgroup of G that acts transitively on Ω.*

Proof. Let $g_1, \ldots, g_t$ be random subproducts made from S (using an arbitrary but fixed ordering of S), where $t = c \ln |\Omega|$ for some sufficiently large constant c. We can suppose that $c \ge 45$. For $i \in [0, t]$, let $G_i := \langle g_1, \ldots, g_i \rangle$, and let N_i denote the number of orbits of G_i on Ω. We claim that, for all $i \in [1, t]$,

$$\text{if } N_{i-1} > 1 \text{ then } \text{Prob}\big(N_i \le \tfrac{7}{8} N_{i-1}\big) \ge \tfrac{1}{3}. \tag{2.8}$$

Indeed, let $\Delta_1, \ldots, \Delta_k$, $k = N_{i-1} > 1$, be the orbits of G_{i-1} on Ω. For each $j \in [1, k]$, let X_j be a 0–1 valued random variable with $X_j = 1$ if $\Delta_j^{g_i} \ne \Delta_j$, and $X_j = 0$ otherwise. By Lemma 2.3.1, $\text{Prob}(X_j = 1) \ge 1/2$, and so for the random variable $X = \sum_{j=1}^k X_j$ we have $E(X) \ge k/2$. (As usual, $E(A)$ denotes the expected value of a random variable A.) Let $p := \text{Prob}(X \le k/4)$. Then $pk/4 + (1-p)k \ge E(X) \ge k/2$, implying $p \le 2/3$. Hence, with probability at least 1/3, at least $k/4$ orbits of G_{i-1} are proper subsets of an orbit of G_i. However, this means that at least $k/8$ orbit of G_i are the union of more than one orbit of G_{i-1}, implying (2.8).

Let $Y_1, \ldots, Y_t$ be 0–1 valued random variables, with $Y_i = 0$ if $N_{i-1} > 1$ and $N_i > 7N_{i-1}/8$, and $Y_i = 1$ otherwise. By (2.8), we have $\text{Prob}(Y_i = 1) \ge 1/3$, and $\sum_{i=1}^t Y_i \ge \ln |\Omega| / \ln(8/7)$ implies that G_t is transitive. Applying Lemma 2.3.3

with the parameters $\varepsilon = 1 - 3/(c\ln(8/7))$, $p = 1/3$, and $t = c\ln|\Omega|$, we obtain

$$\operatorname{Prob}\left(\sum_{i=1}^{t} Y_i \le \frac{\ln|\Omega|}{\ln(8/7)}\right) \le e^{-\frac{1}{2}\left(1-\frac{3}{c\ln(8/7)}\right)^2 \frac{1}{3}c\ln|\Omega|} \le e^{-c\ln|\Omega|/24}. \quad (2.9)$$

In the last inequality, we have used our assumption that $c \ge 45$. □

2.3.3. Closure Algorithms without Membership Testing

As promised in Section 2.1.2, we give versions of the G-closure algorithm without using membership testing. The situation we consider is the following. A black-box group $G = \langle S\rangle$ acts on a (not necessarily different) black-box group H. We suppose that given any $h \in H$ and $g \in G$, we can compute the image $h^g \in H$. Also, we suppose that an upper bound l_H is known for the length of subgroup chains in H. Given $A \subseteq H$, our goal is to find generators for $\langle A^G\rangle \le H$.

The breadth-first-search technique of Section 2.1.2 found generators for a subgroup chain $\langle A\rangle = \langle U_0\rangle \le \langle U_1\rangle \le \cdots \le \langle U_{l_H}\rangle = \langle A^G\rangle$, where U_i generated the group $\langle U_{i-1}, \{u^g \mid u \in U_{i-1}, g \in S\}\rangle$. Membership testing was used to discard redundant generators and ensure that $|U_i| \le l_H$ for all i.

Without the benefit of membership testing, a possibility is to start with $U_0 := A$ and define sets T_i, U_i recursively. Given U_{i-1}, let $T_i := U_{i-1} \cup \{u^g \mid u \in U_{i-1}, g \in S\}$ and let U_i be a set of size $O(l_H)$ obtained via the algorithm in the proof of Lemma 2.3.4 or Theorem 2.3.6 such that, with high probability, $\langle U_i\rangle = \langle T_i\rangle$. Clearly, $\langle A\rangle = \langle U_0\rangle \le \langle U_1\rangle \le \langle U_2\rangle \le \cdots \le \langle A^G\rangle$ and $\langle U_i\rangle < \langle U_{i+1}\rangle$ if $\langle U_i\rangle \ne \langle A^G\rangle$. Since a subgroup chain cannot grow more than l_H times, $\langle U_{l_H}\rangle$ must be closed for the G-action and so $\langle U_{l_H}\rangle = \langle A^G\rangle$.

A faster algorithm, which avoids the construction of small generating sets for the intermediate groups in the procedure described in the previous paragraph, is given in [Cooperman and Finkelstein, 1993]. It is based on the following lemma.

Lemma 2.3.8. *Suppose that $G = \langle S\rangle$ acts on H and let $U \subseteq H$. Let g be a random subproduct of S and u be a random subproduct of U. If $\langle U\rangle$ is not closed for the G-action then* $\operatorname{Prob}(u^g \notin \langle U\rangle) \ge 1/4$.

Proof. By hypothesis, $K := \{k \in G \mid \langle U\rangle^k = \langle U\rangle\} \lneqq G$. Hence, by Lemma 2.3.2, $\operatorname{Prob}(g \notin K) \ge 1/2$. If $g \notin K$ then $X := \langle U\rangle^{g^{-1}} \cap \langle U\rangle \ne \langle U\rangle$. Thus, by Lemma 2.3.2, $\operatorname{Prob}(u \notin X) \ge 1/2$. Combining the two probabilities, we obtain that $\operatorname{Prob}(u^g \notin \langle U\rangle) = \operatorname{Prob}(u \notin X \mid g \notin K)\operatorname{Prob}(g \notin K) \ge 1/4$. □

Theorem 2.3.9. *Suppose that $G = \langle S \rangle$ acts on a group H, an upper bound l_H for the length of subgroup chains in H is given, and $\delta > 0$ is an arbitrary constant. Then there is a Monte Carlo algorithm that, with probability at least $1 - \delta$, constructs $O(l_H + |A|)$ generators for the G-closure of some $A \subseteq H$, using $O(l_H(|A| + |S| + l_H))$ group operations.*

Proof. The algorithm is a straightforward consequence of Lemma 2.3.8. Initialize $U := A$. Given U, let u be a random subproduct of U and let g be a random subproduct of S. Add u^g to U. An application of Lemma 2.3.3 of basic type shows that increasing U $O(l_H)$ times creates a generating set for $\langle A^G \rangle$ with arbitrarily small constant error. □

2.3.4. Derived and Lower Central Series

As pointed out in [Babai et al., 1995], given a black-box group $G = \langle S \rangle$ and an upper bound l_G for the length of subgroup chains in G, there is a Monte Carlo algorithm that, with arbitrarily small constant error probability, constructs a generating set of size $O(l_G)$ for the derived subgroup G' of G. Namely, we compute the set $A := \{[a, b] \mid a, b \in S\}$, then we construct a generating set of size $O(|S|^2 + l_G)$ for $G' = \langle A^G \rangle$ by the algorithm in the proof of Theorem 2.3.9, and finally we reduce the number of generators for G' to $O(l_G)$ by the algorithm in the proof of Lemma 2.3.4 or Theorem 2.3.6. This process can be iterated to obtain generators for the groups in the derived series and lower central series of G.

Note that it is possible to test solvability and nilpotence of G without membership testing: all we have to do is to check whether the l_Gth member of the appropriate series is the identity subgroup of G. Also, often bounds sharper than l_G for the length of the derived series are available. For example, if G is a permutation group of degree n then the length of the derived series is $O(\log n)$ (cf. [Dixon, 1968]) whereas G may have subgroup chains of length $\Theta(n)$.

However, we cannot test subnormality of some $H \leq G$ without membership testing. Repeated applications of the algorithm in the proof of Theorem 2.3.9 yield a candidate subnormal series between H and G, but eventually we have to decide whether two sets of group elements generate the same subgroup.

In the rest of this section, we describe a more practical algorithm for computing commutator subgroups that avoids the blowup in the number of generators. We start with an identity about the commutator of products.

Lemma 2.3.10.

$$[a_1 a_2 \cdots a_n, b_1 b_2 \cdots b_m] = \prod_{i=1}^{n} \prod_{j=m}^{1} [a_i, b_j]^{b_{j+1} \cdots b_m a_{i+1} \cdots a_n}.$$

Proof. By induction, it is easy to check that

$$[a, b_1 b_2 \cdots b_m] = \prod_{j=m}^{1} [a, b_j]^{b_{j+1}\cdots b_m}$$

and

$$[a_1 a_2 \cdots a_n, b] = \prod_{i=1}^{n} [a_i, b]^{a_{i+1}\cdots a_n}.$$

The combinination of these two identities proves the lemma. □

Lemma 2.3.11. *Let $H = \langle U \rangle$ and $K = \langle V \rangle$ be two subgroups of a common parent group and let u and v be random subproducts from the sets U and V, respectively. Moreover, let $N \leq [H, K]$, $N \lhd \langle H, K \rangle$. If $N \neq [H, K]$ then* $\mathrm{Prob}([u, v] \notin N) \geq 1/4$.

Proof. Let $U = \{u_1, \ldots, u_n\}$, $V = \{v_1, \ldots, v_m\}$, $u = u_1^{\varepsilon_1} \cdots u_n^{\varepsilon_n}$, and $v = v_1^{\nu_1} \cdots v_m^{\nu_m}$. If $N \neq [H, K]$ then there exist $u_i \in U$, $v_j \in V$ such that $[u_i, v_j] \notin N$. Fix the last such pair u_i, v_j with respect to the ordering implied by Lemma 2.3.10: $(i_1, j_1) \prec (i_2, j_2)$ if either $i_1 < i_2$, or $i_1 = i_2$ and $j_1 > j_2$. Then, by Lemma 2.3.10, $[u, v]$ can be written in the form $[u, v] = x[u_i^{\varepsilon_i}, v_j^{\nu_j}]^y z$ for some $x \in [H, K]$, $y \in \langle H, K \rangle$, and $z \in N$. So

$$\begin{aligned}\mathrm{Prob}([u, v] \notin N) &\geq \mathrm{Prob}(\varepsilon_i \nu_j = 1 | x \in N)\,\mathrm{Prob}(x \in N) \\ &\quad + \mathrm{Prob}(\varepsilon_i \nu_j = 0 | x \notin N)\,\mathrm{Prob}(x \notin N) \\ &\geq (1/4)\,\mathrm{Prob}(x \in N) + (3/4)\,\mathrm{Prob}(x \notin N) \geq 1/4.\end{aligned}$$ □

Theorem 2.3.12. *Suppose that $H = \langle U \rangle$ and $K = \langle V \rangle$ are subgroups of G and an upper bound l_G for the length of subgroup chains in G is known. Let $\delta > 0$ be an arbitrary constant. Then there is a Monte Carlo algorithm that, with probability at least $1 - \delta$, constructs $O(l_G)$ generators for $[H, K]$, using $O(l_G(|U| + |V| + l_G))$ group operations.*

Proof. We construct a generating set for $[H, K]$ in two phases. In the first phase, we place cl_G commutators $[u, v]$ into a set T, where u and v are random subproducts from the sets U and V, respectively. Based on Lemma 2.3.11, an application of Lemma 2.3.3 of basic type shows that, with error probability at most $e^{-c' l_G}$, the normal closure of $\langle T \rangle$ generates $[H, K]$. In the second phase, using the algorithm in the proof of Theorem 2.3.9, we compute generators for $\langle T^{\langle H, K \rangle} \rangle$. □

Remark 2.3.13. During the computation of the derived or lower central series of G, we apply the algorithm in the proof of Theorem 2.3.12 l_G times, and the error probabilities of all rounds have to be added. This causes no problem, because all error estimates are of the form $e^{-c' l_G}$ for some constant c' and $l_G e^{-c' l_G}$ is still exponentially small as a function of l_G.

Lemma 2.3.11 also can be applied to test whether a given group $G = \langle S \rangle$ is abelian. Note that the straightforward deterministic method, checking whether each pair of generators commutes, requires $\Theta(|S|^2)$ group operations.

Lemma 2.3.14. *Given a black-box group $G = \langle S \rangle$ and an arbitrary constant $\delta > 0$, there is a Monte Carlo algorithm that, with probability at least $1 - \delta$, decides whether G is abelian, using $O(|S|)$ group operations.*

Proof. We apply Lemma 2.3.11 with $H = K = G, U = V = S$, and $N = 1$. If G is not abelian then two random subproducts on S do not commute with probability at least $1/4$. So, if c pairs of random subproducts commute then Prob(G is abelian) $\geq 1 - (3/4)^c$. □

2.4. Random Prefixes

The material in this section is reproduced from the paper [Babai et al., 1995], copyright © 1995 by Academic Press; reprinted by permission of the publisher.

2.4.1. Definition and Basic Properties

The paper by [Babai et al., 1995] also introduces the method of random prefixes, which is an alternative to the applications of random subproducts. Random prefixes can be used to reduce generating sets roughly with the asymptotic efficiency as the algorithm in the proof of Theorem 2.3.6 and to speed up asymptotically the computation of algebraic closures and commutator subgroups (cf. Theorems 2.3.9 and 2.3.12).

Given a sequence of group elements $g_1, g_2, \ldots, g_k$, a *random prefix* of this sequence is the product of a randomly chosen initial segment of a random ordering of the g_i. To apply random prefixes, we need an analogue of Lemma 2.3.2 to ensure that a random prefix of generators has a fair chance to avoid any given proper subgroup. Our first goal is the deduction of such an analogue.

We start with some technical results concerning *spreader permutations*. Let $M = \{1, 2, \ldots, m\}$, $\bar{M} = M \cup \{0, m+1\}$, $H \subseteq M$, and $\bar{H} = H \cup \{0, m+1\}$.

For $x \in H$, let

$$\delta_H(x) = \min_{y \in \bar{H} \setminus \{x\}} |x - y|$$

(i.e., $\delta_H(x)$ is the distance of x to its nearest neighbor in $\bar{H}$). Define $\delta(H) := \sum_{x \in H} \delta_H(x)$ and $\text{spread}(H) := \delta(H)/(m+1)$. It is easy to see that $0 < \text{spread}(H) < 1$ (in fact, $|H|/(m+1) \leq \text{spread}(H) \leq 1 - 1/(|H|+1)$). This quantity, the *spreading factor of* H, measures how evenly the set H is distributed within M. We say that H is ε-*spread* if $\text{spread}(H) \geq \varepsilon$.

Let $P = \{p_1, \ldots, p_r\}$ be a set of permutations of M. We call P an (ε, m)-*spreader* if for every nonempty $H \subseteq M$, at least one of the sets $H^{p_j} (1 \leq j \leq r)$ is ε-spread.

Lemma 2.4.1. *Let H be a random subset of size $h \neq 0$ of $M = \{1, \ldots, m\}$ and $1/30 \leq \varepsilon < 1/15$. Then* $\text{Prob}(\text{spread}(H) \geq \varepsilon) > 1 - (15\varepsilon)^{h/4}$.

Proof. If $|H| \geq \varepsilon(m+1)$ then $\text{spread}(H) \geq |H|/(m+1) \geq \varepsilon$ with probability 1.

If $1 \leq h \leq 5$ and $h < \varepsilon(m+1)$ then we compute the ratio q of h-element subsets $a_1 < a_2 < \cdots < a_h$ of M such that $a_i - a_{i-1} \geq \lceil \varepsilon(m+1)/h \rceil$ for $1 \leq i \leq h+1$, with $a_0 := 0$ and $a_{h+1} := m+1$. For such sets H, $\text{spread}(H) \geq h\lceil \varepsilon(m+1)/h \rceil/(m+1) \geq \varepsilon$. Introducing the notation $d := \lceil \varepsilon(m+1)/h \rceil - 1$, we have

$$\begin{aligned}
&\text{Prob}(\text{spread}(H) \geq \varepsilon) \geq q \\
&= \binom{m-(h+1)d}{h} \Big/ \binom{m}{h} > \left(\frac{m-h+1-(h+1)d}{m-h+1}\right)^h \\
&= \left(1 - \frac{(h+1)d}{m-h+1}\right)^h > 1 - h\frac{(h+1)d}{m-h+1} > 1 - \frac{(h+1)\varepsilon(m+1)}{m-h+1} \\
&> 1 - \frac{(h+1)\varepsilon(m+1)}{(1-\varepsilon)(m+1)} > 1 - (h+1)\varepsilon(15/14) > 1 - (15\varepsilon)^{h/4}.
\end{aligned}$$

In the case $6 \leq h < \varepsilon(m+1)$, we consider H as the range of an injective function f from $H_0 := \{1, 2, \ldots, h\}$ to M. We also define $\overline{H_0} := H_0 \cup \{0, h+1\}$ and extend f by setting $f(0) := 0$ and $f(h+1) := m+1$.

Let $k := \lfloor h/2 \rfloor$. For $0 < c < 1$, we estimate the probability of the event $A(c)$ that there exist k distinct elements $H_0^* \subset \overline{H_0}$, $H_0^* = \{a_{i,j} \mid 1 \leq i \leq l, 1 \leq j \leq k_i\}$, where $2 \leq k_i \leq 3$ for $1 \leq i \leq l$ and $\sum k_i = k$, satisfying the following property: For all $1 \leq i \leq l$ and $1 \leq j \leq k_i - 1$, $f(a_{i,j}) < f(a_{i,j+1}) < f(a_{i,j}) +$

$c(m+1)$. Such k elements can be chosen

$$\binom{h+2}{l}\binom{l}{k-2l}\frac{(h+2-l)!}{(h+2-k)!}$$

ways, since the elements $a_{1,1}, a_{2,1}, \ldots, a_{l,1}$ are from $\overline{H_0}$; from this set, we have to choose a $k-2l$-element subset to determine those $a_{i,1}$ which have $k_i = 3$; finally, fixing the two subsets of the $a_{i,1}$, there are $(h+2-l)!/(h+2-k)!$ choices for the remaining $a_{i,j}$.

Fixing the set H_0^*, we can define f by first choosing the (random) values $f(a_{i,1})$ for $1 \le i \le l$ and then continuing with the choice of $f(a_{i,j})$ for $j \ge 2$. The probability that $f(a_{i,j+1})$ falls between $f(a_{i,j})$ and $f(a_{i,j})+c(m+1)$ is at most $c(m+1)/(m+1-(k-1))$. (Note that this estimate is valid even in the case $a_{i,j+1} = h+1$, although the value of $f(h+1)$ is not chosen randomly: In this case, consider the probability that $f(a_{i,j})$ falls between $(1-c)(m+1)$ and $m+1$.) Therefore,

$$\mathrm{Prob}(A(c)) < \binom{h+2}{l}\binom{l}{k-2l}\frac{(h+2-l)!}{(h+2-k)!}\left(\frac{c(m+1)}{m+1-k}\right)^{k-l}.$$

Since $k \le h/2 \le \varepsilon(m+1)/2 \le (m+1)/30$, we have $(m+1)/(m+1-k) \le 30/29$. We claim that

$$\binom{h+2}{l}\binom{l}{k-2l}\frac{(h+2-l)!}{(h+2-k)!} < \left(\frac{29h}{4}\right)^{k-l}, \tag{2.10}$$

and so $\mathrm{Prob}(A(c)) < (15ch/2)^{k-l}$.

For $6 \le h \le 200$, (2.10) can be checked by computer. (GAP needs about 10 seconds on a SparcStation2 to do that.) For larger values, we use Stirling's formula. Note that the sequence $\sqrt{2\pi n}(n/e)^n/n!$, $n = 0, 1, \ldots$ is monotone increasing with limit 1, so $\sqrt{2\pi n}(n/e)^n < n! < (10/9)\sqrt{2\pi n}(n/e)^n$ for all $n \ge 1$. Substituting the appropriate $\sqrt{2\pi n}(n/e)^n$ for the factorials in the denominator and $(10/9)\sqrt{2\pi n}(n/e)^n$ in the numerator gives an upper estimate for the left-hand side of (2.10).

Using that for $1/3 < x < 1/2$

$$(3x-1)^{3x-1}(1-2x)^{1-2x} > 4^x(29e/8)^{x-1}, \tag{2.11}$$

in the case $k/3 < l < k/2$ the term $(3l-k)^{3l-k}(k-2l)^{k-2l}$ can be estimated from below by $k^l 4^l (29e/8)^{l-k}$. Also, $(h+2-k)^{h+2-k} > e^2(h/2)^{h+2-k}$. Using these estimates, it is straightforward to check that (2.10) holds. If $l \in \{k/3, k/2\}$ then $\binom{l}{k-2l} = 1$, and Stirling's formula easily yields (2.10).

What does it mean that $A(c)$ fails? Let $A = (a_0, a_1, \ldots, a_{h+1})$ be the sequence of elements of $H \cup \{0, m+1\}$, listed in increasing order. Then let $A_1, \ldots, A_n$ be the maximal subsequences of consecutive elements with difference less than $c(m+1)$; that is, $a_{i+1} - a_i < c(m+1)$ if a_i and a_{i+1} are in the same A_j, but $\min(A_{j+1}) - \max(A_j) \geq c(m+1)$. Each A_j of size greater than 1 can be partitioned into subsequences of length two and three, which can be considered as the f-images of the set H_0^*. Therefore, if $A(c)$ fails then $\bigcup_{j\in J} |A_j| < k$, where the summation runs over the A_j with $|A_j| > 1$. Hence more than $h-k$ elements $x \in H$ are in one-element sets A_j, which means that $\delta_H(x) \geq c(m+1)$. In particular, if h is even then we choose $c := 2\varepsilon/h$ and obtain that $\mathrm{spread}(H) \geq (h-k)c = \varepsilon$ with probability greater than $1-(15\varepsilon)^{k-l} \geq 1-(15\varepsilon)^{h/4}$. If h is odd then we choose $c := 2\varepsilon/(h+1)$ and obtain that $\mathrm{spread}(H) \geq (h-k)c = \varepsilon$ with probability greater than $1-(15\varepsilon h/(h+1))^{k-l} \geq 1-(15\varepsilon h/(h+1))^{(h-1)/4} > 1-(15\varepsilon)^{h/4}$. (The last inequality is the only point in the proof of case $h \geq 6$ where we used the lower bound $\varepsilon \geq 1/30$.) □

Theorem 2.4.2. *Let* $1/30 \leq \varepsilon < 1/15$, $C > 0$, *and* $r > (4+4(C+1)\log m)/\log(1/(15\varepsilon))$. *Furthermore, let* $P = \{p_1, \ldots, p_r\}$ *be a set of* r *random permutations of* $\{1, 2, \ldots, m\}$. *Then* P *is an* (ε, m)*-spreader with probability at least* $1 - m^{-C}$.

Proof. Let H be a nonempty subset of $\{1, 2, \ldots, m\}$ and $h := |H|$. By Lemma 2.4.1, the probability that $\mathrm{spread}(H) < \varepsilon$ for all permutations in P is at most $(15\varepsilon)^{hr/4}$. Hence the probability that not all nonempty subsets are ε-spread is at most

$$\sum_{h=1}^{m} \binom{m}{h} (15\varepsilon)^{hr/4} = \left(1 + (15\varepsilon)^{r/4}\right)^m - 1 < 2(15\varepsilon)^{r/4} m < m^{-C}.$$

Here, the first inequality is true because the condition imposed on r implies that $(15\varepsilon)^{r/4} < 1/(me)$, and $(1+x)^m < 1+2xm$ for $0 < x < 1/(me)$. The second inequality is a straightforward consequence of the condition imposed on r. □

Now we are in position to connect spreader permutations, random prefixes, and expansion of subgroups.

Lemma 2.4.3. *Let $G = \langle S \rangle$ be a black-box group and $P = \{p_1, \dots, p_r\}$ be an $(\varepsilon, |S|)$-spreader, and let $K \lneqq G$. Moreover, let g be a random prefix of S, obtained from a randomly chosen element of P. Then* $\text{Prob}(g \notin K) \geq \varepsilon/(2r)$.

Proof. Let $H := \{s \in S \mid s \notin K\}$ and let $a_1 < a_2 < \cdots < a_h$ denote the positions corresponding to the elements of H in the randomly chosen permutation $p \in P$. We also define $a_0 := 0$ and $a_{h+1} := |S| + 1$ and let g_j denote the random prefix of S corresponding to the first j elements of p. Then, for a fixed $i \in [1, h]$, either $g_j \notin K$ for all j with $a_{i-1} \leq j < a_i$ or $g_j \notin K$ for all j with $a_i \leq j < a_{i+1}$ (it is also possible that both of these events occur). So $\text{Prob}(g \notin K) \geq \text{spread}(\{a_1, \dots, a_h\})/2$ and, since P is an $(\varepsilon, |S|)$-spreader, $\text{spread}(\{a_1, \dots, a_h\}) \geq \varepsilon$ with probability at least $1/r$. □

2.4.2. Applications

We shall apply random prefixes to G-closure computations. We consider the same situation as in Section 2.3.3, namely, $G = \langle S \rangle$ acts on H, we can compute $h^g \in H$ for any $h \in H$, $g \in G$, and an upper bound l_H is known for the length of subgroup chains in H.

Lemma 2.4.4. *Suppose that $G = \langle S \rangle$ acts on H and $\langle U \rangle \leq H$ is not closed for the G-action. Let $P = \{p_1, \dots, p_{r_1}\}$ be an $(\varepsilon, |S|)$-spreader and $Q = \{q_1, \dots, q_{r_2}\}$ be an $(\varepsilon, |U|)$-spreader. Moreover, let g and u be random prefixes on the sets S and U, respectively, obtained from randomly chosen elements of P and Q. Then* $\text{Prob}(u^g \notin \langle U \rangle) \geq \varepsilon^2/(4r_1r_2)$.

Proof. By hypothesis, $K := \{k \in G \mid \langle U \rangle^k = \langle U \rangle\} \lneqq G$. Hence, by Lemma 2.4.3, $\text{Prob}(g \notin K) \geq \varepsilon/(2r_1)$. If $g \notin K$ then $X := \langle U \rangle^{g^{-1}} \cap \langle U \rangle \neq \langle U \rangle$. Thus, by Lemma 2.4.3, $\text{Prob}(u \notin X) \geq \varepsilon/(2r_2)$. Combining the two probabilities, we obtain $\text{Prob}(u^g \notin \langle U \rangle) \geq \text{Prob}(u \notin X | g \notin K)\text{Prob}(g \notin K) \geq \varepsilon^2/(4r_1r_2)$. □

Theorem 2.4.5. *Suppose that $G = \langle S \rangle$ acting on a group H, an upper bound l_H for the length of subgroup chains in H, a subset $A \subseteq H$, and a constant $\delta > 0$ are given. Then there is a Monte Carlo algorithm that, with probability at least $1 - \delta$, constructs $O(l_H \log |S| (\log l_H + \log \log |S|))$ generators for $\langle A^G \rangle$, using*

$$O(l_H \log |S| (\log l_H + \log \log |S|)^3 + |A| \log l_H + |S| \log |S|)$$

group operations.

Proof. We can suppose that $|A| \in O(l_H)$ because if $|A|$ is too big then we construct $O(l_H)$ generators for $\langle A \rangle$. By Theorem 2.3.6, this can be done by $O(|A| \log l_H)$ group operations.

We would like to apply Lemma 2.4.4. We can fix a value ε in the interval $[1/30, 1/15)$, and there is no problem concerning the set S: We can construct an $(\varepsilon, |S|)$-spreader $P = \{p_1, \ldots, p_{r_1}\}$ of size $O(\log |S|)$ and collect in a set T_G all products of elements of S corresponding to initial segments of the permutations in P. For the construction of random permutations in P, see Exercise 2.1. The construction of T_G requires $O(|S| \log |S|)$ group operations.

However, the set U containing the already constructed generators for $\langle A^G \rangle$ changes during the algorithm, so we cannot precompute the product of initial segments in random permutations of U. Computing spreader permutations of increasing degree as $|U|$ changes would take too much time. Therefore, we proceed in the following way.

Let $m = cl_H \log |S| (\log l_H + \log \log |S|)$ for a sufficiently large constant c and let $Q = \{q_1, \ldots, q_{r_2}\}$ be an (ε, m)-spreader of size $O(\log m) = O(\log l_H + \log \log |S|)$. We can suppose that m is a power of 2. We use an m-long array U to store generators for $\langle A^G \rangle$. Initially, U contains the elements of A and the rest of U is padded with the identity element. As new generators x for $\langle A^G \rangle$ are constructed, we replace one of the identity elements in U by x. Now comes the idea that speeds up the computation. Instead of products of initial segments of the U^{q_i}, we store the product of elements in positions $l2^j + 1, l2^j + 2, \ldots, (l+1)2^j$ in U^{q_i} for all $i \in [1, r_2]$, $j \in [0, \log m]$, and $l \in [0, m/2^j - 1]$. This requires the storage of less than $2mr_2$ group elements.

From this data structure T_H, the product of the first k elements of U^{q_i} can be computed with $O(\log m) = O(\log l_H + \log \log |S|)$ group operations, by taking segments of lengths corresponding to the terms in the binary expansion of k. Also, after replacing an element of U, the data structure can be updated by at most $r_2 \log m \in O((\log l_H + \log \log |S|)^2)$ group operations, since after updating segments of length 2^j, the r_2 segments of length 2^{j+1} containing the new element can be updated by one multiplication per segment.

This gives us the following algorithm: Replace the identity elements in U by group elements of the form u^g as described in Lemma 2.4.4, always updating the data structure T_H. If the construction of the spreaders P, Q was successful then Lemma 2.4.4 and an application of the basic type of Lemma 2.3.3 imply that by the time all $m - |A| \in O(l_H r_1 r_2)$ identity elements are replaced, U generates $\langle A^G \rangle$ with probability greater than $1 - e^{-c' l_H}$ for some constant $c' > 0$. The number of group operations required is $O(mr_2 \log m) = O(l_H \log |S| (\log l_H + \log \log |S|)^3)$. □

Corollary 2.4.6. *Suppose that an upper bound l_G is known for the length of subgroup chains in a group G and that $O(l_G)$ generators are given for G and for some $H \leq G$. Let $\delta > 0$. Then there is a Monte Carlo algorithm that, with probability at least $1 - \delta$, constructs $O(l_G)$ generators for the normal closure $\langle H^G \rangle$, using $O(l_G \log^4 l_G)$ group operations.*

Proof. By Theorem 2.4.5, $O(l_G \log^2 l_G)$ generators can be obtained by the indicated number of group operations. By Theorem 2.3.6, this generating set can be reduced to one of size $O(l_G)$ with $O(l_G \log^3 l_G)$ further group operations. □

Our last application is the asymptotic speedup of commutator subgroup computations.

Lemma 2.4.7. *Let $H = \langle U \rangle$ and $K = \langle V \rangle$ be two subgroups of a common parent group, and let $N \lneqq [H, K]$, $N \lhd \langle H, K \rangle$. Moreover, let $P = \{p_1, \ldots, p_{r_1}\}$ be an $(\varepsilon, |U|)$-spreader and $Q = \{q_1, \ldots, q_{r_2}\}$ be an $(\varepsilon, |V|)$-spreader, and let u and v be random prefixes on the sets U and V, respectively, obtained from randomly chosen elements of P and Q. Then* $\text{Prob}([u, v] \notin N) \geq \varepsilon^2/(4r_1r_2)$.

Proof. By Lemma 2.3.10, $X := \{h \in H \mid [h, K] \leq N\}$ is a subgroup of H and, by hypothesis, $X \neq H$. So Lemma 2.4.3 implies that $\text{Prob}(u \notin X) \geq \varepsilon/(2r_1)$. If $u \notin X$ then $Y := \{k \in K \mid [u, k] \in N\} \neq K$, so, by Lemma 2.4.3, $\text{Prob}([u, v] \notin N) \geq \varepsilon/(2r_2)$. Combining the two probabilities, we obtain $\text{Prob}([u, v] \notin N) \geq \text{Prob}([u, v] \notin N | u \notin X)\text{Prob}(u \notin X) \geq \varepsilon^2/(4r_1r_2)$. □

Theorem 2.4.8. *Suppose that $H = \langle U \rangle$ and $K = \langle V \rangle$ are subgroups of G and an upper bound l_G for the length of subgroup chains in G is known. Let $\delta > 0$. Then there is a Monte Carlo algorithm that, with probability at least $1 - \delta$, constructs $O(l_G)$ generators for $[H, K]$, using $O(l_G \log^4 l_G + (|U| + |V|) \log l_G)$ group operations.*

Proof. With $O((|U| + |V|) \log l_G)$ group operations, we can construct generating sets of size $O(l_G)$ for H and K. So we suppose that $|U| \in O(l_G)$ and $|V| \in O(l_G)$.

A generating set for $[H, K]$ is obtained in two phases. In the first phase, we construct an $(\varepsilon, |U|)$-spreader $P = \{p_1, \ldots, p_{r_1}\}$ and an $(\varepsilon, |V|)$-spreader $Q = \{q_1, \ldots, q_{r_2}\}$, and we place the products of initial segments of U^{p_i}, V^{q_i} into the sets T_U and T_V, respectively. This requires $O(|U| \log |U| + |V| \log |V|) = O(l_G \log l_G)$ group operations. Then, for a sufficiently large constant c, we place $c l_G r_1 r_2 \in O(l_G \log^2 l_G)$ commutators $[u, v]$ into a set T, where u and v

are randomly chosen elements of T_U and T_V, respectively. After that, using $O(l_G \log^3 l_G)$ group operations, we construct $O(l_G)$ generators for $\langle T \rangle$.

By Lemma 2.4.7, the normal closure of $\langle T \rangle$ in $\langle H, K \rangle$ is $[H, K]$ with high probability. In the second phase, we compute generators for this normal closure. By Corollary 2.4.6, this requires $O(l_G \log^4 l_G)$ group operations. □

Exercises

2.1. Design an algorithm that constructs a uniformly distributed random element of S_n in $O(n)$ time. (You can suppose that a random element of a list can be chosen in $O(1)$ time.) *Hint:* We have to construct an injective function $f : [1, n] \to [1, n]$. When $f(1), \dots, f(k)$ are already defined, store the remaining $n - k$ possible function values in an array of length $n - k$. How should this array be modified when $f(k + 1)$ is defined?

2.2. Design an algorithm that constructs a uniformly distributed random element of A_n.

2.3. Let M be a finite state Markov chain, with transition probability matrix P. Prove that $p_{ij}^{(k)}$ is the (i, j)-entry in P^k.

2.4. Let M be a Markov chain and suppose that states u, v can be reached from each other. Prove that u and v are both aperiodic or they have the same period.

2.5. Prove that the stationary distribution of a finite, irreducible, aperiodic Markov chain is the uniform one if and only if the column sums of the transition probability matrix are all 1.

2.6. Prove the inequality (2.11).

2.7. [Beals and Babai, 1993] Suppose that G is a nonabelian black-box group and $N \lhd G, N \neq G$. Suppose also that a subset $A := \{g_1, \dots, g_k\} \subseteq G \backslash \{1\}$ is given, and we know that $A \cap N \neq \emptyset$. Design a Monte Carlo algorithm that, with high probability, computes a nontrivial element of a proper normal subgroup of G. *Hint:* If $\{a, b\} \cap N \neq \emptyset$ then $[a, b] \in N$. What can we do if a and b commute?

3

Permutation Groups

A Complexity Overview

In this chapter, we start the main topic of this book with an overview of permutation group algorithms.

3.1. Polynomial-Time Algorithms

In theoretical computer science, a universally accepted measure of efficiency is *polynomial-time computation*. In the case of permutation group algorithms, groups are input by a list of generators. Given $G = \langle S \rangle \le S_n$, the input is of length $|S|n$ and a polynomial-time algorithm should run in $O((|S|n)^c)$ for some fixed constant c. In practice, $|S|$ is usually small: Many interesting groups, including all finite simple groups, can be generated by two elements, and it is rare that in a practical computation a permutation group is given by more than ten generators. On the theoretical side, any $G \le S_n$ can be generated by at most $n/2$ permutations (cf. [McIver and Neumann, 1987]). Moreover, any generating set S can be easily reduced to less than n^2 generators in $O(|S|n^2)$ time by a deterministic algorithm (cf. Exercise 4.1), and in Theorem 10.1.3 we shall describe how to construct at most $n - 1$ generators for any $G \le S_n$. Hence, we require that the running time of a polynomial-time algorithm is $O(n^c + |S|n^2)$ for some constant c.

In this book, we promote a slightly different measure of complexity involving n, $|S|$, and $\log |G|$ (cf. Section 3.2), which better reflects the practical performance of permutation group algorithms. However, "traditional" polynomial time is still very important for us: Experience shows that a lot of ideas developed in the polynomial-time context are later incorporated in practical algorithms; conversely, procedures performing well in practice often have versions with polynomial running time.

Nice surveys on polynomial-time computations can be found in [Luks, 1993] and [Kantor and Luks, 1990]. The latter paper also contains a comprehensive polynomial-time toolkit. The following list is a part of that toolkit; we overview those tasks that can be done by a polynomial-time algorithm for all inputs $G = \langle S \rangle \le S_n$. For special classes of groups (mostly for solvable groups or, slightly more generally, for groups with nonabelian composition factors of bounded order), there is a significant extension of this list. We mention some of the additional problems, which can be handled in polynomial time if the input group has bounded nonabelian composition factors, in Section 3.3. Also, as a general rule, it seems to be easier to compute a targeted subgroup if it is *normal* in G: For example, $C_G(H)$ is computable in polynomial time if $H \lhd G$ (since $H \lhd G$ implies $C_G(H) \lhd G$) but no polynomial-time algorithm is known for arbitrary inputs $H \le G$.

Let $G = \langle S \rangle \le \mathrm{Sym}(\Omega)$. Then the following tasks can be computed by deterministic polynomial-time algorithms:

(a) Find orbits and blocks of imprimitivity of G.
(b) Given $h \in \mathrm{Sym}(\Omega)$, test whether $h \in G$.
(c) Find the order of G.
(d) Find a generator–relator presentation for G.
(e) Given $\Delta \subseteq \Omega$, find the pointwise stabilizer of Δ in G.
(f) Compute the kernel of the homomorphism $\pi : G \to \mathrm{Sym}(\Delta)$. ($\pi$ is defined by specifying the images of generators.)
(g) (i) Given $T \subseteq G$, find the normal closure $\langle T^G \rangle$.
  (ii) Given $H \le G$, test whether H is subnormal in G; and, if so, find a sequence $H = L_0 \lhd L_1 \lhd \cdots \lhd L_m = G$.
  (iii) Find the derived series and lower central series of G (and hence test solvability and nilpotence).
(h) Find the center of G.
(i) Find a composition series $1 = H_0 \lhd H_1 \lhd \cdots \lhd H_m = G$ for G, and find a faithful permutation representation for each of the composition factors H_i/H_{i-1}, specifically, a homomorphism $\pi_i : H_i \to \mathrm{Sym}(\Delta_i)$ with $\ker(\pi_i) = H_{i-1}$ and $|\Delta_i| \le |\Omega|$.
(j) If G is simple, identify the isomorphism type of G.
(k) (i) Given $H \lhd G$, test whether H is minimal normal in G. If it is not, then find $N \lhd G$ such that $1 \lneqq N \lneqq H$.
  (ii) Find a chief series for G.
(l) (i) If p is a prime, find a Sylow p-subgroup of G containing a given p-subgroup P of G.
  (ii) Given Sylow p-subgroups P_1, P_2 of G, find $g \in G$ such that $P_1^g = P_2$.
  (iii) Given a Sylow p-subgroup P of L where $L \lhd G$, find $N_G(P)$.

(m) Given $H \leq G$, find $\mathrm{Core}_G(H)$. (For the definitions used in (m) – (p), see Section 1.2.1.)
(n) Find the socle of G.
(o) Find the intersection of all maximal normal subgroups of G.
(p) For any collection Σ of simple groups, find $O_\Sigma(G)$ and $O^\Sigma(G)$.
(q) Find all of the above in quotient groups of G.

We give historical remarks and indicate some of the group theory involved in the different algorithms.

Item (a) is essentially all we can do using only the given generators of G (although there are some exceptions, e.g., testing whether a permutation group is regular). The first algorithm for computing blocks of imprimitivity appeared in [Atkinson, 1975]. For all other tasks, we need to compute *bases* and *strong generating sets* (cf. Section 4.1), the basic data structures introduced in the seminal works [Sims, 1970, 1971a]. It was first analyzed in [Furst et al., 1980] that Sims's algorithm for computing strong generating sets runs in polynomial time. Items (b)–(g) are straightforward consequences of the strong generator construction. Concerning (d), we note that the presentation obtained from such a construction usually contains redundant defining relations; the first method to delete some of the redundant relators is in [Cannon, 1973]. Sims's strong generating set construction uses only basic group theory: Lagrange's theorem and Schreier's lemma (cf. Lemma 4.2.1).

Computing the center is also elementary (cf. [Luks, 1987]); in fact, it can be reduced to a point stabilizer construction (cf. [Cooperman et al., 1989]). The preliminary version of [Luks, 1987] for finding a composition series, circulated since 1981, was the first algorithm based on consequences of the classification of finite simple groups. [Neumann, 1986] also contains an algorithm for computing a composition series, which runs in polynomial time for groups with some restrictions of the primitive groups involved in them. Problem (j), the identification of simple groups, is solved in [Kantor, 1985b].

The nontrivial case of (k)(i) is when H is abelian. This was resolved in [Rónyai, 1990], as an application of methods handling associative algebras. (k)(ii) is a direct consequence of the first part.

Item (l), the work on the algorithmic Sylow machinery in [Kantor, 1985b, 1990] is the most complicated algorithm in our list: In addition to consequences of the classification of finite simple groups, it requires a case-by-case study of the classical finite simple groups. Considering the elementary nature of the Sylow theorems and their important role in group theory, it would be worthwhile to find simpler algorithms to deal with Sylow subgroups.

Problem (m) is solved in [Kantor and Luks, 1990]. The algorithm uses (l) and so the classification of finite simple groups. The nonabelian part of the socle is first computed in [Babai et al., 1983] whereas the abelian part, based on [Rónyai, 1990], is obtained in [Kantor and Luks, 1990].

The solution of (o) and computing $O^{\Sigma}(G)$ is implicit in [Babai et al., 1987]. The subgroups $O_p(G)$ are obtained in [Kantor, 1985b] and [Neumann, 1986] and the general case of $O_{\Sigma}(G)$ is handled in [Kantor and Luks, 1990].

The paper [Kantor and Luks, 1990] generalized the entire polynomial-time toolkit to work with quotient groups G/K of $G \leq S_n$. The trivial approach, namely finding a permutation representation of the factor group, does not work: In Exercise 6.6, we shall give examples $K \lhd G \leq S_n$ from [Neumann, 1986] for an infinite sequence of n values so that no faithful permutation representation of G/K acts on less than $2^{n/4}$ points. The quotient group machinery required a radically new approach and it is based on Sylow subgroup computations. Therefore, it is much more difficult to compute, for example, the upper central series of G than to compute $Z(G)$.

We shall present algorithms for most of the tasks listed in our toolkit. However, in most cases, these will *not* be the deterministic polynomial-time versions. Rather, we shall describe faster, randomized algorithms, which we define in the next section.

3.2. Nearly Linear-Time Algorithms

A lot of interesting groups do not have small-degree permutation representations; however, in their permutation representations as $G \leq S_n$, $\log |G|$ is small compared to n. This phenomenon can be formalized in the following way. We call an (infinite) family $\mathcal{G}$ of permutation groups *small-base groups* if each $G \in \mathcal{G}$ of degree n satisfies $\log |G| < \log^c n$ for some fixed constant c. Important families of groups, including all permutation representations of all finite simple groups except the alternating ones, belong to this category (with $c = 2$). Primitive groups not containing alternating composition factors in their socle are also small-base groups.

Permutation group algorithms belong to the class of procedures where the running time may differ significantly on inputs of the same size, because it usually depends both on the degree and on the order of the group the given permutations generate. The running time of most of the polynomial-time algorithms surveyed in Section 3.1 is at least $\Omega(n^2)$. However, on tens of thousands of points, even a $\Theta(n^2)$ algorithm may be prohibitively slow (and using $\Theta(n^2)$ memory is currently out of the question). Therefore, to handle small-base groups of large degree, we have to find alternatives to the polynomial-time methods that are more efficient on small-base inputs. We are interested in algorithms with running time estimates of the form $O(n|S| \log^c |G|)$; we call such procedures *nearly linear-time algorithms* because, for small-base groups, they run in *nearly linear*, $O(n|S|\log^{c'}(n|S|))$, time of the input length. Using the soft version of the big-O notation (cf. Section 1.2), the time bound of nearly linear-time algorithms on small-base input groups is $O^{\sim}(n|S|)$.

In Chapters 4–6, we shall see that there are nearly linear-time versions for a significant part of the polynomial-time library: [Beals, 1993a], [Schönert and Seress, 1994] for (a), [Babai et al., 1991] for (b)–(g), [Beals and Seress, 1992] for (h) and (i), [Morje, 1995] for (j) and for (l) in groups with no composition factors of exceptional Lie type, [Holt and Rees, 1994] and [Ivanyos and Lux, 2000] for (k), and [Luks and Seress, 1997] for $O_{\Sigma}(G)$ in the cases when Σ consists of one cyclic group or it is the family of all cyclic groups. The price we pay for the speedup is that, with a few exceptions, the algorithms are randomized Monte Carlo (cf. Section 1.3 for the definition), and so they may return incorrect answers. Also, these algorithms may not have the lowest time complexity on general inputs. For example, the version of basic permutation group manipulation from [Babai et al., 1991], which we present in Section 4.5, runs in $O(n\log^4|G|\log n+|S|n\log|G|)$ time, which is $O^{\sim}(n^5+|S|n^2)$ in the worst case. However, there are $O^{\sim}(n^4+|S|n^2)$ deterministic (cf. [Babai et al., 1997b]) and $O^{\sim}(n^3+|S|n)$ Monte Carlo (cf. [Babai et al., 1995]) algorithms for the same task. We mention, however, that usually some of the $\log|G|$ factors in the running time estimates can be replaced by a bound on the length of subgroup chains, which may be significantly smaller than $\log|G|$.

A recent direction of research is to upgrade the Monte Carlo nearly linear-time algorithms to Las Vegas–type algorithms. The papers [Kantor and Seress, 1999, 2001] establish a general framework for the upgrade, which we shall discuss in Section 8.3.

Most of the nearly linear-time algorithms listed above are implemented in GAP and are available in the library of the standard GAP distribution. Practical computations almost exclusively deal with small-base input groups. The nearly linear-time algorithms usually run faster than quadratic ones even in the individual cases when the factor $\log^c|G|$ occurring in their running time estimate is larger than n. Because of their practical importance, we devote a significant portion of this book to nearly linear-time algorithms.

3.3. Non-Polynomial-Time Methods

Some important problems do not have (at present) polynomial-time solutions. They are interesting both in theory and practice: On one hand, the graph isomorphism problem is reducible to them in polynomial time; on the other hand, they are important tools in the study of groups.

Examples of such problems include instances of the following, all of which are reducible to each other in polynomial time (cf. [Luks, 1993]):

(a) Given $\Delta \subseteq \Omega$, compute the setwise stabilizer $G_\Delta = \{g \in G \mid \Delta^g = \Delta\}$.
(b) Given $H, G \le \mathrm{Sym}(\Omega)$, compute $C_G(H)$.
(c) Given $H, G \le \mathrm{Sym}(\Omega)$, compute $H \cap G$.
(d) Given $H, G \le \mathrm{Sym}(\Omega)$ and $x_1, x_2 \in \mathrm{Sym}(\Omega)$, are the double cosets Hx_1G and Hx_2G equal?
(e) Given $x_1, x_2 \in G$, decide whether they are conjugate.

Although all known algorithms for these problems have exponential worst-case complexity, they are not considered difficult in practice. Further research is needed to find the reason for this phenomenon. It is improbable that the decision problems (d) and (e) are NP-complete; otherwise the polynomial-time hierarchy of computational problems would collapse (see [Babai and Moran, 1988]; we refer the reader to [Garey and Johnson, 1979] for the definitions regarding complexity classes). Moreover, it is concievable that there may be polynomial-time algorithms (at least for the classes of groups occurring in practice) to solve them; or the *average case* running time of the implemented methods may be polynomial. However, it is not even clear from which probability distribution the "average" input should be chosen. In the uniform distribution on the subgroups of S_n, a typical permutation group $G \le S_n$ has large order: $\log |G|$ is $\Omega(n)$ (see [Pyber, 1993]). However, some heuristics indicate that almost all $G \le S_n$ are nilpotent, and for nilpotent groups, all instances of the above list are solvable in polynomial time. More generally, there are polynomial-time solutions if the non-abelian composition factors of G have bounded order. [Luks, 1993, Section 6] contains a unified treatment of all cases (a)–(e).

An important problem, which is considered more challenging than those mentioned in (a)–(e) both in theory and practice, is the computation of normalizers. It is not even known whether $N_{\mathrm{Sym}(\Omega)}(G)$ is computable in polynomial time; for comparison, we mention that $C_{\mathrm{Sym}(\Omega)}(G)$ is easily obtainable (cf. Section 6.1.2). Also, it is an open issue whether the normalizer problem is polynomially equivalent to (a)–(e).

The practical way to attack problems (a)–(e) and the normalizer problem is by *backtrack methods* (cf. Chapter 9). These are systematic considerations of all elements of G, in which only those that satisfy the desired property are chosen. Various tricks are built-in to eliminate large parts (entire cosets of certain subgroups) of G at once, by establishing that the coset cannot contain any of the desired elements or that we already constructed all such elements.

Backtrack methods for permutation groups were first described in [Sims, 1971b] where the construction of centralizers was presented. Since then, a large library of algorithms has been developed; we mention [Butler, 1982] for a general description of backtrack searches, [Butler, 1983] and [Holt, 1991] for computing normalizers, and [Butler and Cannon, 1989] for computing Sylow subgroups. A second generation of backtrack algorithms is developed in [Leon, 1991], where ideas from the graph isomorphism program *nauty* from [McKay, 1981] are utilized. The thesis [Theißen, 1997] extends Leon's method to obtain a very efficient normalizer algorithm.

4

Bases and Strong Generating Sets

4.1. Basic Definitions

In this section, we define the fundamental data structures in [Sims, 1970, 1971a] for efficient permutation group manipulation. Suppose that $G \leq \mathrm{Sym}(\Omega)$ and $|\Omega| = n$; in fact, in this chapter we can identify Ω with $\{1, 2, \ldots, n\}$.

A sequence of elements $B = (\beta_1, \ldots, \beta_m)$ from Ω is called a *base* for G if the only element of G to fix B pointwise is the identity. The sequence B defines a subgroup chain

$$G = G^{[1]} \geq G^{[2]} \geq \cdots \geq G^{[m]} \geq G^{[m+1]} = 1, \tag{4.1}$$

where $G^{[i]} := G_{(\beta_1,\ldots,\beta_{i-1})}$ is the pointwise stabilizer of $\{\beta_1, \ldots, \beta_{i-1}\}$. The base is called *nonredundant* if $G^{[i+1]}$ is a proper subgroup of $G^{[i]}$ for all $i \in [1, m]$. Different nonredundant bases can have different size (cf. Exercise 4.2).

By repeated applications of Lagrange's theorem, we obtain

$$|G| = \prod_{i=1}^{m} \left|G^{[i]} : G^{[i+1]}\right|. \tag{4.2}$$

Because the cosets of $G^{[i]}$ mod $G^{[i+1]}$ correspond to the elements of the orbit $\beta_i^{G^{[i]}}$, we obtain $|G^{[i]} : G^{[i+1]}| = |\beta_i^{G^{[i]}}| \leq n$ for all $i \in [1, m]$. Moreover, if B is nonredundant then $|G^{[i]} : G^{[i+1]}| \geq 2$. These inequalities, combined with (4.2), immediately yield $2^{|B|} \leq |G| \leq n^{|B|}$, or $\log|G|/\log n \leq |B| \leq \log|G|$. The last inequality justifies the name "small-base group" introduced in Section 3.2: When $|G|$ is small, so is $|B|$.

A *strong generating set* (SGS) for G relative to B is a generating set S for G with the property that

$$\left\langle S \cap G^{[i]}\right\rangle = G^{[i]}, \text{ for } 1 \leq i \leq m+1. \tag{4.3}$$

As a simple example, consider the group $G = S_4$, in its natural action on the set $[1,4]=\{1,2,3,4\}$. The sequence $B=(1,2,3)$ is a nonredundant base for G, since $G^{[1]}=\mathrm{Sym}([1,4])\gneqq G^{[2]}=\mathrm{Sym}([2,4])\gneqq G^{[3]}=\mathrm{Sym}([3,4])\gneqq G^{[4]}=1$. The sets $S=\{(1,2,3,4),(3,4)\}$ and $T=\{(1,2,3,4),(2,3,4),(3,4)\}$ generate G, but S is *not* a strong generating set relative to B since $\langle S\cap G^{[2]}\rangle = \mathrm{Sym}([3,4]) \neq G^{[2]} = \mathrm{Sym}([2,4])$. In contrast, T is an SGS relative to B.

Given an SGS, the orbits $\beta_i^{G^{[i]}}$ can be easily computed (cf. Theorem 2.1.1(i)). These orbits are called the *fundamental orbits* of G. By (4.2), from the orbit sizes we immediately get $|G|$. Also, keeping track of elements of $G^{[i]}$ in the orbit algorithm that carry β_i to points in $\beta_i^{G^{[i]}}$, we obtain transversals R_i for $G^{[i]}$ mod $G^{[i+1]}$. We always require that the representative of the coset $G^{[i+1]}\cdot 1 = G^{[i+1]}$ is the identity.

The Sifting Procedure

By Lagrange's theorem, every $g\in G$ can be written uniquely in the form $g = r_m r_{m-1}\cdots r_1$ with $r_i\in R_i$. This decomposition can be done algorithmically: Given $g\in G$, find the coset representative $r_1\in R_1$ such that $\beta_1^g=\beta_1^{r_1}$. Then compute $g_2:=gr_1^{-1}\in G^{[2]}$; find $r_2\in R_2$ such that $\beta_2^{g_2}=\beta_2^{r_2}$; compute $g_3:=g_2r_2^{-1}\in G^{[3]}$; etc. This factorization procedure is called *sifting* and may be considered as a permutation group version of Gaussian elimination.

Sifting can also be used for *testing membership* in G. Given $h\in\mathrm{Sym}(\Omega)$, we attempt to factor h as a product of coset representatives. If the factorization is successful then $h\in G$. Two things may go awry: It is possible that, for some $i\le m$, the ratio $h_i:=hr_1^{-1}r_2^{-1}\cdots r_{i-1}^{-1}$ computed by the sifting procedure carries β_i out of the orbit $\beta_i^{G^{[i]}}$; the other possibility is that $h_{m+1}:=hr_1^{-1}r_2^{-1}\cdots r_{m-1}^{-1}r_m^{-1}\neq 1$. In both cases, obviously $h\notin G$. The ratio h_i with the largest index i ($i\le m+1$) computed by the sifting procedure is called the *siftee* of h.

Computing and storing a transversal explicitly may require $\Theta(n^2)$ time and memory. To avoid that, transversals can be stored in Schreier trees. A *Schreier tree data structure* for G is a sequence of pairs (S_i,T_i) called *Schreier trees*, one for each base point β_i, $1\le i\le m$. Here T_i is a directed labeled tree, with all edges directed toward the root β_i and edge labels from a set $S_i\subseteq G^{[i]}$. The vertices of T_i are the points of the fundamental orbit $\beta_i^{G^{[i]}}$. The labels satisfy the condition that for each directed edge from γ to δ with label h, $\gamma^h=\delta$. If γ is a vertex of T_i then the sequence of the edge labels along the unique path from γ to β_i in T_i is a word in the elements of S_i such that the product of these permutations moves γ to β_i. Thus each Schreier tree (S_i,T_i) defines *inverses* of a set of coset representatives for $G^{[i+1]}$ in $G^{[i]}$.

We store inverses of coset representatives in the Schreier trees because sifting requires the inverses of these transversal elements.

Besides the $O(|S_i|n)$ memory requirement to store the permutations in S_i, storing T_i requires only $O(n)$ additional memory, because T_i can be stored in an array V_i of length n. The γth entry of V_i is defined if and only if $\gamma \in \beta_i^{G^{[i]}}$, and in this case $V_i[\gamma]$ is a pointer to the element of S_i that is the label of the unique directed edge of T_i starting at γ. Because of this representation as an array, Sims originally called Schreier trees *Schreier vectors*.

Continuing our example $G = S_4$ with base $B = (1, 2, 3)$ and SGS $T = \{(1, 2, 3, 4), (2, 3, 4), (3, 4)\}$, let us construct Schreier trees for G using the label sets $S_i := T \cap G^{[i]}$. The trees T_i can be constructed as the breadth-first-search trees, which compute the orbits $\beta_i^{G^{[i]}}$ (cf. Section 2.1.1) but, since the edges of the trees must be directed toward the roots, we have to use the inverses of the elements of S_i in the construction of the T_i. The label set S_i determines uniquely only the levels of the tree T_i, because the vertices on level j may be the images of more vertices on level $j-1$, under more permutations. For example, in T_1, level 0 contains the point 1, level 1 contains only the point 4 (since 4 is the only point that is the image of 1 under the inverse of some element of S_1), and $(1, 2, 3, 4)$ is the only possible label for the edge $\overrightarrow{(4, 1)}$. Level 2 consists only of the point 3, but we have three possibilities for defining the label of $\overrightarrow{(3, 4)}$ because the inverses of $(1, 2, 3, 4)$, $(2, 3, 4)$, and $(3, 4)$ all map 4 to 3. Therefore, the label of the edge $\overrightarrow{(3, 4)}$ depends on the order in which the inverses of the elements of S_1 are applied to take the image of 4. One possibility for the three Schreier trees is coded in the arrays $((), (2, 3, 4), (2, 3, 4), (1, 2, 3, 4))$, $(*, (), (2, 3, 4), (2, 3, 4))$, and $(*, *, (), (3, 4))$. Here $()$ denotes the identity permutation and $*$ denotes that the appropriate entry of the array is not defined because the corresponding point is not in the fundamental orbit of β_i. If, for example, we need a transversal element carrying the first base point 1 to 3 then from the first array we obtain that $(2, 3, 4) \cdot (1, 2, 3, 4) = (1, 2, 4, 3)$ maps 3 to 1, and its inverse is the desired transversal element.

4.2. The Schreier–Sims Algorithm

Strong generating set constructions are based on the following lemma by O. Schreier (hence giving rise to the name Schreier–Sims algorithm). Recall that if $H \leq G$ and R is a right transversal for G mod H, then for $g \in G$ we denote the unique element of $Hg \cap R$ by $\bar{g}$.

Lemma 4.2.1. *Let $H \leq G = \langle S \rangle$ and let R be a right transversal for G mod H, with $1 \in R$. Then the set*

$$T = \{rs(\overline{rs})^{-1} \mid r \in R, s \in S\}$$

generates H.

The elements of T are called *Schreier generators* for H.

Proof. By definition, the elements of T are in H, so it is enough to show that $T \cup T^{-1}$ generates H. Note that $T^{-1} = \{rs(\overline{rs})^{-1} \mid r \in R, s \in S^{-1}\}$.

Let $h \in H$ be arbitrary. Since $H \leq G$, h can be written in the form $h = s_1 s_2 \cdots s_k$ for some nonnegative integer k and $s_i \in S \cup S^{-1}$ for $i \leq k$. We define a sequence $h_0, h_1, \ldots, h_k$ of group elements such that

$$h_j = t_1 t_2 \cdots t_j r_{j+1} s_{j+1} s_{j+2} \cdots s_k \tag{4.4}$$

with $t_i \in T \cup T^{-1}$ for $i \leq j$, $r_{j+1} \in R$, and $h_j = h$. Let $h_0 := 1 s_1 s_2 \cdots s_k$. Recursively, if h_j is already defined then let $t_{j+1} := r_{j+1} s_{j+1} (\overline{r_{j+1} s_{j+1}})^{-1}$ and $r_{j+2} := \overline{r_{j+1} s_{j+1}}$. Clearly, $h_{j+1} = h_j = h$, and it has the form required in (4.4).

We have $h = h_k = t_1 t_2 \cdots t_k r_{k+1}$. Since $h \in H$ and $t_1 t_2 \cdots t_k \in \langle T \rangle \leq H$, we must have $r_{k+1} \in H \cap R = \{1\}$. Hence $h \in \langle T \rangle$. □

Remark 4.2.2. In this book we deal only with finite groups, and so every element h of a given group $G = \langle S \rangle$ can be written as a product $h = s_1 s_2 \cdots s_k$ of generators and we do not have to deal with the possibility that some s_i is the inverse of a generator. In the proof of Lemma 4.2.1, we included the possibility $s_i \in S^{-1}$ since this lemma is valid for infinite groups as well, and in an infinite group we may need the inverses of generators to write every group element as a finite product.

In the analysis of the Schreier–Sims algorithm, we shall also use the following observation from [Sims, 1971a].

Lemma 4.2.3. *Let* $\{\beta_1, \ldots, \beta_k\} \subseteq \Omega$ *and* $G \leq \mathrm{Sym}(\Omega)$. *For* $1 \leq j \leq k+1$, *let* $S_j \subseteq G_{(\beta_1, \ldots, \beta_{j-1})}$ *such that* $\langle S_j \rangle \geq \langle S_{j+1} \rangle$ *holds for* $j \leq k$. *If* $G = \langle S_1 \rangle$, $S_{k+1} = \emptyset$, *and*

$$\langle S_j \rangle_{\beta_j} = \langle S_{j+1} \rangle \tag{4.5}$$

holds for all $1 \leq j \leq k$ *then* $B = (\beta_1, \ldots, \beta_k)$ *is a base for* G *and* $S = \bigcup_{1 \leq j \leq k} S_j$ *is an SGS for* G *relative to* B.

Proof. We use induction on $|\Omega|$. Our inductive hypothesis is that $S^* := \bigcup_{2 \leq j \leq k} S_j$ is an SGS for $\langle S_2 \rangle$, relative to the base $B^* := (\beta_2, \ldots, \beta_k)$. Let $G^{[i]} := G_{(\beta_1, \ldots, \beta_{i-1})}$; we have to check that (4.3) holds for $2 \leq i \leq k+1$. For $i = 2$, (4.3) holds since applying (4.5) with $j = 1$, we obtain $G_{\beta_1} = \langle S_2 \rangle \leq \langle S \cap G_{\beta_1} \rangle$. The reverse containment is obvious. For $i > 2$, (4.3) follows from the fact that $S^* \cap G_{(\beta_1,\ldots,\beta_{i-1})}$ generates $\langle S_2\rangle_{(\beta_2,\ldots,\beta_{i-1})}$ by the inductive hypothesis, and so $G^{[i]} \geq \langle S \cap G_{(\beta_1,\ldots,\beta_{i-1})}\rangle \geq \langle S^* \cap G_{(\beta_1,\ldots,\beta_{i-1})}\rangle = \langle S_2\rangle_{(\beta_2,\ldots,\beta_{i-1})} = (G_{\beta_1})_{(\beta_2,\ldots,\beta_{i-1})} = G^{[i]}$. □

The Schreier–Sims Algorithm

Given $G = \langle T\rangle$, we can construct an SGS in the following way. We maintain a list $B = (\beta_1, \ldots, \beta_m)$ of already known elements of a nonredundant base and an approximation S_i for a generator set of the stabilizer $G_{(\beta_1,\ldots,\beta_{i-1})}$, for $1 \leq i \leq m$. During the construction, we always maintain the property that, for all i, $\langle S_i\rangle \geq \langle S_{i+1}\rangle$. We say that the data structure is *up to date below level* j if (4.5) holds for all i satisfying $j < i \leq m$.

In the case when the data structure is up to date below level j, we compute a transversal R_j for $\langle S_j\rangle$ mod $\langle S_j\rangle_{\beta_j}$. Then we test whether (4.5) holds for $i = j$. By Lemma 4.2.1, this can be done by sifting the Schreier generators obtained from R_j and S_j in the group $\langle S_{j+1}\rangle$. (In this group, membership testing is possible, because Lemma 4.2.3 implies that we have a strong generating set for $\langle S_{j+1}\rangle$.) If all Schreier generators are in $\langle S_{j+1}\rangle$ then the data structure is up to date below level $j-1$; otherwise, we add a nontrivial siftee h of a Schreier generator to S_{j+1} and obtain a data structure that is up to date below level $j+1$. In the case $j = m$, we also choose a new base point β_{m+1} from supp(h).

We start the algorithm by choosing $\beta_1 \in \Omega$ that is moved by at least one generator in T and setting $S_1 := T$. At that moment, the data structure is up to date below level 1; the algorithm terminates when the data structure becomes up to date below level 0. Lemma 4.2.3 immediately implies correctness.

As an example, let us follow the construction of an SGS for the input $G = \langle (2,3), (1,2,4)\rangle \leq \mathrm{Sym}([1,4])$. We start by choosing 2 as the first base point, as the first element of $\Omega = [1,4]$ moved by the first generator of G. (Note that although generating sets and orbits are sets, in the computer they are stored as sequences and so we may talk about their first element.) We define $S_1 := \{(2,3), (1,2,4)\}$ and compute the Schreier tree coded by the array $((1,2,4), (), (2,3), (1,2,4))$. At that moment, the data structure is up to date below level 1. Next, we have to check whether the Schreier generators for the point stabilizer G_2 are in the trivial group, which is the current data structure below level 1. The first Schreier generator is composed from the coset representative $r = (1,4,2)$, which carries the base point 2 to 1, and from the generator $s = (2,3)$. So $rs = (1,4,3,2)$, $\overline{rs} = (1,4,2)$, and $rs(\overline{rs})^{-1} = (3,4)$. The siftee of $(3,4)$ in the trivial group is obviously itself and not the identity, so we define the new base point 3 (as the first element moved by the siftee). We also define $S_2 := \{(3,4)\}$ and compute the Schreier tree coded by $(*, *, (), (3,4))$.

At that moment, the data structure is up to date below level 2. Checking the Schreier generators reveals that we have a correct SGS for the group $\langle S_2 \rangle$, so we can step back a level and the data structure is up to date below level 1. The next Schreier generator on level 1 with nontrivial siftee is composed from $r = (2, 3)$, which carries the base point 2 to 3, and from the generator $s = (1, 2, 4)$. Now $rs = (1, 2, 3, 4)$, $\overline{rs} = (2, 3)$, and $rs(\overline{rs})^{-1} = (1, 3, 4)$. The siftee of $(1, 3, 4)$ in $\langle S_2 \rangle$ is $(1, 4)$, so we have to redefine S_2 as $S_2 := \{(3, 4), (1, 4)\}$ and recompute the second Schreier tree. The new tree is coded by the array $((1, 4), *, (), (3, 4))$. At that moment, the data structure is up to date below level 2. For the coset representative $r = ((1, 4) \cdot (3, 4))^{-1} = (3, 1, 4)$ carrying the base point 3 to 1 and for $s = ((3, 4) \in S_2$, we obtain $rs = (1, 3)$, $\overline{rs} = (3, 1, 4)$, and $rs(\overline{rs})^{-1} = (1, 4)$. The siftee of $(1, 4)$ is itself, so we add 1 to the base, define $S_3 := \{(1, 4)\}$, and compute the third Schreier tree coded in the array $((), *, *, (1, 4))$. At that moment, the data structure is up to date below level 3. After that, we check that all Schreier generators composed on levels 3, 2, 1 have trivial siftees, and so the data structure is up to date below level 0 and we can terminate the algorithm. The output is the base $(2, 3, 1)$ and the SGS $\{(2, 3), (1, 2, 4), (3, 4), (1, 4)\}$.

Because of the fundamental importance of SGS constructions for permutation group computations, we do a detailed complexity analysis. We analyze two versions. In the first one, we store transversals explicitly; in the second one, transversals are stored in Schreier trees.

The number of base points is at most $\log |G|$. For a fixed base point β_k, the set S_k changes at most $\log |G|$ times during the algorithm, since the group $\langle S_k \rangle$ must increase each time we add an element to S_k. We have to recompute the transversal R_k after each change of S_k. This can be done by an orbit computation as in Theorem 2.1.1(i) (using the inverses of the elements of S_k, since the edges in the Schreier tree are directed toward the root, and now we proceed in the opposite direction). However, we have to find the image of a point in the orbit of β_k under an element of S_k *only once* during the entire algorithm; hence the time spent in image computations (for all base points, during the entire algorithm) is $O(|B| \log |G| n + |T| n)$, which is $O(n \log^2 |G| + |T| n)$.

In the first version, we also have to compute transversals explicitly. For a fixed base point, this may require $O(n)$ permutation multiplications at cost $O(n)$ each, so the total time spent in transversal computations is $O(n^2 \log |G|)$. In the second version, the bookkeeping necessary to set up the Schreier trees can be done in $O(n \log |G|)$ total time.

Next, we estimate the time needed to sift Schreier generators. Again, we observe that although the sets R_k and S_k change, they are always only augmented and therefore any elements of R_k and S_k must be combined to a

Schreier generator only once. The total number of Schreier generators is $\sum_k |R_k||S_k| \in O(n\log^2|G| + |T|n)$. In the first version, sifting a Schreier generator can be done by $O(\log|G|)$ permutation multiplications, so the total cost is $O(n^2\log^3|G| + |T|n^2\log|G|)$. In the second version, recovering a coset representative from a Schreier tree may require $O(n)$ permutation multiplications, so the cost of one sift is $O(n^2\log|G|)$; the total cost is $O(n^3\log^3|G| + |T|n^3\log|G|)$.

The first version stores $\sum_k |S_k| \in O(\log^2|G|)$ strong generators and $\sum |R_k| \in O(n\log|G|)$ coset representatives; therefore it requires $O(n^2\log|G| + |T|n)$ memory, whereas the second one needs only $O(n\log^2|G| + |T|n)$. Note that we combined the terms $\log^2|G|$ and $n\log|G|$ to $n\log|G|$, which is incorrect if $\log|G| \notin O(n)$. However, the stated memory requirements are valid even in this case, since $|S_k|$ can also be estimated from above by the length of the longest subgroup chain in G. This length is always less than $3n/2$ (cf. [Cameron et al., 1989]), so $\sum_k |S_k| \in O(n\log|G|)$.

Hence we have proved the following theorem.

Theorem 4.2.4. *Given $G = \langle T\rangle$, a strong generating set for G can be computed by deterministic algorithms in $O(n^2\log^3|G| + |T|n^2\log|G|)$ time using $O(n^2\log|G| + |T|n)$ memory or in $O(n^3\log^3|G| + |T|n^3\log|G|)$ time using $O(n\log^2|G| + |T|n)$ memory.*

Some remarks are in order. First, we note that in both versions a factor n can be shaved off from the term including $|T|$ in the time analysis (Exercise 4.4).

The time–space tradeoff in complexity demonstrated in Theorem 4.2.4 is quite common in computational group theory. Because of hardware restrictions, we usually choose versions with smaller memory usage.

In the analysis of the second version, we estimated the depth of Schreier trees by n. As a cyclic group with one generator shows, this bound can be achieved; however, practical experience indicates that, for most groups and generating sets, a breadth-first-search Schreier tree will have significantly smaller depth. Nevertheless, even in such nicely behaving groups, Schreier trees with large depth may occur! During the construction, when we define a new base point, the last group in our stabilizer chain is temporarily a cyclic group with one generator. If the group contains elements of large order, this cyclic group may define a Schreier tree of large depth. When a second generator is added to this level, the breadth-first Schreier tree defined by the two generators may have small depth; however, if we always augment previous Schreier trees, a long branch remains in the tree forever. (This situation occurs for example in the group $\mathrm{GL}_d(q)$, acting as a permutation group on the q^d points of the underlying vector space.) Hence, it may be useful to recompute the ith Schreier tree from scratch each time a new generator is added to S_i. Depending on the types of groups with which we work, this recomputation may mean savings in the sifting part of the algorithm or it may be just a waste of time because the augmented tree works just as well. We brought up this issue to indicate the nature of problems we may face at implementation, in addition to the complexity considerations.

For arbitrary inputs, the worst-case complexity of the first version is $O(n^5)$ (as in the memory estimates, the extra $\log n$ factors can be avoided by using stronger bounds than the trivial $\log(n!)$ on the length of subgroup chains). [Knuth, 1991] gives a family of generators for S_n for which the running time of the algorithm is indeed $\Theta(n^5)$. Based on experiments, we believe that, using random generators for S_n, the expected running time is $\Theta(n^4)$. In Section 10.2, we shall describe faster algorithms for the recognition and effective handling of symmetric and alternating groups.

SGS constructions are essential parts of almost all permutation group algorithms since strong generating sets provide the only known means to test membership in groups. Therefore, it is of utmost importance to have efficient ways to construct strong generating sets. This importance is the reason why we deal at length with the various attempts to speed up the Schreier–Sims algorithm.

4.3. The Power of Randomization

The time-critical part of the Schreier–Sims procedure is the sifting of Schreier generators. Therefore, a possible way to speed up the algorithm is to find a fast solution of the following problem:

$$\text{Given } G = \langle S \rangle, \text{ a transversal } R \text{ for } G \bmod G_\beta, \text{ and } H \leq G_\beta, \text{ decide whether } H = G_\beta. \tag{4.6}$$

Of course, we want to apply the solution in the case when $G = \langle S_i \rangle$ and $H = \langle S_{i+1} \rangle$ for two consecutive levels during the SGS construction.

Practical experience shows that if $H \neq G_\beta$ then *usually* at least half of the Schreier generators constructed from R and S are in $G_\beta \backslash H$. Therefore, testing only a random sample of Schreier generators, we obtain a *heuristic* solution for (4.6), and thus a heuristic randomized algorithm for SGS construction.

We caution that, although testing a small number of randomly chosen Schreier generators usually suffices, it is easy to construct examples when there is *only one* generator $g \in S$ such that Schreier generators obtained from g and R may show $H \neq G_\beta$, because for all $k \in S \backslash \{g\}$ and $r \in R$, $rk(\overline{rk})^{-1} \in H$ (cf. Exercise 4.5). Nevertheless, it is possible to extend the heuristics above to an algorithm with rigorous probability analysis, leading to a fast Monte Carlo SGS construction. We shall describe this extension in Section 4.5.

Another possibility for a randomized SGS construction is based on the fact that if an alleged SGS is not correct then, with probability at least 1/2, a uniformly distributed random element of G does not sift through the corresponding transversal system. More precisely, let $G = \langle S \rangle$, $B = (\beta_1, \ldots, \beta_m) \subseteq \Omega$ such that no element of S fixes B pointwise, and let $S_i := S \cap G_{(\beta_1,\ldots,\beta_{i-i})}$ for $1 \leq i \leq m+1$. For $1 \leq i \leq m$, we can compute the orbits $\beta_i^{\langle S_i \rangle}$ and sets of group elements $R_i \subseteq \langle S_i \rangle$ such that for all $\gamma \in \beta_i^{\langle S_i \rangle}$ there is a unique $r_\gamma \in R_i$ with $\beta_i^{r_\gamma} = \gamma$. (Note that R_i may be only a proper subset of a transversal for $\langle S_i \rangle$ mod $\langle S_{i+1} \rangle$ if S is not an SGS for G. The R_i may be computed explicitly or stored in a Schreier tree data structure.) Given $h \in \mathrm{Sym}(\Omega)$, we can apply the sifting procedure to attempt to factor h in the form $h = r_m \cdots r_1$ with $r_i \in R_i$. We say that h *sifts through* if this factorization procedure succeeds.

Lemma 4.3.1. *Let* $G = \langle S \rangle$, $B = (\beta_1, \ldots, \beta_m)$, *and* S_i, *for* $1 \leq i \leq m+1$, *be as in the previous paragraph. If B is not a base or S is not a strong generating set relative to B then the probability that a uniformly distributed random* $g \in G$ *does not sift through the transversal system built from S is at least* $1/2$.

Proof. Let i be the largest integer such that $\langle S_i \rangle_{\beta_i} \gneqq \langle S_{i+1} \rangle$. By Lemma 4.2.3, such an i exists and we have a correct SGS for $\langle S_{i+1} \rangle$. In particular, we can test membership in the group $\langle S_{i+1} \rangle$.

Let p be the probability that a given uniformly distributed random $g \in G$ sifts through the first i levels of our data structure and, in the case when g sifts through, let $\bar{g}$ denote the unique element of $R_i \cdots R_2 R_1$, which is the product of transversal elements computed by the sifting procedure. Then $g(\bar{g})^{-1} \in G_{(\beta_1,\ldots,\beta_i)}$, and every element of $G_{(\beta_1,\ldots,\beta_i)}$ has the same chance to occur as $g(\bar{g})^{-1}$ for some $g \in G$. Hence $\mathrm{Prob}(g(\bar{g})^{-1} \in \langle S_{i+1} \rangle) = 1/|G_{(\beta_1,\ldots,\beta_i)} : \langle S_{i+1} \rangle|$. We know that $|G_{(\beta_1,\ldots,\beta_i)} : \langle S_{i+1} \rangle| \geq 2$, since $G_{(\beta_1,\ldots,\beta_i)} \geq \langle S_i \rangle_{\beta_i} \gneqq \langle S_{i+1} \rangle$; we also know that $g(\bar{g})^{-1} \in \langle S_{i+1} \rangle$ if and only if it sifts through the transversal system of $\langle S_{i+1} \rangle$. In summary, g does not sift through with probability $(1-p) + (1 - 1/|G_{(\beta_1,\ldots,\beta_i)} : \langle S_{i+1} \rangle|)p \geq 1 - p + p/2 \geq 1/2$. □

To get a better approximation of an SGS, a nontrivial siftee can be used to augment S. Hence, *if* uniformly distributed random elements of G are available, we have a Monte Carlo algorithm that, with arbitrarily small error probability prescribed by the user, constructs an SGS. A version of this procedure is described and analyzed in more detail in Lemma 5.4.1.

The Random Schreier–Sims Algorithm

If uniformly distributed random elements are not available then we have to resort to one of the methods for random element generation described in Section 2.2. The first practical implementation is described in [Leon, 1980b, Section 7]. Random elements are constructed via the first method described in Section 2.2, as the vertices traversed during a random walk on the Cayley graph $\Gamma(G, T)$ defined by the input $G = \langle T \rangle$. Leon uses the stopping criterion that twenty consecutively constructed random elements of G sift through. The term "random Schreier–Sims algorithm" in the literature usually refers to Leon's algorithm.

Similar SGS computations based on the other two methods for random element selection described in Section 2.2 are not implemented. In recent releases of GAP, a heuristic speedup of the nearly linear-time SGS construction of Section 4.5 is used to replace the random Schreier–Sims algorithm.

In practice, fast heuristic SGS constructions are used in conjunction with independent *strong generating tests* to check the correctness of the result. The simplest such test, which can be applied quite frequently in practice, is when the order of G is known in advance. Because the methods never put permutations in the SGS that are not elements of the input group, the possible error is always one-sided: Namely, the group order computed by (4.2) from the alleged SGS is never greater than the true value. Therefore, if the computed and the true values are the same then the construction is correct.

We postpone the description of further strong generating tests to Chapter 8, when more algorithmic machinery will be available to us.

4.4. Shallow Schreier Trees

For large values of n, storing transversals for a point stabilizer chain of some $G \le S_n$ explicitly requires too much memory. However, if transversals are stored in Schreier trees then the sifting process slows down proportionally to the depth of trees. Hence we are interested in methods for constructing Schreier trees that have both small depth and use only a small number of edge labels.

In this section, we present two constructions: A deterministic one from [Babai et al., 1991] and a randomized one from [Cooperman et al., 1990]. Both methods manipulate *subsets* of groups, which themselves are not subgroups.

Given a sequence of elements $(g_1, \ldots, g_k) \subseteq G \le \mathrm{Sym}(\Omega)$, $|\Omega| = n$, the *cube* C_k of these elements is defined as

$$C_k = \left\{ g_1^{\varepsilon_1} g_2^{\varepsilon_2} \cdots g_k^{\varepsilon_k} \;\middle|\; \varepsilon_i \in \{0, 1\} \right\}. \tag{4.7}$$

Also, let $C_k^{-1} := \{g \in G \mid g^{-1} \in C_k\}$. Clearly, $|C_k| \le 2^k$; we say that the cube is *nondegenerate* if equality is attained. Cubes were introduced in [Babai and Szemerédi, 1984] in a highly nonconstructive setting: They were used to establish that testing membership in black-box groups is in the complexity class NP.

Nondegenerate cubes are usually obtained by repeated applications of the following simple observation.

Lemma 4.4.1. *Let* $g_1, \ldots, g_k, g_{k+1} \in G$. *Then* $|C_{k+1}| = 2|C_k|$ *if and only if* $g_{k+1} \notin C_k^{-1}C_k$.

Proof. $|C_{k+1}| = 2|C_k| \iff C_k g_{k+1} \cap C_k = \emptyset \iff g_{k+1} \notin C_k^{-1}C_k$. □

Testing membership in $C_k^{-1}C_k$ is difficult and no method other than essentially listing all elements is known. However, as observed in [Babai et al., 1991], it is easy to check the following sufficient condition for nonmembership: If, for some $\alpha \in \Omega$, $\alpha^g \notin \alpha^{C_k^{-1}C_k}$ then $g \notin C_k^{-1}C_k$. The set $\alpha^{C_k^{-1}C_k}$ is computable in $O(kn)$ time: Define $\Delta_0 := \{\alpha\}$ and recursively $\Delta_i := \Delta_{i-1} \cup \Delta_{i-1}^{h_i}$, where h_i is the ith member in the sequence $g_k^{-1}, \ldots, g_1^{-1}, g_1, \ldots, g_k$. Then $\Delta_{2k} = \alpha^{C_k^{-1}C_k}$. Also, we can keep track how points in $\alpha^{C_k^{-1}C_k}$ were first reached, thereby defining a breadth-first-search tree structure on $\alpha^{C_k^{-1}C_k}$ (cf. Section 2.1.1 for the definition of breadth-first-search trees). Based on this observation, a shallow Schreier tree for the transversal of a point stabilizer can be built.

Lemma 4.4.2. *Given* $G = \langle S \rangle \le \mathrm{Sym}(\Omega)$ *and* $\alpha \in \Omega$, *a Schreier tree of depth at most* $2\log|G|$ *for the transversal* G *mod* G_α *can be built in* $O(n \log^2 |G| + |S|n)$ *time, by a deterministic algorithm.*

Proof. We define elements $g_1, \ldots, g_k$ of G such that their cube is nondegenerate and $\alpha^{C_k^{-1}C_k} = \alpha^G$. Since C_k is nondegenerate, $k \le \log|G|$ and the Schreier tree with labels $\{g_i, g_i^{-1} \mid i \le k\}$ has depth at most $2\log|G|$.

Let g_1 be an arbitrary (nonidentity) element of S. If $g_1, \ldots, g_i$ are already defined and $\alpha^{C_i^{-1}C_i} \neq \alpha^G$ then there exists $\gamma_i \in \alpha^{C_i^{-1}C_i}$ and $u_i \in S$ such that $\gamma_i^{u_i} \notin \alpha^{C_i^{-1}C_i}$. Let g be the product of labels along the path from γ_i to α in the tree on $\alpha^{C_i^{-1}C_i}$ and let $g_{i+1} := g^{-1}u_i$.

The orbit α^G is computed in $O(|S|n)$ time. For fixed $i \le k$, the set $\alpha^{C_i^{-1}C_i}$ can be obtained in $O(n \log|G|)$ time and g_{i+1} is computed using $O(\log|G|)$ group multiplications, requiring also $O(n\log|G|)$ time. Finally, we observe that, for each $\beta \in \alpha^G$ and $s \in S$, we have to compute β^s at most once during the entire algorithm when searching for appropriate γ_i and u_i values, since if $\beta^s \in \alpha^{C_i^{-1}C_i}$

then $\beta^s \in \alpha^{C_j^{-1}C_j}$ for all $j > i$ as well. Therefore, the total cost of all image computations β^s during the algorithm is $O(|S|n)$. □

Remark 4.4.3. In implementations, we use the following slight modification of the algorithm described in Lemma 4.4.2. Using the notation introduced there, suppose that $g_1, \ldots, g_i$ are already defined. Let Δ be the union of the first $2i$ levels of the breadth-first-search tree rooted at α, in the graph defined by the edge labels $\{g_1, \ldots, g_i, g_1^{-1}, \ldots, g_i^{-1}\}$. If $\Delta \neq \alpha^G$ then we construct g_{i+1} such that $\alpha^{g_{i+1}} \notin \Delta$. Clearly, $\alpha^{C_i^{-1}C_i} \subseteq \Delta$, so we have the same theoretical guarantee for the depth of the output Schreier tree as in Lemma 4.4.2. In practice, the depth is usually less than $\log|\alpha^G|$.

The randomized method of [Cooperman et al., 1990], from which the following two lemmas are taken, constructs a Schreier tree of depth $O(\log|\alpha^G|)$ for G mod G_α. It assumes the availability of uniformly distributed random elements of G. A variant of Lemma 4.4.4 seems to have first appeared in [Erdős and Rényi, 1965].

For a random variable A, the expected value of A is denoted by $E(A)$.

Lemma 4.4.4. *Let $G \leq \mathrm{Sym}(\Omega)$, let Δ be an orbit of G, with $|\Delta| = m$, and let Γ be an arbitrary subset of Δ. If g is a uniformly distributed random element of G then $E(|\Gamma^g \backslash \Gamma|) = |\Gamma|(m - |\Gamma|)/m$.*

Proof. Since G acts transitively on Δ, for any $\gamma \in \Gamma$ and $\delta \in \Delta \backslash \Gamma$ there are $|G|/m$ elements $h \in G$ such that $\gamma^h = \delta$. Thus $\mathrm{Prob}(\delta \in \Gamma^g) = |\Gamma|/m$ for any $\delta \in \Delta \backslash \Gamma$ and $E(|\Gamma^g \backslash \Gamma|) = |\Gamma|(m - |\Gamma|)/m$. □

Lemma 4.4.5. *Let $G \leq \mathrm{Sym}(\Omega)$, let Δ be an orbit of G, with $|\Delta| = m$, and let $\emptyset \neq \Gamma \subset \Delta$. Moreover, let $0 < p < 1/2$, and let g be a uniformly distributed random element of G. Then*

(i) *if $|\Gamma| \leq m/2$ then*

$$\mathrm{Prob}\left(|\Gamma^g \cup \Gamma| \geq |\Gamma|\left(2 - \frac{|\Gamma|}{(1-p)m}\right)\right) \geq p;$$

(ii) *if $|\Gamma| \geq m/2$ then*

$$\mathrm{Prob}\left(|\Delta \backslash (\Gamma^g \cup \Gamma)| \leq (m - |\Gamma|)\frac{m - |\Gamma|}{(1-p)m}\right) \geq p.$$

Proof. (i) The condition $|\Gamma^g \cup \Gamma| \geq |\Gamma|(2 - |\Gamma|/((1-p)m))$ is equivalent to

$|\Gamma^g \backslash \Gamma| \ge c|\Gamma|$, for $c = 1 - |\Gamma|/((1-p)m)$. Let $x = \mathrm{Prob}(|\Gamma^g \backslash \Gamma| \ge c|\Gamma|)$. Then $x|\Gamma| + (1-x)c|\Gamma| \ge E(|\Gamma^g \backslash \Gamma|)$. By Lemma 4.4.4, this inequality is equivalent to $x \ge p$.

(ii) In this case, the condition $|\Delta \backslash (\Gamma^g \cup \Gamma)| \le (m - |\Gamma|)^2/((1-p)m)$ is equivalent to $|\Gamma^g \backslash \Gamma| \ge (1-c)(m-|\Gamma|)$, for $c = (m-|\Gamma|)/((1-p)m)$. Let $x = \mathrm{Prob}(|\Gamma^g \backslash \Gamma| \ge (1-c)(m-|\Gamma|))$. Then $x(m-|\Gamma|)+(1-x)(1-c)(m-|\Gamma|) \ge E(|\Gamma^g \backslash \Gamma|)$. By Lemma 4.4.4, this inequality is equivalent to $x \ge p$. □

The following theorem, with weaker constants, is the main result of [Cooperman et al., 1990].

Theorem 4.4.6. *Let* $G \le \mathrm{Sym}(\Omega)$, *let* Δ *be an orbit of* G, *with* $|\Delta| = m$, *and let* $\delta \in \Delta$. *Moreover, let* $(g_1, g_2, \ldots, g_t)$, $t \ge 8\log m + 16$, *be a sequence of uniformly distributed random elements of* G. *Then, with probability at least* $1 - m^{-0.29}$, $\delta^{\langle g_1, g_2, \ldots, g_t \rangle} = \Delta$ *and the breadth-first-search Schreier tree for* G *mod* G_δ *using the labels* $g_1^{-1}, g_2^{-1}, \ldots, g_t^{-1}$ *has depth at most* $2\log m + 4$. *For* $c > 1$, *using* $c(8\log m + 16)$ *random elements ensures a Schreier tree of depth at most* $2\log m + 4$ *with probability at least* $1 - m^{-0.29c}$.

Proof. The basic idea of the proof is the following. Let D_j denote the cube of the first j random elements, and let $\Delta_j := \delta^{D_j}$. We also fix some $0 < p < 1/2$. Then $\Delta_{j+1} = \Delta_j \cup \Delta_j^{g_{j+1}}$, and Lemma 4.4.5 implies that, with probability at least p, the size of the Δ_j increase by at least a constant factor when $|\Delta_j| \le m/2$, and the size of the $\Delta \backslash \Delta_j$ decrease by at least a constant factor when $|\Delta_j| > m/2$. Hence, after $O(\log m)$ rounds, $\delta^{D_j} = \Delta$. Also, the vertices in δ^{D_j} are on the first j levels of the Schreier tree (note that the Schreier tree uses the inverses of the g_i as labels, since edges are directed toward the root δ).

We can obtain the estimate for the depth of the Schreier tree described in the statement of the theorem by a refinement of the argument in the previous paragraph. Let us fix $p := 0.46$. We define a subsequence $H = (h_1, h_2, \ldots, h_s)$ of $(g_1, g_2, \ldots, g_t)$, consisting of those g_i for which the already constructed part of Δ increases by the amount indicated in Lemma 4.4.5. Formally, suppose that $(h_1, h_2, \ldots, h_j)$ is already constructed, where $h_i = g_{k_i}$ for some $k_1 < k_2 < \cdots < k_j$. Let C_j denote the cube of $(h_1, h_2, \ldots, h_j)$, and let $\Gamma_j := \delta^{C_j}$. If $|\Gamma_j| \le m/2$ then we define h_{j+1} as the g_i with the smallest index $i > k_j$ satisfying

$$|\Gamma_j^{g_i} \cup \Gamma_j| \ge |\Gamma_j| \left(2 - \frac{|\Gamma_j|}{0.54m}\right),$$

whereas if $|\Gamma_j| > m/2$ then we define h_{j+1} as the g_i with the smallest index

$i > k_j$ satisfying

$$m - \left|\Gamma_j^{g_i} \cup \Gamma_j\right| \leq \frac{(m - |\Gamma_j|)^2}{0.54m}.$$

We have to show that

(a) $s \geq \lfloor 2\log m \rfloor + 4$, with probability at least $1 - m^{-0.29}$ if $t \geq 8\log m + 16$, and with probability at least $1 - m^{-0.29c}$ if $t \geq c(8\log m + 16)$; and
(b) $\delta^{C_{\lfloor 2\log m\rfloor+4}} = \Delta$.

(a) and (b) imply that the breadth-first-search tree using the first $\lfloor 2\log m \rfloor + 4$ of the h_i as generators already has depth at most $\lfloor 2\log m \rfloor + 4$ and, of course, the depth of the breadth-first-search tree using all g_i cannot be greater.

(a) is an easy consequence of Lemma 2.3.3, applied with the parameters $\varepsilon = 0.4, t = \lceil 8\log m + 16 \rceil, p = 0.46$ and $\varepsilon = 1 - 0.6/c, t = \lceil c(8\log m + 16)\rceil, p = 0.46$ respectively. To prove (b), consider the function

$$f_m(x) = \begin{cases} \lceil x(2 - x/0.54m)\rceil & \text{if } 0 < x \leq 0.5m \\ m - \lfloor (m-x)^2/0.54m \rfloor & \text{if } 0.5m < x < m \end{cases}$$

defined for the integers $1 \leq x \leq m$. We use the notation $f_m^{\{k\}}(x) := f_m(f_m(f_m \ldots f_m(x)\ldots))$($k$ iterations). Since $f_m(x)$ is monotone increasing, it is enough to show that

$$f_m^{\{\lfloor 2\log m\rfloor+4\}}(1) = m. \tag{4.8}$$

For $m < 2^8/0.54$, (4.8) can be checked by a small computer calculation. For $m \geq 2^8/0.54$, we can argue as follows. If $k > 1.36\log m - 3.35 > \log_{5/3} 0.18m$ then $f_m^{\{k\}}(1) \geq 0.18m$, since $f_m(x) \geq 5x/3$ for $x \leq 0.18m$. If $x \geq 0.18m$ then $f_m^{\{5\}}(x) > 0.625m$. Finally, if $x > 0.625m$ and $k > 0.93 + \log\log 0.54m$ then $f_m^{\{k\}}(x) = m$. To see this, observe that for such an x and k,

$$m - f_m^{\{k\}}(x) \leq \frac{(m-x)^{2^k}}{(0.54m)^{(2^k-1)}} < 1.$$

Since $f_m^{\{k\}}(x)$ is an integer, it must be equal to m. To finish the proof, we check that $1.36\log m - 3.35 + 5 + 0.93 + \log\log 0.54m < 2\log m + 4$ if $m \geq 2^8/0.54$. □

Remark 4.4.7. A possible variant is to generate the random elements one by one and to discard immediately those that do not extend the already constructed part of Δ by the amounts indicated in the statement of Lemma 4.4.5. In this way, we never store more than $\lfloor 2\log m \rfloor + 4$ group elements and have the same theoretical guarantee for success (number of random elements to construct, depth of tree, and error estimate) as in Theorem 4.4.6.

The cutoff point $0.18m$ and the probability $p = 0.46$ in the previous analysis are quite arbitrary. Using a cutoff point closer to 0 and probability closer to 0.5, the same argument as in the proof of Theorem 4.4.6 shows that, for each fixed $\varepsilon > 0$, the breadth-first-search tree of $(2+\varepsilon)\log m + c$ random group elements has depth at most $(1+\varepsilon)\log m + c$ with probability at least $1 - m^{-C}$. Here c and C are positive constants, depending on ε. In practice, the breadth-first-search tree of $\log m$ random elements usually has depth less than $\log m$.

Given an arbitrary strong generating set for a group G, a slight modification of the idea of Lemma 4.4.2 produces an SGS of size at most $\log |G|$, which also defines shallow Schreier trees. The following lemma is from [Cooperman and Finkelstein, 1992].

Lemma 4.4.8. *Let $G = \langle S \rangle$ and suppose that S is a strong generating set relative to the base $B = (\beta_1, \ldots, \beta_m)$ with the corresponding point stabilizer chain $G = G^{[1]} \geq G^{[2]} \geq \cdots \geq G^{[m+1]} = 1$. Then an SGS R relative to B can be computed in nearly linear time by a deterministic algorithm, such that $|R| \leq \log |G|$ and the breadth-first-search Schreier tree defined by $(R \cup R^{-1}) \cap G^{[i]}$ for $G^{[i]}$ mod $G^{[i+1]}$ has depth at most $2\log |G^{[i]}|$, for $1 \leq i \leq m$.*

Proof. We construct an SGS as a sequence of group elements $R := (r_1, r_2, \ldots, r_k)$ such that their cube is nondegenerate. This will ensure that $k \leq \log |G|$.

We work in a "bottom-up" manner, constructing an SGS for $G^{[m]}$, $G^{[m-1]}$, etc. Suppose that an initial segment $R_j := (r_1, \ldots, r_j)$ of R is already defined such that $R_j \subseteq G^{[i]}$ and $R_j \cap G^{[i+1]}$ is an SGS for $G^{[i+1]}$ with Schreier trees of depth as stated. Let $\Delta \subseteq \beta_i^{G^{[i]}}$ be the set of points reachable from β_i via a path of length at most $2j$ using the edges defined by $R_j \cup R_j^{-1}$. If $\Delta \neq \beta_i^{G^{[i]}}$ then, as in the proof of Lemma 4.4.2, we construct $r_{j+1} \in G^{[i]}$ such that $\beta_i^{r_{j+1}} \notin \Delta$. If $\Delta = \beta_i^{G^{[i]}}$ then we decrease i by 1. The procedure stops when i reaches 0. $\square$

We remark that [Sims, 1971a] contains an even faster method, which constructs an SGS of size at most $\log |G|$ as a *subset* of the given SGS (cf. Exercise 4.7). However, there is no theoretical guarantee better than the orbit sizes on the depths of the new Schreier trees.

4.5. Strong Generators in Nearly Linear Time

In this section, we describe a nearly linear-time Monte Carlo SGS construction from [Babai et al., 1991]. In Section 5.2, we shall give a version of this algorithm that, with high probability, constructs an SGS in small-base groups even faster.

We follow the outline of the second version of the Schreier–Sims procedure analyzed in Section 4.2 (where we store coset representatives implicitly in Schreier trees). In this second version, there are two nonlinear bottlenecks: We have to guarantee that the Schreier trees are of small depth, and we have to find a fast solution for the problem presented in (4.6) in Section 4.3 (i.e., we have to decide whether a subgroup $H \leq G_\beta$ is proper, or $H = G_\beta$).

The input is $G = \langle T \rangle \leq \mathrm{Sym}(\Omega)$, with $|\Omega| = n$. Suppose that an initial segment $B = (\beta_1, \ldots, \beta_k) \subseteq \Omega$ of a nonredundant base and an approximation S_i for a generator set of the stabilizer $G^{[i]} := G_{(\beta_1,\ldots,\beta_{i-1})}$ for $1 \leq i \leq k$ are already computed. As in the original Schreier–Sims algorithm, we always maintain the property that $\langle S_i \rangle \geq \langle S_{i+1} \rangle$ for all i. In this section, we use a slight modification of the definition of an up-to-date SGS data structure, which was originally given in Section 4.2. We say that the data structure is *up to date below level* j if $\langle S_i \rangle_{\beta_i} = \langle S_{i+1} \rangle$ holds for all i with $j < i \leq k$ *and* the sum of depths of the Schreier trees for levels $j+1, \ldots, k$ is at most $6 \log |\langle S_{j+1} \rangle|$.

The SGS Construction

In the case when the data structure is up to date below level j, we use the algorithm described in the proof of Lemma 4.4.2 to compute a Schreier tree of depth at most $2 \log |\langle S_j \rangle|$ for $\beta_j^{\langle S_j \rangle}$. After that, we apply a Monte Carlo solution of (4.6), to be described later, that either returns some $g \in \langle S_j \rangle_{\beta_j} \backslash \langle S_{j+1} \rangle$ or reports that $\langle S_j \rangle_{\beta_j} = \langle S_{j+1} \rangle$. In the former case, we add g to S_{j+1} and declare that the data structure is up to date below level $j+1$. If $\langle S_j \rangle_{\beta_j} = \langle S_{j+1} \rangle$ then we have an SGS for $\langle S_j \rangle$, and so we can construct uniformly distributed random elements of $\langle S_j \rangle$. We compute a Schreier tree for $\langle S_j \rangle$ mod $\langle S_{j+1} \rangle$ as described in Remark 4.4.7, using a sequence $(g_1, \ldots, g_s)$ of random elements as labels, with $s := \lfloor 2 \log |\beta_j^{\langle S_j \rangle}| + 4 \rfloor \leq 6 \log |\beta_j^{\langle S_j \rangle}|$. If this Schreier tree construction succeeds then we set $S_j := S_{j+1} \cup \{g_1, \ldots, g_s\}$ and obtain a data structure that is up to date below level $j-1$.

We want to handle the base point β_1 in the same manner as the later base points. To this end, in the spirit of Exercise 4.4, we start the algorithm by defining $S_0 := T$, corresponding to an imaginary base point $\beta_0 \notin \Omega$ (we can think of β_0 as a point added to the permutation domain, which is fixed by every element of G, and so $G = G_{\beta_0}$ and the Schreier tree for G mod G_{β_0} is trivial).

Then we say that the data structure is up to date below level 0; the algorithm terminates when the data structure becomes up to date below level -1.

Lemma 4.5.1. *The Schreier tree computations of the SGS construction require $O(n \log n \log^4 |G|)$ total time and $O(n \log |G|)$ memory. This part of the SGS construction algorithm is of Las Vegas type.*

Proof. We use the notation introduced at the beginning of this section. There are at most $\log |G|$ base points, and each base point β_i is processed at most $\log |G|$ times, since the subgroup $\langle S_i \rangle$ increases at each processing. Hence the algorithms in Lemma 4.4.2 and Remark 4.4.7 are invoked at most $\log^2 |G|$ times. Lemma 4.4.2 is deterministic. Therefore, to obtain a fixed, but arbitrarily small, constant error probability for the entire algorithm, we have to ensure that each invocation of the algorithm in Remark 4.4.7 fails with probability at most $c'/\log^2 |G|$, for some constant c' prescribed by the overall error requirement. This can be achieved by choosing the value of c in the statement of Theorem 4.4.6 as $c = c'' \log n$ for an appropriate constant c''. In fact, using a large enough but constant c'', we can reduce the error probability to less than $1/n^d$, since $\log^2 |G| < (n \log n)^2$, and so the algorithm in Remark 4.4.7 fails with probability less than $2^{-0.29c'' \log n} < 1/(\log^2 |G| n^d)$. Note that, when invoking this algorithm on level j, during the execution of the algorithm we actually check whether the vertex set of the Schreier tree we construct is $\beta_j^{\langle S_j \rangle}$; therefore the Schreier tree computation part of the SGS algorithm is of Las Vegas type.

Concerning the time requirement, note that $|S_i|$ is $O(\log |G|)$ during the entire algorithm, since S_i is obtained by adding $O(\log |\beta_i^{\langle S_i \rangle}|)$ elements to S_{i+1}. Hence one call of the algorithm in Lemma 4.4.2 runs in $O(n \log^2 |G|)$ time.

On level i, one call of the algorithm in Remark 4.4.7 runs in $O(n \log |G| \cdot \log n \log |G^{[i]} : G^{[i+1]}|)$ time, since our requirement on the depths of Schreier trees ensures that a random element can be generated by $O(\log |G|)$ multiplications and, as discussed at the error estimate part above, we have to generate $O(\log n \log |G^{[i]} : G^{[i+1]}|)$ random elements.

Since both kinds of Schreier tree computations are called at most $\log^2 |G|$ times, the total time requirement is as stated. □

Remark 4.5.2. With a time–space tradeoff, it is possible to achieve Schreier tree computations that run in $O(n \log n \log^3 |G|)$ time but use $O(n \log^2 |G|)$ memory. The applications of the algorithm in Remark 4.4.7 already run within this tighter time bound and we claim that, for fixed i, the at most $\log |G|$ invocations of Lemma 4.4.2 on level i run in $O(n \log^2 |G|)$ total time. This can be achieved by saving the sequence $R_i := (r_1, r_2, \ldots, r_{j_i})$ from elements of $G^{[i]}$ with nondegenerate cube computed at the first invocation of Lemma 4.4.2 on level i, and letting the algorithm augment R_i at later calls of Lemma 4.4.2. The generating set S_i changes at each call at level i, so the orbit $\beta_i^{\langle S_i \rangle}$ has to be computed at each call, but the other steps of the algorithm are executed only once, as if the $O(\log^2 |G|)$ elements of all the different S_i were given together, as one generating set.

What is left is to describe the solution for (4.6). This solution is based on the following *local expansion lemma* from [Babai, 1992].

Lemma 4.5.3. *Let S denote a set of generators of the group G and set $T = S \cup S^{-1} \cup \{1\}$. Let D be any finite subset of T^t, the set of t-term products of members of T (in any order). Suppose that there exists q with $0 < q \leq 1/(2t+1)$ such that $|D| \leq (1 - 2qt)|G|$. Then, for at least one generator $g \in S$,*

$$|D \backslash Dg| \geq q|D|.$$

Proof. Suppose, on the contrary, that $|D \backslash Dg| < q|D|$ for all $g \in S$. Using this assumption, we first prove that $G = T^{2t}$.

Observe that, for each $g \in S$, $|D \backslash Dg^{-1}| = |Dg \backslash D| = |D \backslash Dg| < q|D|$ and for any $x, y \in G$,

$$D \backslash Dxy \subseteq (D \backslash Dy) \cup (D \backslash Dx)y. \tag{4.9}$$

From these, an easy induction on k shows that for any $u \in T^k$, we have $|D \backslash Du| < kq|D|$. As long as $kq \leq 1$, this implies that $u \in D^{-1}D$. Since $q \leq 1/(2t + 1)$, we obtain $T^{2t+1} \subseteq D^{-1}D \subseteq T^{2t}$ and so $T^{2t+1} = T^{2t}$. Moreover, since S generates G, we have $G = \bigcup_{k \geq 0} T^k$; from this we obtain $G = T^{2t}$.

To finish the proof, we count the number of pairs (x, u) with $x \in D$, $u \in G$, and $xu \in D$ two different ways. On one hand, for each fixed $x \in D$, the number of $u \in G$ with $xu \in D$ is exactly $|D|$, so the number of pairs (x, u) is $|D|^2$. On the other hand, fixing $u \in G = T^{2t}$ we have $|D \backslash Du| < 2qt|D|$, so the number of $x \in D$ with $xu \in D$ is greater than $(1 - 2qt)|D|$ and the number of pairs (x, u) is greater than $(1 - 2qt)|D||G|$. However, this contradicts the condition $|D| \leq (1 - 2qt)|G|$. □

The Algorithm Solving (4.6)

We apply the local expansion lemma with the following parameters. Using the notation of (4.6), let S be the union of labels used in the Schreier tree of R and in the Schreier tree data structure of H. Moreover, let t be the sum of depths of all of the trees mentioned above and let $q = 1/4t$. First, we sift the input generators of G through HR. If one of them has a nontrivial siftee then $H \neq G_\beta$. Hence it is enough to consider the case when all input generators sift through, and so S generates G.

Let $D = HR$; then $D \subseteq (S \cup S^{-1} \cup \{1\})^t$. If $H \neq G_\beta$ then $|G_\beta : H| \geq 2$, $|D| \leq |G|/2$, and all conditions of Lemma 4.5.3 are satisfied.

Note that although D is not necessarily a group, just a *subset* of a group, it is defined as a product of transversals. Consequently, it is possible to test membership in D and to generate uniformly distributed random elements of D. By Lemma 4.5.3, if $H \neq G_\beta$ then, for a randomly chosen $d \in D$,

$$\mathrm{Prob}(\exists g \in S(dg \notin D)) \geq 1/4t. \tag{4.10}$$

Hence, if $4ct$ random elements $d \in D$ are chosen and $dg \in D$ for all such d and all $g \in S$ then $\mathrm{Prob}(H = G_\beta) > 1 - e^{-c}$.

How can we test that $dg \in D$? Observe that each $d \in D$ is factored uniquely in the form $d = hr$ for some $h \in H$ and $r \in R$. Moreover, $dg = hrg \in D$ if and only if $hrg(\overline{hrg})^{-1} \in H$. (As usual, $\bar{k}$ denotes the element of R such that $k \in G_\beta \bar{k}$.) Now h fixes β, so $\overline{hrg} = \overline{rg}$. Also, $hrg(\overline{hrg})^{-1} \in H$ if and only if $rg(\overline{hrg})^{-1} \in H$. Summarizing, we obtain that $dg \in D$ if and only if the Schreier generator $rg(\overline{rg})^{-1} \in H$. Consequently, instead of generating random elements of D, it is enough to check whether Schreier generators constructed by combining randomly chosen elements of R with all elements of S are in H. Hence the solution of (4.6) we obtained can be considered as a theoretical justification (with error estimate) of a version of the first heuristic described in Section 4.3.

The algorithm, as just described, requires that, after choosing some random $r \in R$, we check the Schreier generators $rg(\overline{rg})^{-1}$ for all $g \in S$. Our next goal is to present an improvement that avoids checking all elements of S. Picking random elements of S may not work well; this is the essence of Exercise 4.5. Instead, we use an enhancement of the random subproduct method (cf. Section 2.3). Let k' be an integer chosen randomly from the uniform distribution on $[0, |S| - 1]$ and let $k = \lfloor k'/2 \rfloor$. Moreover, let w_1 be a random subproduct of a random ordering of S and let w_2 be a random subproduct of a random ordering of a random subset S' of size k of S (cf. Exercise 2.1 for the construction of random orderings). Finally, for $g \in G$, $D \subseteq G$, and integer a, let $P(g, a)$ denote the proposition that $|\{d \in D \mid dg \notin D\}| \geq a$. Note that

$$P(g, a) \wedge \neg P(h, b) \Longrightarrow P(gh, a - b) \wedge P(hg, a - b), \tag{4.11}$$

where $\neg P$ denotes the negation of the proposition P (cf. Exercise 4.8).

Lemma 4.5.4. *Suppose that $P(g, a)$ holds for some $g \in S$. Then, with probability at least $1/4$, $P(w_i, a/4)$ holds for at least one $i \in \{1, 2\}$.*

Proof. Let us fix $g \in S$ such that $P(g, a)$ holds. First, we observe that $\mathrm{Prob}(g \in S') \le 1/2$, since $\binom{|S|-1}{k}/\binom{|S|}{k} = (|S| - k)/|S| \ge 1/2$.

We have $w_1 = ug^{\varepsilon}v$, where $\varepsilon \in \{0, 1\}$ and u, v are random subproducts of two subsets of S that partition $S \backslash \{g\}$. Assume that v is the random subproduct on the set that is not larger than the other one; the other case can be handled similarly.

Let p denote the probability of the event that $P(h, a/4)$ holds when h is a random subproduct of a random ordering of a random subset of size k of $S \backslash \{g\}$. (The number $k = \lfloor k'/2 \rfloor$ is also a random variable.) Because the distribution of v is the same as the distribution of h in this definition, we have $\mathrm{Prob}(P(v, a/4)) = p$.

Now we prove that the conditional probability $\mathrm{Prob}(P(ug^{\varepsilon}, a/2)| \neg P(v, a/4))$ is at least $1/2$. This is true because ε is independent of u and v. Therefore, by (4.11),

$$\begin{aligned}
&\mathrm{Prob}(P(ug^{\varepsilon}, a/2)|\neg P(v, a/4)) \\
&\quad = \mathrm{Prob}(P(ug^{\varepsilon}, a/2)|P(u, a/2) \wedge \neg P(v, a/4))\,\mathrm{Prob}(P(u, a/2)) \\
&\qquad + \mathrm{Prob}(P(ug^{\varepsilon}, a/2)|\neg P(u, a/2) \wedge \neg P(v, a/4))\,\mathrm{Prob}(\neg P(u, a/2)) \\
&\quad \ge \mathrm{Prob}(\varepsilon = 0|P(u, a/2) \wedge \neg P(v, a/4))\,\mathrm{Prob}(P(u, a/2)) \\
&\qquad + \mathrm{Prob}(\varepsilon = 1|\neg P(u, a/2) \wedge \neg P(v, a/4))\,\mathrm{Prob}(\neg P(u, a/2)) = 1/2.
\end{aligned}$$

Finally, using (4.11) again,

$$\begin{aligned}
\mathrm{Prob}(P(w_1, a/4)) &\ge \mathrm{Prob}(P(ug^{\varepsilon}, a/2)|\neg P(v, a/4))\,\mathrm{Prob}(\neg P(v, a/4)) \\
&\ge (1 - p)/2
\end{aligned}$$

and

$$\mathrm{Prob}(P(w_2, a/4)) \ge \mathrm{Prob}(P(w_2, a/4)|g \notin S')\,\mathrm{Prob}(g \notin S') \ge p/2.$$

Since $\max\{(1 - p)/2, p/2\} \ge 1/4$, $P(w_i, a/4)$ holds with probability at least $1/4$ for $i = 1$ or 2. □

In the application to solve (4.6), we have $D = HR$. Using the notation introduced after Lemma 4.5.3, (4.10) gives that if $H \ne G_\beta$ then $P(g, |D|/4t)$ holds for some $g \in S$. Hence Lemma 4.5.4 implies that, for a randomly chosen $r \in R$

and for the random subproducts w_1, w_2 as just described,

$$\text{Prob}(rw_1(\overline{rw_1})^{-1} \notin H \text{ or } rw_2(\overline{rw_2})^{-1} \notin H) \geq \frac{1}{64t}. \tag{4.12}$$

During the SGS construction, we call the subroutine solving (4.6) $O(\log^2 |G|)$ times, so each call can fail with probability at most $c'/\log^2 |G|$, for some constant c'. In these applications, we have $t \leq 6 \log |G|$, so (4.12) and $\log |G| < n \log n$ imply that testing $c''t \log n \in O(\log n \log |G|)$ pairs $rw_1(\overline{rw_1})^{-1}$, $rw_2(\overline{rw_2})^{-1}$ achieves the desired error probability. In fact, as discussed in the proof of Lemma 4.5.1, at the error estimate of the Schreier tree construction part of the SGS construction algorithm, choosing a large enough constant c'' ensures that the overall probability of failure is less than $1/n^d$.

Constructing $rw_i(\overline{rw_i})^{-1}$ and sifting it in H requires $O(n \log |G|)$ time. Hence the overall time requirement of calls to solve (4.6) is $O(n \log n \log^4 |G|)$. Combining this result with Lemma 4.5.1, we obtain the following:

Theorem 4.5.5. *Suppose that $G = \langle T \rangle \leq \text{Sym}(\Omega)$ and $|\Omega| = n$. There is a Monte Carlo algorithm that, with error probability less than $1/n^d$ for a constant d prescribed by the user, constructs an SGS for G in $O(n \log n \log^4 |G| + |T| n \log |G|)$ time. The memory requirement of the algorithm is $O(n \log |G| + |T| n)$.*

The term $|T| n \log |G|$ in the running time stems from the handling of level 0. Recall that we stored the original generators in a set S_0. When the data structure is up to date below level 0, we have to sift the elements of S_0 in $\langle S_1 \rangle$ to ensure that $G = \langle S_1 \rangle$.

By Remark 4.5.2, the Schreier tree computations run in $O(n \log n \log^3 |G|)$ time. In Section 5.2 we shall give a version of the algorithm that, with high probability, runs in $O(n \log n \log^3 |G|)$ time for small-base group inputs.

4.5.1. Implementation

A version of the nearly linear-time Monte Carlo SGS construction was first implemented in the C language (see [Seress and Weisz, 1993]). The main purpose was to experiment with the different Schreier tree constructions and to get a feeling for the true probabilities in (4.12), which turn out to be much higher (usually, at least 1/3) than the estimate $1/64t$ obtained in (4.12).

It turns out that almost always the Schreier tree construction described in Remark 4.4.3 already produces at most $\log |\alpha^G|$ generators, such that these generators and their inverses define a Schreier tree of depth at most $\log |\alpha^G|$ for

G mod G_α, and applying the algorithm in Remark 4.4.7 does not improve the depth significantly. Hence, in the GAP implementation, we use the algorithm in Remark 4.4.3, and no additional work for Schreier tree computations is necessary.

As indicated already, the practical performance of the algorithm is far better than our probability estimate in (4.12). One reason may be that the Schreier tree construction of Lemma 4.4.2 "randomizes" the coset representatives in the sense that by writing out a coset representative as a product of original generators, the length of the word may be exponential compared to the distance from the root of the Schreier tree. Also, the usage of the random subproducts w_1 and w_2 provides further random mixing of the generators. So, although we could guarantee only with probability $\Omega(1/\log|G|)$ that a Schreier generator obtained from a random coset and the w_is witnesses that $H \neq G_\beta$, the practical performance is probably closer to the case when the Schreier generator is constructed using a random element as a generator. Such a Schreier generator has a chance of at least $1 - 1/|G_\beta : H| \geq 1/2$ to witness that $H \neq G_\beta$.

Based on that experience, the following very fast heuristic version is implemented in GAP. While working on level i of the data structure, we test *only one* Scheier generator obtained from a random coset and (the long) random subproduct w_1. If this Schreier generator sifts through $\langle S_{i+1}\rangle$ then we declare that $\langle S_i\rangle_{\beta_i} = \langle S_{i+1}\rangle$ and that the data structure is up to date below level $i-1$. Although such a declaration may be wrong with a significant probability, individual errors seem to be corrected as the Schreier–Sims procedure moves along the base points. We mention that the implementation uses a further heuristic, replacing the sifting of Schreier generators; we shall describe this further heuristic at the end of Section 5.2.

To ensure the correctness of the output, the user has two options in GAP. The default value is a deterministic test, which decides with certainty whether the SGS construction is correct. The user also has the option to choose a randomized test, which declares that, with probability at least $1 - \delta$, the SGS construction is correct; the error bound δ is specified by the user.

In this section we describe only the randomized SGS construction test, which is based on the following lemma. The deterministic strong generating set used in GAP is the so-called Verify routine by Sims, which we shall discuss in Section 8.2. In Chapter 8, we shall present other strong generating tests as well.

Lemma 4.5.6. *Let S be an alleged strong generating set of $G = \langle S\rangle \leq \mathrm{Sym}(\Omega)$ relative to the base $(\beta_1, \ldots, \beta_m)$. For $1 \leq i \leq m$, let $S_i = S \cap \mathrm{Sym}(\Omega)_{(\beta_1,\ldots,\beta_{i-1})}$ and let R_i be the transversal built from S_i for $G_{(\beta_1,\ldots,\beta_{i-1})}$ mod $G_{(\beta_1,\ldots,\beta_i)}$. Finally, let $D = R_m R_{m-1} \cdots R_1$, let g be a uniformly distributed random element of*

D, t be the sum of the depths of Schreier trees coding the R_i, and w_1, w_2 be a pair of random subproducts built from S, as described in the algorithm solving (4.6). If S is not a strong generating set then $\text{Prob}(gw_j \notin D) \geq 1/64t$ *for at least one $j \in \{1, 2\}$.*

Proof. By Lemma 4.2.3, if S is not a strong generating set then $\langle S_i \rangle_{\beta_i} \gneqq \langle S_{i+1} \rangle$ for at least one $i \in [1, m]$. This implies that $|D| \leq |G|/2$ and Lemma 4.5.3 can be applied with $q = 1/4t$. We obtain that $\text{Prob}(dh \notin D) \geq 1/4t$ for some $h \in S$ and so, by Lemma 4.5.4, $\text{Prob}(gw_j \notin D) \geq 1/64t$. □

It is clear that if $64t \ln(1/\delta)$ pairs gw_1, gw_2 sift through D then the probability that S is indeed an SGS is at least $1 - \delta$.

The advantages of separating the construction of an SGS and the checking of the correctness of the construction are twofold. On one hand, in the oft occurring case when $|G|$ is known in advance, it is very easy to check correctness by comparing $|D|$ and $|G|$ and, in the case $|D| = |G|$, we can terminate already after the fast heuristic construction. On the other hand, even if $|G|$ is not known in advance, we have to apply the expensive sifting of Schreier generators about a factor $\log^2 |G|$ less times than in the original algorithm, which checks whether $\langle S_i \rangle_{\beta_i} = \langle S_{i+1} \rangle$ each time a level i is processed.

If the input SGS is not correct then the probability that $gw_j \notin D$ seems to be much higher in practice than the estimate $1/64t$ in Lemma 4.5.6, and errors are detected very quickly. In the case when $gw_j \notin D$ is detected, the siftee of gw_j can be added to S and we can return to the construction phase of the algorithm. Hence, the vast majority of time in an SGS construction is spent checking the already correct final result. However, since the algorithm is Monte Carlo, we cannot terminate and guarantee the correctness of the answer with the prescribed probability earlier than the number of checks implied by the estimate $1/64t$ in Lemma 4.5.6.

Exercises

4.1. Given $G = \langle S \rangle$, construct less than n^2 generators for G in $O(|S|n^2)$ time. *Hint:* Sift the elements of S.

4.2. Give an example of a permutation group $G \leq S_n$ with two nonredundant bases: One of size $\lfloor \log |G| \rfloor$ and one of size $\lceil \log |G| / \log n \rceil$. *Hint:* G can be chosen cyclic.

4.3. Prove Lemma 4.2.1 without the assumption that $1 \in R$.

4.4. Design a version of the Schreier–Sims algorithm that handles all base points uniformly (i.e., do not place all input generators immediately into the generating set for $G^{[1]}$) and that achieves the time savings indicated following Theorem 4.2.4.

4.5. (Luks) Let $H = \{h_1, \ldots, h_{n-2}\}$ be a regular group acting on $\{1, \ldots, n-2\}$, and let $G = \langle h_1, \ldots, h_{n-2}, (n-1, n)\rangle$. Show that $B = (1, n-1)$ is a nonredundant base for G but the probability that a randomly chosen Schreier generator detects that $G_1 \neq 1$ is only $1/(n-1)$.

4.6. Determine the time requirement of the algorithm presented in the proof of Lemma 4.4.8.

4.7. [Sims, 1971a] Design and analyze an algorithm that computes an SGS of size at most $\log |G|$ as a subset of a given SGS S. *Hint:* Let S be an SGS relative to the base $(\beta_1, \ldots, \beta_m)$, let $G = G^{[1]} \geq \cdots \geq G^{[m+1]} = 1$ be the corresponding point stabilizer chain, and let $S_i = S \cap G^{[i]}$. List the elements of S such that S_{i+1} comes before $S_i \backslash S_{i+1}$, for all $i \in [1, m]$. Do an orbit computation to decide whether the jth element s_j of the list is in the subgroup generated by the first $j-1$ elements; if it is then discard s_j.

4.8. Prove the implication (4.11) in Section 4.5. *Hint:* Use that

$$(Dg \backslash D)h \subseteq (Dgh \backslash D) \cup (D \backslash Dh)$$

and

$$D \backslash Dg \subseteq (D \backslash Dhg) \cup (Dh \backslash D)g.$$

5

Further Low-Level Algorithms

In this chapter, we collect some basic algorithms that serve as building blocks to higher level problems. Frequently, the efficient implementation of these low-level algorithms is the key to the practicality of the more glamorous, high-level problems that use them as subroutines.

5.1. Consequences of the Schreier–Sims Method

The major applications of the Schreier–Sims SGS construction are membership testing in groups and finding the order of groups, but an additional benefit of the algorithm is that its methodology can be applied to solve a host of other problems in basic permutation group manipulation. We list the most important applications in this section. The running times depend on which version of the Schreier–Sims algorithm we use; in particular, all tasks listed in this section can be performed by nearly linear-time Monte Carlo algorithms. For use in Chapter 6, we also point out that if we already have a nonredundant base and SGS for the input group then the algorithms presented in Sections 5.1.1–5.1.3 are all Las Vegas. Concerning the closure algorithms in Section 5.1.4, see Exercise 5.1.

5.1.1. Pointwise Stabilizers

Any version of the Schreier–Sims method presented in Chapter 4 can be easily modified to yield the pointwise stabilizer of some subset of the permutation domain. Suppose that $G = \langle S \rangle \leq \mathrm{Sym}(\Omega)$ and $\Delta \subseteq \Omega$ are given, and we need generators for $G_{(\Delta)}$. To this end, we fix an ordering of Ω in which the elements of Δ come first and construct a base and strong generating set such that the base elements are ordered according to that chosen ordering. Then there is an initial segment of the base, say the first k elements, that contains the base points in Δ and so the $(k+1)$st group in the stabilizer chain is $G_{(\Delta)}$.

The only modification in the Schreier–Sims algorithm is that we temporarily lift the restriction that the constructed base must be nonredundant and we declare every element of Ω to be a base point (in the chosen ordering). After the strong generating construction is finished, we can simply delete the redundant base points. Note that on levels with $|G^{[i]} : G^{[i+1]}| > 1$, the created strong generating set and Schreier trees do not have to be changed. Also, the time complexity of the algorithm remains the same as it was originally, since sifting through a level with fundamental orbit of size 1 can be done in $O(1)$ time.

5.1.2. Homomorphisms

The same idea as in pointwise stabilizer computations can be used to compute kernels of actions. Given $G = \langle S \rangle \leq \mathrm{Sym}(\Omega)$ and a homomorphism $\varphi : G \to \mathrm{Sym}(\Delta)$ (defined by the images of the elements of S), we can consider G as a permutation group acting on $\Omega \cup \Delta$. Namely, for $g \in S$, denote by $(g, \varphi(g))$ the element of $\mathrm{Sym}(\Omega) \times \mathrm{Sym}(\Delta)$ whose restrictions to Ω and Δ are g and $\varphi(g)$, respectively. Then $\bar{G} := \{(g, \varphi(g)) \mid g \in G\}$ is isomorphic to G, and $\ker(\varphi)$ is the restriction of the pointwise stabilizer $\bar{G}_{(\Delta)}$ to Ω. Similarly, we can determine whether a map $\varphi : S \to \mathrm{Sym}(\Delta)$ defines a homomorphism. The map φ defines a homomorphism if and only if the pointwise stabilizer of Ω in the group $\langle (g, \varphi(g)) \mid g \in S \rangle$ is the identity.

Specifying the images of the generators at a homomorphism $\varphi : G = \langle S \rangle \to \mathrm{Sym}(\Delta)$ is not sufficient for the effective handling of φ. We also need to be able to compute $\varphi(g)$ for any $g \in G$ and a representative from the coset $\varphi^{-1}(g)$ for any $g \in \varphi(G)$. To this end, we store two strong generating sets S_1, S_2 with the corresponding Schreier tree data structures for the group $\bar{G} = \langle (g, \varphi(g)) \mid g \in S \rangle \leq \mathrm{Sym}(\Omega \cup \Delta)$. The first SGS S_1 is relative to a base $B_1 = (\beta_1, \ldots, \beta_m)$ such that an initial segment $(\beta_1, \ldots, \beta_k)$ of B_1 consists of points of Ω and $\bar{G}_{(\beta_1, \ldots, \beta_k)}$ fixes Ω pointwise; S_2 is relative to a base B_2 such that an initial segment of B_2 consists of points in Δ and the pointwise stabilizer of this initial segment is $\bar{G}_{(\Delta)}$. Given $g \in G$, $\varphi(g)$ can be computed by sifting $(g, 1)$ in the Schreier tree data structure corresponding to S_1 and then restricting the inverse of the siftee of $(g, 1)$ to Δ. Similarly, given $g \in \varphi(G)$, a representative of the preimage $\varphi^{-1}(g)$ can be obtained by sifting $(1, g)$ in the Schreier tree data structure corresponding to S_2 and restricting the inverse of the siftee to Ω.

In implementations, permutation groups of degree n are represented acting on the set $[1, n]$. In particular, $\varphi : G \to \mathrm{Sym}(\Delta)$ is specified by pairs of permutations

$(g, \varphi(g))$, where g acts on $[1, n]$ for $n := |\Omega|$ and $\varphi(g)$ acts on $[1, k]$ for $k := |\Delta|$. Therefore, when computing in $\mathrm{Sym}(\Omega) \times \mathrm{Sym}(\Delta)$, we have to fix a permutation c acting on $[1, n + k]$ such that $[1, k]^c = [n + 1, n + k]$ and use c and c^{-1} to conjugate the elements of $\mathrm{Sym}(\Delta)$ to $\mathrm{Sym}([n + 1, n + k])$ and vice versa.

In practice, we often know an SGS S for $G \leq \mathrm{Sym}(\Omega)$ before the homomorphism φ is defined, and φ is specified by the images of the elements of the SGS. In this case, we have at hand the SGS S_1 defined in this section without any additional work. Note also that restricting the elements of S_2 to Δ, we obtain an SGS for the image group $\varphi(G)$.

5.1.3. Transitive Constituent and Block Homomorphisms

Two oft occurring special cases of group homomorphisms, which can be handled faster than the general case of homomorphisms specified by the images of generators, are transitive constituent homomorphisms and block homomorphisms. A *transitive constituent homomorphism* is the restriction of $G \leq \mathrm{Sym}(\Omega)$ to an orbit $\Delta \subseteq \Omega$. In this case, we do not have to extend the G-action to a domain of size $|\Omega| + |\Delta|$; we only have to compute a strong generating set according to an ordering of Ω starting with Δ. The kernel of the homomorphism is the pointwise stabilizer of Δ, and preimages can be computed as in the general case. Moreover, the image of a group element $g \in G$ can be computed without sifting, just restricting g to Δ. Therefore, we do not have to compute a second SGS for G as in the case of general homomorphisms.

In implementations, similarly to the general case of homomorphisms, we need to construct a permutation $c \in \mathrm{Sym}([1, |\Omega|])$ that maps Δ to $[1, |\Delta|]$. Then c and c^{-1} can be used to define the image group of the transitive constituent homomorphism as a permutation group acting on $[1, |\Delta|]$ and to convert the elements of this image group to permutations acting on Δ.

The other special case of homomorphisms is block homomorphism. Suppose that a not necessarily transitive group $G \leq \mathrm{Sym}(\Omega)$ is given, and let $\Delta := \{\Delta_1, \ldots, \Delta_k\}$ be a partition of Ω with the property that G permutes the pairwise disjoint subsets $\Delta_1, \ldots, \Delta_k$ of Ω. The elements of Δ are called *blocks* for G. A *block homomorphism* $\varphi : G \to S_k$ is defined by the permutation action of G on Δ.

An efficient method for computing with block homomorphisms is described in [Butler, 1985]. The goal is to avoid working with permutations on $\Omega \cup \Delta$ and use only the permutation representation of G on Ω. The algorithm assumes that given a base and SGS for some $H \leq \mathrm{Sym}(\Omega)$, we can quickly compute another base $\bar{B}$ and SGS for H, where $\bar{B}$ starts with some specified $\omega \in \Omega$. This can be achieved by a base change algorithm (cf. Section 5.4). This application of base change is much faster than the computation of an SGS from scratch.

The Block Homomorphism Algorithm

Computing the image of some $g \in G$ at the block homomorphism φ is easy: We store an $|\Omega|$-long list L, where the ith entry of L records which block the ith element of Ω belongs to. Since g permutes the blocks, it is enough to pick a representative $\delta_j \in \Delta_j$ from each block and look up in L the block containing δ_j^g. This defines the image $j^{\varphi(g)}$ for all $j \in [1, k]$.

Next, we describe the computation of $\ker(\varphi)$. Since $\ker(\varphi)$ consists of those elements of G that fix the blocks setwise, we compute recursively generators for the subgroups $G^{[j]} := G_{(\Delta_1,\ldots,\Delta_{j-1})}$, $j = 1, 2, \ldots, k+1$ (i.e., $G^{[j]}$ stabilizes the first $j-1$ blocks but can permute elements of a block among themselves). Then we can set $\ker(\varphi) := G^{[k+1]}$. Suppose that an SGS for $G^{[j]}$ is already known, and let us fix some $\delta_j \in \Delta_j$. We apply a base change to get a base B_j for $G^{[j]}$ starting with δ_j and the corresponding SGS. In particular, we have generators X'_{j+1} for the subgroup $(G^{[j]})_{\delta_j}$ and a transversal R_j (coded in a Schreier tree) for $G^{[j]} \bmod (G^{[j]})_{\delta_j}$. Our goal is to add some elements $r_{j_1}, r_{j_2}, \ldots$ of R_j to X'_{j+1} so that $X_{j+1} := X'_{j+1} \cup \{r_{j_1}, r_{j_2}, \ldots\}$ generates $G^{[j+1]}$.

For any $g \in G^{[j]}$, $g \in G^{[j+1]}$ if and only if $\delta_j^g \in \Delta_j$. Based on this observation, we choose $r_{j_1}, r_{j_2}, \ldots \in R_j$ recursively so that $\delta_j^{r_{j_i}} \in \Delta_j$ and the orbits $\delta_j^{\langle X'_{j+1} \cup \{r_{j_1},\ldots,r_{j_i}\}\rangle}$ increase as i increases. At the addition of each r_{j_i}, the size of the orbit of δ_j at least doubles, since the groups $\langle X'_{j+1} \cup \{r_{j_1}, \ldots, r_{j_i}\}\rangle$ define a strictly increasing sequence of groups with the property that the stabilizer of δ_j does not increase. Hence, with the addition of at most $\log|\Delta_j|$ elements from R_j, we can achieve that the set $X_{j+1} = X'_{j+1} \cup \{r_{j_1}, r_{j_2}, \ldots\}$ satisfies $\delta_j^{\langle X_{j+1}\rangle} = \delta_j^{G^{[j]}} \cap \Delta_j$. This means that $\langle X_{j+1}\rangle = G^{[j+1]}$. Note that X_{j+1} is an SGS for $G^{[j+1]}$, relative to the base B_j.

The kernel computation just described has an additional advantage: In the process, we get strong generating sets for the subgroups in the chain $G = G^{[1]} \geq G^{[2]} \geq \cdots \geq G^{[k+1]} = \ker(\varphi)$, which is the point stabilizer chain of the image group $\varphi(G)$ on the permutation domain Δ. More precisely, for each $j \in [1, k]$, we get two SGS for $G^{[j]}$: The first SGS comes from the recursion, and it is discarded during the base change, when the second SGS relative to the base B_j is constructed. After the SGS X_{j+1} for $G^{[j+1]}$ is constructed, we discard the second SGS for $G^{[j]}$ as well, but we store the Schreier tree (S_j, T_j), which codes the transversal R_j for $G^{[j]} \bmod (G^{[j]})_{\delta_j}$. Hence, the output of the kernel computation is an SGS X_{k+1} for $\ker(\varphi)$ and a sequence of Schreier trees (S_j, T_j)

for $1 \leq j \leq k$. The set

$$\bigcup_{j=1}^{k} \varphi(S_j)$$

is an SGS for $\varphi(G)$. Note that we do not compute and store $\varphi(S_j)$ explicitly; we can work with the elements of S_j, which are permutations of Ω.

The last basic task we have to perform is to find a preimage for some given $g \in \varphi(G)$. This can be accomplished by a slight modification of the usual sifting procedure, using the transversals R_j. The only concern is that $g \in \mathrm{Sym}(\Delta) \cong S_k$ and $R_j \subseteq \mathrm{Sym}(\Omega)$, so we cannot take the product of g with inverses of transversal elements, as required in sifting. Instead, we construct recursively $r_j \in R_j$ for $j = 1, 2, \ldots$ so that $g\varphi(r_1)^{-1} \cdots \varphi(r_{j-1})^{-1} \in \varphi(G^{[j]})$. Suppose that $r_1, \ldots, r_{j-1}$ and the product $p_j := r_1^{-1} \cdots r_{j-1}^{-1}$ of their inverses are already constructed. Let $l := j^g$, and suppose that $\delta_l^{p_j}$ is in the block $\Delta_{l'}$. Then we pick $r_j \in R_j$ such that $\delta_j^{r_j} \in \Delta_{l'}$. Following this procedure, $p_{k+1} := r_1^{-1} \cdots r_k^{-1}$ is the inverse of a preimage of g. Note that, as customary with sifting, the same procedure can be used to test membership in $\varphi(G)$. Given $g \in S_k$, either the procedure constructs a preimage of g, or we notice for some j that no element of R_j carries δ_j to the block $\Delta_{l'}$, as required in the sifting procedure above.

5.1.4. Closures and Normal Closures

Suppose that we have already computed a base $B = (\beta_1, \ldots, \beta_m)$ and a strong generating set S relative to B for some group $G \leq \mathrm{Sym}(\Omega)$, and now we need an SGS for the group $H = \langle S \cup T \rangle$, for some $T \subseteq \mathrm{Sym}(\Omega)$. Following the terminology in GAP, we call H the *closure* of G by T. The incremental nature of the Schreier–Sims procedure enables us to compute an SGS for H as an augmentation of the SGS for G, instead of starting the construction from scratch. Namely, using the terminology of Section 4.2, we add T to the generating set of G, recompute the first fundamental orbit β_1^H and the transversal for H mod H_{β_1}, and declare that our data structure is up to date below level 1. When the data structure becomes up to date below level 0, we have an SGS for H relative to a base that contains B as a (not necessarily proper) initial segment.

We can also compute normal closures and G-closures. Suppose that $H = \langle T \rangle \leq \mathrm{Sym}(\Omega)$ and $G = \langle S \rangle \leq \mathrm{Sym}(\Omega)$ are given, T is an SGS for H, and we need to compute the G-closure $\langle H^G \rangle$. Since a strong generating set enables us to test membership in a permutation group, the method of Section 2.1.2 can be applied. According to that algorithm, when we detect that $h^g \notin H$ for some $h \in T$ and $g \in S$, we replace H by the closure $\langle H, h^g \rangle$. However, it was observed in practice that frequent calls of the closure algorithm are quite slow. The reason is that the algorithm spends a significant amount of time for SGS computations in subgroups of $\langle H^G \rangle$ that will be replaced later by larger subgroups of $\langle H^G \rangle$. Therefere, to reduce the number of calls of the closure algorithm, it is beneficial to collect all h^g with $h \in T, g \in S, h^g \notin H$ in a list L and call the closure algorithm for $\langle H, L \rangle$. Of course, it is possible that $\langle H, L \rangle$ is not yet the G-closure of H and the procedure has to be iterated; however, the number of calls of the closure algorithm usually decreases significantly, and the G-closure algorithm becomes faster.

In GAP, this observation is combined with the random subproduct method. Given $H, G \leq \mathrm{Sym}(\Omega)$ as in the previous paragraph, in the list L we collect permutations h^g, where h is a random subproduct of the generators of H and g is a random subproduct of the generators of G. In this way, we can avoid working with long lists of permutations in the case when $|S||T|$ is too large. In the GAP implementation, L is chosen to be of length 10. Also, when computing the random subproduct h, we also use the previously constructed elements of L as generators. With this strategy, the G-closure is almost always obtained by computing at most two lists L of length 10, and so calling the closure algorithm at most twice. After the alleged G-closure $\bar{H}$ of H is computed, we may check the correctness of the result deterministically, by confirming that the conjugates of the generators of $\bar{H}$ by S are in $\bar{H}$. Alternatively, by Lemma 2.3.8, if k random subproduct conjugates h^g fail to increase $\bar{H}$ then $\bar{H} = \langle \bar{H}^G \rangle$ with probability at least $1 - (3/4)^k$.

Normal closure computations enable us to get generators for commutator subgroups and hence for the elements of the derived and lower central series. In particular, we can test solvability and nilpotence. We note, however, that the methods of Chapter 7 test solvability and nilpotence more quickly.

5.2. Working with Base Images

Given a base B for some $G \leq \mathrm{Sym}(\Omega)$, the images of base points uniquely determine the elements of G. Namely, if $B^g = B^h$ for some $g, h \in G$ then gh^{-1} fixes B pointwise and so $g = h$. Therefore, a set $A \subseteq G$ may be described by storing only the images of B under the elements of A. If, as usually happens in practical situations, $|B|$ is significantly smaller than $|\Omega|$ then this method saves memory compared to the storage of permutations.

We need, however, a method to recover the elements of G from the base images. Given an SGS for G relative to B, this task can be readily accomplished.

Lemma 5.2.1. *Let S be an SGS for $G \le \mathrm{Sym}(\Omega)$ relative to B and let t denote the sum of depths of Schreier trees coding the coset representative sets along the point stabilizer chain of G. Then, given an injection $f : B \to \Omega$, it is possible to find a permutation $g \in G$ with $B^g = f(B)$ or determine that no such element of G exists in $O(t|\Omega|)$ time, by a deterministic algorithm.*

Proof. Let $B = (\beta_1, \ldots, \beta_m)$ be the base of G and let $G = G^{[1]} \ge \cdots \ge G^{[m+1]} = 1$ be the corresponding point stabilizer chain. The element $g \in G$ with the required property can be obtained by a modification of the standard sifting procedure for permutations. If $f(\beta_1)$ is in the orbit $\beta_1^{G^{[1]}}$ then, taking the product of edge labels along the path from $f(\beta_1)$ to β_1 in the first Schreier tree, we obtain a permutation $r_1 \in G$ such that $f(\beta_1)^{r_1} = \beta_1$. Hence, the function $f_2 : B \to \Omega$ defined by $f_2(\beta_i) := f(\beta_i)^{r_1}$ fixes β_1. If $f_2(\beta_2) \in \beta_2^{G^{[2]}}$ then, as a product of edge labels in the second Schreier tree, we obtain $r_2 \in G$ such that the function $f_3 : B \to \Omega$ defined by $f_3(\beta_i) := f(\beta_i)^{r_1 r_2}$ fixes the first two base points. Continuing this process, we obtain some $g = r_1 r_2 \cdots r_m \in G$ such that $f(B) = B^{(g^{-1})}$. The number of permutation multiplications in the procedure is at most t. If, for some $i \in [1, m]$, we have $f_i(\beta_i) \notin \beta_i^{G^{[i]}}$ then we conclude that there is no $g \in G$ with $f(B) = B^g$. □

A possible application of this method is when we have to store numerous elements of G or generators for numerous subgroups of G. For example, this is the case when we need to store conjugacy class representatives of G or the subgroup lattice of G. It is enough to store the base images of the conjugacy class representatives or the base images of the generators of the subgroups and, when the group elements are actually needed, Lemma 5.2.1 can be invoked. We shall describe another applications of Lemma 5.2.1 in Sections 6.1.3 and 6.1.4.

Lemma 5.2.1 can be sped up significantly in the case when we do not need the permutation g with $B^g = f(B)$ explicitly but only need to establish the existence of such g or it is enough to have g written as a word in the strong generators. For simplicity, let us assume that the SGS S of G is closed for inverses (i.e., $S = S^{-1}$). This assumption is automatically satisfied if, as in GAP, the Schreier trees are constructed by the algorithm in the proof of Lemma 4.4.2.

Lemma 5.2.2. *Given an injection $f : B \to \Omega$, it is possible to either find a word w of length at most t in the strong generators whose product is the unique $g \in G$ with $B^g = f(B)$ or to determine that no such element of G exists, in $O(|B|t)$ time.*

Proof. We modify the algorithm described in the proof of Lemma 5.2.1 by not computing the inverses $r_1, \ldots, r_m$ of transversal elements explicitly, but by writing them only as words in the strong generators. Namely, if $f(\beta_1)$ is in the orbit $\beta_1^{G^{[1]}}$ then we take the sequence of edge labels along the path from $f(\beta_1)$ to β_1 in the first Schreier tree. In this way, we obtain a word w_1 in the strong generators such that $f(\beta_1)^{w_1} = \beta_1$. Then we define the function $f_2 : B \to \Omega$ by $f_2(\beta_i) := f(\beta_i)^{w_1}$. Iterating this process, eventually we obtain a word $w_1 w_2 \cdots w_m$ such that $w := (w_1 w_2 \cdots w_m)^{-1}$ satisfies $f(B) = B^w$. Since $S = S^{-1}$, we can write w as a word in the strong generators as well. As before, if $f_i(\beta_i) \notin \beta_i^{G^{[i]}}$ for some $i \in [1, m]$ then we conclude that there is no $g \in G$ with $f(B) = B^g$. □

We shall refer to the procedure described in Lemma 5.2.2 as *sifting as a word*. The first important application of sifting as words is the following theorem.

Theorem 5.2.3. *Given a base B for some $G = \langle S \rangle \leq \mathrm{Sym}(\Omega)$, with $|\Omega| = n$, an SGS for G can be computed in $O(n|B|^2|S| \log^3 |G|)$ time, by a deterministic algorithm. In particular, if a nonredundant base is known then an SGS can be obtained by a nearly linear-time deterministic algorithm.*

Proof. We modify the basic Schreier–Sims algorithm, described in Section 4.2, as follows: Coset representatives for the stabilizer chain of G are stored in Schreier trees, computed by the algorithm in the proof of Lemma 4.4.2. This ensures that the depth of each tree is at most $2 \log |G|$. The time-critical part of the procedure is the sifting of Schreier generators. We construct and sift Schreier generators as words. If the sifting produces a nontrivial residue then we compute this residue as a permutation of Ω and add it to the strong generating set. Nontrivial residues occur at most $|B| \log |G|$ times.

We leave for the reader to check the details that the modified procedure satisfies the claimed time bound. □

In practice, it occurs frequently that we know a base in advance. In a number of cases, we compute an SGS for some group G, and then the bulk of the computation deals with various subgroups of G. In these cases, the base of G can be used as a base for all of its subgroups. This will play an important role in Chapter 6. Of course, a known base can be used in a combination with the randomized methods of Chapter 4 as well.

Base image computations can be utilized at the construction of an SGS even in the case when a base for G is not known in advance. We finish this section with such an application from [Babai et al., 1991], which, with high probability, constructs an SGS of a small-base group faster than the version described in Theorem 4.5.5.

The following lemma is folklore; it seems to appear first in print in [Cameron et al., 1984].

Lemma 5.2.4. *Let $G \le \mathrm{Sym}(\Omega)$, with $|\Omega| = n$, be transitive. Suppose that the minimal base size of G is b. Then, for all $g \in G\backslash\{1\}$, we have $|\mathrm{supp}(g)| \ge n/b$.*

Proof. Let B be a minimal base for G (considered as a *subset* of Ω, not a sequence) and let Δ be the support of some $g \neq 1$, $|\Delta| = d$. We define two hypergraphs on Ω with edge sets $\mathcal{B} = \{B^h \mid h \in G\}$ and $\mathcal{D} = \{\Delta^h \mid h \in G\}$, respectively. Both hypergraphs are uniform and, since G is transitive, each element of Ω is contained in the same number of edges of $\mathcal{B}$ and $\mathcal{D}$. Let $\deg(\mathcal{B})$ and $\deg(\mathcal{D})$ denote the valencies of vertices in these hypergraphs. Counting the number of pairs (B', β) with $B' \in \mathcal{B}$ and $\beta \in B'$ two ways, we obtain $\deg(\mathcal{B})n = b|\mathcal{B}|$. Similarly, $\deg(\mathcal{D})n = d|\mathcal{D}|$. Moreover, because the elements of $\mathcal{B}$ are bases for G and the elements of $\mathcal{D}$ are supports of group elements, $B' \cap \Delta' \neq \emptyset$ for all $B' \in \mathcal{B}$, $\Delta' \in \mathcal{D}$. A simple double counting argument gives

$$\deg(\mathcal{B})\deg(\mathcal{D})n = |\{(B', \Delta', \beta) \mid B' \in \mathcal{B}, \Delta' \in \mathcal{D}, \beta \in B' \cap \Delta'\}| \ge |\mathcal{B}||\mathcal{D}|.$$

Substituting $\deg(\mathcal{B}) = b|\mathcal{B}|/n$ and $\deg(\mathcal{D}) = d|\mathcal{D}|/n$ yields $bd \ge n$. □

The speedup of the nearly linear-time SGS construction for small-base groups may proceed as follows. We warn the reader in advance that this is a theoretical exercise: The GAP implementation also uses base image computations, but the implemented version is much simpler than the algorithm presented here. We shall describe how base images are used in the implementation after the proof of Theorem 5.2.6.

We first handle the case of transitive groups. Please note the distinction in the statement of the next theorem between the *maximum* running time, which we never exceed, and the *typical* running time, which happens with high probability. Such distinction occurs frequently in Las Vegas algorithms, where we can check the correctness of the output with certainty, but it is quite unusual in the case of Monte Carlo algorithms.

Theorem 5.2.5. *Suppose that $G = \langle T \rangle \le \mathrm{Sym}(\Omega)$, with $|\Omega| = n$, is transitive, and let $d > 0$ be a constant. There is a Monte Carlo algorithm that constructs an SGS for G in $O(n \log n \log^4 |G| + |T| n \log |G|)$ time. With probability greater than $1 - 1/n^d$, the algorithm terminates in $O(\log^5 |G| \log n + n \log n \log^3 |G| +$*

$|T|n \log |G|)$ *time. In any case, the output is correct with probability greater than* $1 - 1/n^d$. *The memory requirement is* $O(n \log^2 |G| + |T|n)$.

Proof. We assume familiarity with the nearly linear-time SGS construction in Section 4.5, which we briefly summarize here. Recall that we work with an initial segment $B = (\beta_1, \ldots, \beta_k)$ of a nonredundant base and an approximation S_i for a generating set of $G^{[i]} := G_{(\beta_1,\ldots,\beta_{i-1})}$, for $1 \leq i \leq k$. We maintain the property that, for all $i \in [1, k]$, $\langle S_i \rangle \geq \langle S_{i+1} \rangle$, and we say that the data structure is up to date below level j if $\langle S_i \rangle_{\beta_i} = \langle S_{i+1} \rangle$ holds for all i with $j < i \leq k$ *and* the sum of depths of the Schreier trees for levels $j + 1, \ldots, k$ is at most $6 \log |\langle S_{j+1} \rangle|$.

In the case when the data structure is up to date below level j, we compute a Schreier tree for $\langle S_j \rangle$ mod $\langle S_j \rangle_{\beta_j}$ and check whether $\langle S_j \rangle_{\beta_j} = \langle S_{j+1} \rangle$ by sifting Schreier generators composed from random cosets for $\langle S_j \rangle$ mod $\langle S_j \rangle_{\beta_j}$ and short and long random subproducts from S_j. If $\langle S_j \rangle_{\beta_j} = \langle S_{j+1} \rangle$ then we compute a shallower Schreier tree for $\langle S_j \rangle$ mod $\langle S_j \rangle_{\beta_j}$; if not then we add a new generator to S_{j+1}.

As asserted in Remark 4.5.2, the total time spent in Schreier tree computations is $O(n \log n \log^3 |G|)$. Hence, to achieve the running time stated in this theorem, we have to speed up the sifting of Schreier generators.

The idea is that we form the short and long random subproducts only as words, without multiplying out the permutations, and sift the Schreier generators as words in the Schreier tree data structure of $\langle S_{j+1} \rangle$. If a Schreier generator does not sift through as a word then we multiply out the siftee and obtain a permutation in $\langle S_j \rangle_{\beta_j} \backslash \langle S_{j+1} \rangle$. If a Schreier generator sifts through then we have to establish that the siftee w, which is a word fixing B pointwise, is really the identity. We would like to check this property by establishing that some randomly chosen points in Ω are fixed by w.

By Lemma 5.2.4, if b is the minimal base size of G and b randomly chosen points of Ω are fixed by w then $w = 1$ with probability greater than $1 - (1 - 1/b)^b > 1/2$. The problem is that we do not know the value of b in advance. Therefore, we *guess* an upper bound M for $\log |G|$ and run the entire algorithm under the assumption that $\log |G| \leq M$. We shall check each siftee by computing the image of M randomly chosen points. If our guess for M is correct then $b \leq M$, and we detect a nontrivial siftee with probability greater than $1/2$.

We know that G is transitive on Ω and so $|G| \geq n$. Therefore, our initial guess for M is the smallest power of 2 greater than $\log n$, $M = 2^l \geq \log n$. We abort the algorithm immediately if we obtain some evidence that our guess for M is incorrect: Either $\prod_j |\beta_j^{\langle S_j \rangle}| > 2^M$ or the deterministic Schreier tree computation of Lemma 4.4.2 produces a Schreier tree of depth greater than $2M$, or a randomized Schreier tree computation does not return a shallow Schreier tree as required in Remark 4.4.7.

We claim that the running time of the SGS construction with a guess M as an upper bound for $\log |G|$ is

$$O(nM^3 \log n + M^5 \log n + |T|M^2). \tag{5.1}$$

The Schreier tree computations run in $O(nM^3 \log n)$ time, since we do not encounter more than M base points (otherwise the algorithm aborts), each S_j increases at most M times, and by Remark 4.5.2, the invocations of the algorithm in the proof of Lemma 4.4.2 on each fixed level run in $O(nM^2)$ total time. Hence all $\log |G|$ factors in the original timing estimate in Remark 4.5.2 can be replaced by M.

The other part of the algorithm, the sifting of Schreier generators, can be done in $O(nM^3 \log n + M^5 \log n)$ time: We have at most M base points, each S_j increases at most M times, and after each increase of S_j, we sift $O(M \log n)$ Schreier generators as words. Note that we have to sift twice as many Schreier generators as in the version in Section 4.5, since we have to compensate for the fact that we detect a nontrivial siftee only with probability between $1/2$ and 1. The handling of each Schreier generator takes $O(M^2)$ time: This is the time requirement of sifting as a word by Lemma 5.2.2, as well as the required time to compute the images of M randomly chosen points in the siftee. We detect a nontrivial siftee at most M^2 times (at most M times at each of at most M base points) and multiplying out a nontrivial siftee takes $O(Mn)$ time. Finally, the term $O(|T|M^2)$ in (5.1) comes from the sifting of the input generators as words.

If our guess for M is correct then the error probability analysis in Section 4.5 is valid (note again that we have compensated for the fact that not all nontrivial siftees are detected by sifting twice as many Schreier generators as in Section 4.5) and we obtain a correct SGS with probability greater than $1-1/n^d$. However, it is possible that, although the algorithm did not abort, the output is incorrect because our estimate for M was too low and we missed too many nontrivial siftees. Hence, after termination, we use the strong generating test described in Lemma 4.5.6 to check whether the result is correct. If $R_1, \ldots, R_k$ denote the transversals constructed by the algorithm then, by Lemma 4.5.6, sifting $O(M \log n)$ elements of T and of $D := R_k \cdots R_1$ in D detects if the output is incorrect with probability greater than $1 - 1/n^d$. We sift the elements of D and T as permutations, not only as words, and the time requirement for each sift is $O(Mn)$.

If something bad happens (the algorithm aborts or the final strong generating test reveals that the output SGS is incorrect) then we double our guess M and repeat the entire algorithm.

When our guess exceeds the actual value of $\log |G|$ for the first time, the SGS construction succeeds with probability greater than $1 - 1/n^d$ and the algorithm terminates. Since this last guess satisfies $\log |G| \leq M < 2\log |G|$, the running time estimate of the last run is $O(\log^5 |G| \log n + n \log n \log^3 |G| + |T| \log^2 |G|)$. Also, because we double our guesses for M, the running time estimate of the last run dominates the sum of running time estimates for all previous runs.

What happens if, owing to a sequence of unlucky random bits, the algorithm does not terminate when M exceeds $\log |G|$? When the algorithm is run with a guess $M > \log |G|$, the running time is

$$O(n \log^3 |G| \log n + M \log^4 |G| \log n + |T| M \log |G|), \qquad (5.2)$$

since we do not encounter more than $\log |G|$ base points, each S_j increases at most $\log |G|$ times, the sum of depths of Schreier trees is $O(\log |G|)$, and on each level we construct $O(\log |G| \log n)$ Schreier generators (note that this number depended on the sum of depths of trees). Hence all factors M in the estimate (5.1) can be replaced by $\log |G|$, except one: The number of points whose images are traced in the siftees of Schreier generators. Note also that if the guess M exceeds n then the factors M can be replaced by n in (5.2), because we never trace the images of more than n points.

We continue the doubling of our guess M at most until M exceeds $\log(n!)$. Hence the run with the last estimate for M terminates in $O(n \log^4 |G| \log n + |T| n \log |G|)$ time, and this time estimate dominates the sum of running time estimates for all previous runs. □

Theorem 5.2.6. *Given $G = \langle T \rangle \leq \mathrm{Sym}(\Omega)$, with $|\Omega| = n$, and a constant $d > 0$, there is a Monte Carlo algorithm that computes an SGS for G in $O(\log^6 |G| \log n + n \log^4 |G| \log n + |T| n \log |G|)$ time. With probability greater than $1 - 1/n^d$, the algorithm terminates in $O(\log^6 |G| \log n + n \log^3 |G| \log n + |T| n \log |G|)$ time. In any case, the output is correct with probability greater than $1 - 1/n^d$. The memory requirement is $O(n \log^2 |G| + |T| n)$.*

Proof. We start by computing the orbits of G on Ω and choosing an orbit Δ_1 of size greater than 1. We compute a base and SGS for the G-action on Δ_1, by the algorithm described in Theorem 5.2.5. The Schreier tree computations and multiplying out nontrivial siftees as permutations are carried out with permutations of Ω, not only with their restriction to Δ_1. This computation is the base case of the following recursive procedure.

Suppose that a base $B = (\beta_1, \ldots, \beta_k)$ and generating sets S_i and transversals R_i for $1 \le i \le k$ are already computed for the G-action $G|_{\Delta_1 \cup \cdots \cup \Delta_{j-1}}$ on $\Delta_1 \cup \cdots \cup \Delta_{j-1}$, for some orbits $\Delta_1, \ldots, \Delta_{j-1}$ of G. As in the base case, the strong generators and the labels in Schreier trees are computed as permutations of Ω. Then we perform the following three steps:

(a) Decide whether $G_{(\Delta_1 \cup \cdots \cup \Delta_{j-1})} = 1$; if not, then find an orbit Δ_j where $G_{(\Delta_1 \cup \cdots \cup \Delta_{j-1})}$ acts nontrivially.
(b) Compute a nonredundant base for the G-action on Δ_j.
(c) Compute a base and SGS for the G-action on $\Delta_1 \cup \cdots \cup \Delta_j$.

The recursion terminates when Step (a) reports that $G_{(\Delta_1 \cup \cdots \cup \Delta_{j-1})} = 1$, or $\Delta_1 \cup \cdots \cup \Delta_j = \Omega$.

Step (a) is simply calling the strong generating test of Lemma 4.5.6, with the set $D := R_k \cdots R_1$.

If an orbit Δ_j is output by Step (a) then we call the algorithm of Theorem 5.2.5 to compute a nonredundant base $\Gamma := (\gamma_1, \ldots, \gamma_l)$ (and an SGS, which we shall discard) for $G|_{\Delta_j}$. We carry out this computation using only the restrictions of permutations to Δ_j. Since $G|_{\Delta_1 \cup \cdots \cup \Delta_j} \le G|_{\Delta_1 \cup \cdots \cup \Delta_{j-1}} \times G|_{\Delta_j}$, the concatenation of B and Γ is a (possibly redundant) base for $G_{(\Delta_1 \cup \cdots \cup \Delta_j)}$.

In Step (c), we augment the already known SGS for $G|_{(\Delta_1 \cup \cdots \cup \Delta_{j-1})}$, modifying the nearly linear-time algorithm of Section 4.5 by executing all sifts only as words and checking whether a siftee is trivial by computing the images of all points in B and Γ. Initially, we declare that the already known data structure is up to date below level k. Note that, although the Schreier trees at the base points $\beta_1, \ldots, \beta_k$ do not change, we may have to add new generators to some S_i for some $i \le k$.

Now we address the time requirement of the algorithm. The base case, the computation of an SGS for $G|_{\Delta_1}$, runs in $O(|\Delta_1| \log^4 |G| \log n + n \log^3 |G| \log n + |T||\Delta_1| \log |G|)$ maximum time and, with probability greater than $1 - 1/n^{d+2}$, the algorithm terminates in

$$O(\min\{\log |G|, |\Delta_1|\} \log^4 |G| \log n + n \log^3 |G| \log n + |T| \min\{\log |G|, |\Delta_1|\} \log |G|)$$

time. Recursion is called at most $\log |G|$ times, since the order of $G|_{\Delta_1 \cup \cdots \cup \Delta_j}$ is at least twice the order of $G|_{\Delta_1 \cup \cdots \cup \Delta_{j-1}}$.

One call of Step (a) takes $O(nt^2 \log n)$ time, where t is the sum of depths of Schreier trees at the time of the call. This sum is always at most $6 \log |G|$, so the total cost of calls of Step (a) during the entire recursion is $O(n \log^3 |G| \log n)$.

Calling Step (b) on the orbit Δ_j runs in time $O(|\Delta_j| \log^4 |G| \log n + |T||\Delta_j| \log |G|)$ and, with probability greater than $1 - 1/n^{d+2}$, this call terminates in

$$O(\min\{\log |G|, |\Delta_j|\} \log^4 |G| \log n + |\Delta_j| \log^3 |G| \log n$$
$$+ |T| \min\{\log |G|, |\Delta_j|\} \log |G|)$$

time.

In Step (c), the cost of Schreier tree computations and multiplying out nontrivial siftees is $O(n \log^3 |G| \log n)$ during the *entire* algorithm, since the usual estimates (at most $\log |G|$ base points, each S_j increases at most $\log |G|$ times, etc.) are valid and we do not duplicate work at the different calls of Step (c).

Sifting one Schreier generator and deciding whether the siftee is trivial can be done in $O(\log^2 |G|)$ time, since we evaluate siftees on $|B| + |\Gamma| \leq 2 \log |G|$ points. During one call of Step (c), Schreier generators are sifted $O(\log^3 |G| \log n)$ times because there are at most $\log |G|$ base points, each S_j increases at most $\log |G|$ times, and $O(\log |G| \log n)$ generators are sifted at the processing of a level. Consequently, during all calls of Step (c), we spend $O(\log^6 |G| \log n)$ time handling Schreier generators.

If an element of the generating set T sifts through as a word in $G|_{\Delta_1 \cup \cdots \cup \Delta_{j-1}}$ with trivial siftee w (i.e., w fixes all base points in $\Delta_1 \cup \cdots \cup \Delta_{j-1}$), then we store w and sift it along the base points in Δ_j at the next call of Step (c). In this way, we can achieve that the total time requirement of sifting T during calls of Step (c) is $O(|T| \log^2 |G|)$.

Summarizing, we obtain that the maximum time requirement of the entire algorithm is $O(\log^6 |G| \log n + n \log^4 |G| \log n + |T| n \log |G|)$. Note that we ensured that each call of the base case and of Steps (a), (b), and (c) succeeds with probability greater than $1 - 1/n^{d+2}$, so the probability of a correct output is greater than $1 - 3 \log |G|/n^{d+2} > 1 - 1/n^d$ and the algorithm terminates in $O(\log^6 |G| \log n + n \log^3 |G| \log n + |T| n \log |G|)$ time with probability greater than $1 - \log |G|/n^{d+2} > 1 - 1/n^d$. □

Remark 5.2.7. As mentioned in Section 4.5.1, GAP employs a heuristic version of the nearly linear-time Monte Carlo SGS construction. At the processing of a level of the Schreier tree data structure, we form only one Schreier generator g from a random coset and a random subproduct of generators on this level. Then g is sifted as a word, and the siftee is evaluated at each point in orbits of size at most 50, and at five randomly chosen points in orbits of size greater than 50. Although Lemma 5.2.4 cannot be sharpened (cf. Exercise 5.3), five randomly chosen points have a decent chance to detect that a word does not represent the identity on an orbit. By a lemma of Cauchy (widely attributed to Burnside), the average number of fixed points of elements in a transitive permutation group is one (cf. Exercise 5.4).

Groups with numerous orbits do not occur frequently as inputs in practical computations. Usually, we encounter a group H with a lot of orbits as a subgroup of some group G so that when we need an SGS for H, we already have a base B for G. Then $H_B = 1$, so all sifts in H can be done as words and it is enough to evaluate the siftees on B. GAP also contains a version of the algorithm of Lemma 4.5.6, which sifts the elements described in Lemma 4.5.6 as words. This version is utilized if, for example, the user asks for a randomized testing of the correctness of an SGS construction for an input group with known base.

Monte Carlo SGS constructions are applied in GAP for groups acting on more than 100 points. For smaller permutation domains, Sims's original deterministic SGS construction (cf. Section 4.2) is called.

5.3. Permutation Groups as Black-Box Groups

Representing elements of a group $G \leq \mathrm{Sym}(\Omega)$ by storing only the images of a base $B = (\beta_1, \ldots, \beta_m)$ has a serious drawback: From this representation, it is hard to compute the representation of the product of two elements of G. In applications where products have to be computed frequently, invoking Lemma 5.2.1 at each occasion becomes too cumbersome.

Representation as a Black-Box Group

A compromise between storing base images and full permutations is to represent the elements of G by words over an alphabet consisting of the labels S of a Schreier tree data structure of G. Suppose that S is closed for taking inverses. Each $g \in G$ can be written uniquely in the form $g = r_m r_{m-1} \cdots r_1$, where r_i is a coset representative for $G^{[i]} \bmod G^{[i+1]}$. Also, taking the inverses of edge labels along the path from β_i to $\beta_i^{r_i}$ in the ith Schreier tree in the Schreier tree data structure of G, we can decompose r_i as a product of labels. Eventually, g is written as a word in S. We call this word the *standard word* representing g. The length of any standard word is at most the sum t of the depths of the Schreier trees.

Representing the elements of G by standard words defines an isomorphism between G and a black-box group H, where the elements of H are strings of length at most t over the alphabet S. Let $\psi : G \to H$ denote this isomorphism.

Moreover, let μ denote the time requirement in H to compute the product of two given elements of H or the inverse of an element of H, and let ξ be the time requirement to compute a nearly uniformly distributed random element in a given subgroup of H.

Lemma 5.3.1. *There is an absolute constant c (for example, we can choose* $c = 8$*) such that:*

(a) *The quantities* ξ, μ, t *for H are bounded from above by* $\log^c |G|$.
(b) *For any* $g \in G$, $\psi(g)$ *can be computed in* $O(\log^c |G|)$ *time. Conversely, given* $h \in H$, $\psi^{-1}(h)$ *can be computed in* $O(|\Omega| \log^c |G|)$ *time.*

Proof. (a) By Lemma 4.4.2, given any strong generating set of G relative to some nonredundant base B, we can compute Schreier trees of depth at most $2 \log |G|$ for a Schreier tree data structure of G in nearly linear time by a deterministic algorithm. Hence, we can suppose that $t \leq 2|B| \log |G| \leq 2 \log^2 |G|$. (In fact, if the SGS of G was computed by the nearly linear-time Monte Carlo algorithm described in Section 4.5 then $t \in O(\log |G|)$.)

Given $h_1, h_2 \in H$, we compute their products by concatenating them, then follow the images of base points in this concatenated word to obtain a function $f : B \to \Omega$, and use Lemma 5.2.2 to compute the standard word representing $h_1 h_2$. This requires $O(|B|t)$ time. Similarly, the inverse of any $h \in H$ can be computed by taking the inverse of the word h formally, as a word, following the images of base points to obtain a function $f : B \to \Omega$, and again using Lemma 5.2.2. Hence, products and inverses in H can be computed in $O(|B|t)$ time and so $\mu \in O(\log^3 |G|)$. Since the elements of H are represented by standard words in a unique way, comparision of group elements can be done in $O(t)$ time.

Random elements of H can be constructed in $O(t)$ time, by concatenating words representing randomly chosen coset representatives along the point stabilizer subgroup chain of G. In subgroups of H, nearly uniformly distributed random elements can be constructed by the algorithm of Theorem 2.2.4 in $O(\log^8 |H|)$ time.

(b) Given $g \in G$, $\psi(g)$ can be computed in $O(|B|t)$ time by Lemma 5.2.2. Conversely, given $h \in H$, $\psi^{-1}(h)$ is computed by at most t permutation multiplications. □

The advantage of considering G as a black-box group is that we can perform $O(n)$ group operations while still remaining within the nearly linear time frame.

The disadvantage is that we lose the information stored implicitly in the cycle structure of permutations; for example, we cannot compute orders of group elements in polynomial time. Hence, we have to rely on the techniques of Chapter 2.

The idea of considering permutation groups as black-box groups was introduced in [Beals and Seress, 1992]. We shall describe several applications (cf. Sections 6.2, 8.3, and 8.4), but here we only give an illustration of the technique. We describe a nearly linear-time solution from [Morje, 1995] for the following problem, which arose as part of the nearly linear-time computation of Sylow subgroups in groups with no composition factors of exceptional Lie type: Given an arbitrary permutation representation $G \leq \mathrm{Sym}(\Omega)$ for some classical simple group G of Lie type, we want to construct the action of G on one of its natural domains. For example, in the case $G \cong \mathrm{PSL}_d(q)$, we need the action of G on the $(q^d-1)/(q-1)$ points or on the $(q^d-1)/(q-1)$ hyperplanes of the $(d-1)$-dimensional projective space. The permutation representations on the projective points and on the hyperplanes have degree no greater than the original one. Also, in later steps of the Sylow algorithm, the elements of the new permutation domain are identified with the points of the projective space, which enables us to write matrices representing some cleverly chosen generators of G. Then the matrix representation is used for finding Sylow subgroups of G and for conjugating two given Sylow subgroups into each other.

In the following description, our main goal is to demonstrate the use of black-box group techniques, and we use properties of the special linear groups without definitions and proofs. (The basic properties of classical groups of Lie type can be found, for example, in [Taylor, 1992].) The reader unfamiliar with these groups can skip the rest of the section without impeding the understanding of further material. Let $q = p^e$, with p prime. The idea is to find the conjugacy class Σ of transvections and the conjugation action of G on Σ. The set Σ is partitioned into $(q^d - 1)/(q - 1)$ equal parts in two different ways; in the first partition, each part contains the transvections that fix pointwise the same hyperplane, and in the second partition, each part contains the transvections with the same center. The actions of G on these parts are permutation isomorphic to the two desired actions on the projective space.

We may assume that G acts primitively on the original permutation domain Ω, since otherwise we can apply a transitive constituent homomorphism or block homomorphism (cf. Section 5.1.3) to obtain a faithful action of G on a smaller permutation domain. We shall describe in Section 5.5 how can one find a nontrivial block of imprimitivity if G does not act primitively. We shall also assume that $|\Omega| \neq (q^d-1)/(q-1)$ and $|\Omega| \neq (q^d-1)(q^{d-1}-1)/(q^2-1)(q-1)$, which means that the given primitive action of G is not permutation isomorphic to the action on subspaces of dimension one, two, $d-2$, or $d-1$. If the given action is on the one- or $(d-1)$-dimensional subspaces then there is nothing to do; the other two special cases we just excluded are handled by another argument which we do not describe here.

How do we find a transvection? The number of transvections is relatively small (the restrictions in the previous paragraph ensure that $|\Sigma| < |\Omega|$). On one hand, this is good, since we have at least enough time to write down the action of the generators of G as permutations of Σ. On the other hand, we have only a very slim chance of finding a transvection in the most natural way, by taking random elements of G. Hence, we apply an enhancement of this idea: We are searching for some $g \in G$ such that a suitable power of g is a transvection. We consider G as a black-box group H, as described earlier in this section. We choose random elements from H until we find $g \in H$ with order px, where $x|q^{d-2}-1$ but x does not divide $\prod_{i=1}^{d-3}(q^i-1)$. It can be shown that for such a g, the $(q^{d-2}-1)$st power of g is a transvection.

The difficulty with this procedure is that we cannot compute the orders of elements in black-box groups. However, we do not need the exact order of g; rather, we have to check that it satisfies the given divisibility constraints. These can be checked by taking appropriate powers of g, and comparing the result with the identity of H. Powers of g are taken by repeated squaring, as described on the first page of Chapter 2. Once a transvection is found, conjugates are determined by an orbit computation, as in Theorem 2.1.1(iii). The required ordering for this orbit computation may be chosen as the lexicographic ordering of the standard words, with respect to some ordering of the alphabet (i.e., the SGS of G). The action of the generators on Σ is obtained automatically during the orbit computation.

Note that the algorithm we just described is a Las Vegas one. It can be shown that a random element of H satisfies our order restrictions with probability $\Theta(1/(qd))$; hence the running time of the algorithm is better than nearly linear, it is $o(|\Omega|)$. In particular, working in the black-box group, we do not need to write down or read any permutation of Ω.

The basic idea of the algorithm, namely defining a large set X such that an appropriate power of the elements of X falls into a small conjugacy class, is very important. Variations of this idea are used, for example, in [Beals and Babai, 1993], [Hulpke, 1993], and [Beals et al., 2002]. In this book, we shall use the idea in Sections 9.4 and 10.2. There are also applications in nonalgorithmic settings (see, e.g., [Guralnick and Kantor, 2000]).

It turns out that it is possible to construct the matrix representation of black-box classical simple groups without listing any conjugacy classes of group elements (cf. [Kantor and Seress, 2001]). Although we shall state this result formally in Section 8.3, its proof far exceeds the scope of this book both in length and in the required group theoretical background.

5.4. Base Change

An oft occurring problem is that we already have an SGS for a group G relative to some base but we need another base and corresponding SGS that are more suitable for the problem at hand. In Section 5.1.1, we saw that we can construct a base relative to any ordering of the permutation domain; however, more efficient methods exist that exploit the already existing SGS.

The first such base change algorithm was described in [Sims, 1971a, 1971b]. Given an SGS relative to some base $(\beta_1, \ldots, \beta_m)$, Sims's method constructs an SGS relative to the base

$$(\beta_1, \ldots, \beta_{i-1}, \beta_{i+1}, \beta_i, \beta_{i+2}, \ldots, \beta_m).$$

Repeated applications of this procedure enable us to compute an SGS relative to any base $(\alpha_1, \alpha_2, \ldots, \alpha_k)$. Namely, if α_1 does not already occur on the list $(\beta_1, \ldots, \beta_m)$ then we insert it at the earliest possible position (i.e., after the first β_i such that $G_{(\beta_1,\ldots,\beta_i)}$ fixes α) as a redundant base point; after that, with repeated transpositions, we bring α_1 to the first position. Then, working in G_{α_1}, we construct an SGS relative to a base starting with (α_1, α_2), and so on. This incremental procedure is especially efficient when only an initial segment of a new base is prescribed. For example, this is the case when we need generators for the stabilizer of some point $\alpha_1 \in \Omega$.

The Exchange of Two Base Points (Deterministic Version)

Let S be an SGS for G relative to the base $(\beta_1, \ldots, \beta_m)$, let $G = G^{[1]} \geq \cdots \geq G^{[m+1]} = 1$ be the corresponding point stabilizer subgroup chain, and let $R_1, \ldots, R_m$ be the transversals along this subgroup chain. Our goal is to compute an SGS and the transversals $\bar{R}_1, \ldots, \bar{R}_m$ corresponding to the point stabilizer chain $G = \bar{G}^{[1]} \geq \cdots \geq \bar{G}^{[m+1]} = 1$ defined by the new ordering $(\beta_1, \ldots, \beta_{i-1}, \beta_{i+1}, \beta_i, \beta_{i+2}, \ldots, \beta_m)$.

For $1 \leq j < i - 1$ and $i + 1 < j \leq m$, $\bar{R}_j = R_j$. Also, since $\bar{G}^{[i]} = G^{[i]}$, the new ith fundamental orbit $\beta_{i+1}^{\langle S \cap G^{[i]} \rangle}$ and a Schreier tree coding $\bar{R}_i$ are readily obtained. The bulk of the work is to compute $\bar{R}_{i+1}$.

We initialize the new $(i + 1)$st fundamental orbit as $\bar{\Delta}_{i+1} := \{\beta_i\}$. Then we form Schreier generators (only as words, without multiplying out permutations) from $\bar{R}_i$ and $S \cap G^{[i]}$. If, for some word w representing a Schreier generator, the image β_i^w is not in $\bar{\Delta}_{i+1}$ then we multiply out w to obtain a permutation $p \in$

$G_{(\beta_1,\ldots,\beta_{i-1},\beta_{i+1})}$, add p to the set S, and recompute $\bar{\Delta}_{i+1} := \beta_i^{\langle S\cap G_{(\beta_1,\ldots,\beta_{i-1},\beta_{i+1})}\rangle}$. Note that the correct value of $|\bar{\Delta}_{i+1}| = |\bar{R}_{i+1}| = |R_i||R_{i+1}|/|\bar{R}_i|$ is known in advance, so the algorithm may terminate (and usually does) without considering all Schreier generators. We have to add at most $\log|\bar{R}_{i+1}|$ permutations to S, since the already constructed part of $\bar{\Delta}_{i+1}$ at least doubles at each addition. The Schreier tree representing the transversal $\bar{R}_{i+1}$ can be computed from the augmented strong generating set S. If we use the deterministic algorithm in the proof of Lemma 4.4.2 to code transversals in Schreier trees of depth at most $2\log|G|$, then transposing two base points can be done in nearly linear time.

The Exchange of Two Base Points (Las Vegas Version)

The set $\bar{R}_{i+1}$ can also be computed by a randomized nearly linear-time algorithm. Since we have an SGS for $G^{[i]}$, we can construct uniformly distributed random elements g of $G^{[i]}$. Let $\bar{g} \in \bar{R}_i$ so that $g(\bar{g})^{-1} \in G^{[i]}_{\beta_{i+1}}$. As we saw in the proof of Lemma 4.3.1, $g(\bar{g})^{-1}$ is a uniformly distributed random element of $G^{[i]}_{\beta_{i+1}}$, so it has a chance at least $1/2$ to increase the already known part of $\bar{\Delta}_{i+1}$ and $\bar{R}_{i+1}$. Since we know the size of $\bar{R}_{i+1}$ in advance, this construction is Las Vegas type. This randomized construction is usually faster than the deterministic nearly linear one described earlier; an additional advantage is that, by Theorem 4.4.6, the Schreier tree coding $\bar{R}_{i+1}$ will be shallow with high probability. The GAP implementation of base change uses this Las Vegas version.

We note that Sims originally used a different method for computing $\bar{R}_{i+1}$ (cf. Exercise 5.5). This result was generalized in [Brown et al., 1989] and used in a deterministic $O(n^2)$ time algorithm to perform a *cyclic base change*, that is, changing the base from $(\beta_1,\ldots,\beta_m)$ to $(\beta_1,\ldots,\beta_{i-1},\beta_j,\beta_i,\beta_{i+1},\ldots,\beta_{j-1},\beta_{j+1},\ldots,\beta_m)$. In [Cooperman and Finkelstein, 1992], this generalization is combined with Lemma 4.4.8 to obtain a nearly linear-time deterministic cyclic base change algorithm.

Base Change by Conjugation

[Sims, 1971a] also contains the following observation. Suppose that an SGS S relative to the base $(\beta_1,\ldots,\beta_m)$ is known and we need an SGS $\bar{S}$ relative to a base starting with $(\alpha_1,\ldots,\alpha_k)$. If there exists $g \in G$ with $\beta_i^g = \alpha_i$ for $1 \le i \le k$ then $\bar{S} := S^g$ can be chosen, which is usually faster than a sequence of base point transpositions. Therefore, in GAP, the method of transposing base points is combined with the construction of a conjugating permutation c. Initially, we set $c := 1$ and build up c recursively, obtaining permutations that conjugate larger and larger initial segments of the base to the desired points. Suppose that currently, after some base point transpositions, we have an SGS relative to the base $(\gamma_1, \ldots, \gamma_l)$, and $\gamma_i^c = \alpha_i$ for $1 \leq i \leq j-1$. If $\alpha_j^{c^{-1}}$ is in the jth fundamental orbit then we compute the transversal element r from the jth transversal satisfying $\gamma_j^r = \alpha_j^{c^{-1}}$, and we redefine $c := rc$. If $\alpha_j^{c^{-1}}$ is not in the jth fundamental orbit then we bring $\alpha_j^{c^{-1}}$ in the jth position of the base via repeated transpositions and leave c unchanged. In both cases, we get a base $(\gamma_1', \ldots, \gamma_{l'}')$ and c such that $(\gamma_i')^c = \alpha_i$ holds for $i \leq j$. After k steps, we have an SGS S relative to some base such that S^c is an SGS relative to a base starting with $(\alpha_1, \ldots, \alpha_k)$.

If a very long initial segment of a new base is prescribed and it differs significantly from the current base, it may be beneficial to construct the new SGS from scratch, instead of applying a long sequence of transpositions of base points. Since we already have an SGS, uniformly distributed random elements of G are available. Thus the heuristic SGS construction described in Section 4.3, the so-called random Schreier–Sims algorithm, is upgraded to Las Vegas type by Lemma 4.3.1.

Lemma 5.4.1. *Let $G \leq \mathrm{Sym}(\Omega)$, let $\Omega = (\beta_1, \beta_2, \ldots, \beta_n)$ be an ordering of Ω, and suppose that uniformly distributed random elements of G can be constructed in time ξ. Let $d > 0$ be also given. Then there is a Monte Carlo algorithm that constructs an SGS of size $O(d \log n \log |G|)$ for G, relative to the (redundant) base $(\beta_1, \beta_2, \ldots, \beta_n)$. The time requirement of the algorithm is $O(d \log n \log |G|(\xi + nd \log n \log |G|))$ and the probability of error is less than $1/n^d$. If $|G|$ is known in advance then this algorithm is a Las Vegas one.*

Proof. Suppose that we already constructed some subsets $S_1, S_2, \ldots$ of generator sets for the chain of subgroups $G^{[1]} := G, G^{[2]} := G_{\beta_1}, \ldots$ and Schreier trees (S_i, T_i) reaching subsets $\Delta_1, \Delta_2, \ldots$ of the fundamental orbits. During the construction, we maintain the property that, for all i, if the current S_i has k elements then the depth of the Schreier tree T_i is at most k. Because of this restriction, it is possible that Δ_i does not contain all points of $\beta_i^{\langle S_i \rangle}$.

Let g be a uniformly distributed random element of G. We sift g through the partial coset representative system we have already constructed and obtain elements $g_1, g_2, \ldots, g_j, g_i \in G^{[i]}$, up to an index j; the sifting process stops at a subgroup $G^{[j]}$ where the siftee falls into a coset not yet represented in (S_j, T_j). As observed in the proof of Lemma 4.3.1, g_i is a uniformly distributed random element of $G^{[i]}$ for $1 \leq i \leq j$. For $1 \leq i \leq j$ we add g_i to the generator set S_i for $G^{[i]}$, and we add one more level to the tree T_i, by computing $\gamma^{g_i^{-1}}$ for all

$\gamma \in \Delta_i$. Hence the new tree reaches the subset $\Delta_i \cup \Delta_i^{g_i^{-1}}$ of the fundamental orbit. (If $\Delta_i \cup \Delta_i^{g_i^{-1}} = \Delta_i$ then we discard g_i.) The procedure stops when $d \log n$ random elements sift through (or, in the case when $|G|$ is known, $d \log n$ random elements sift through or $\prod_{i=1}^{n} |\Delta_i|$ reaches $|G|$). By Lemma 4.3.1, at termination we have an SGS for G with probability at least $1 - 1/n^d$.

By Theorem 4.4.6, for each i with $G^{[i]} \neq G^{[i+1]}$, the size of $|S_i|$, and hence the depth of the ith Schreier tree, will be $O(d \log n \log |G^{[i]} : G^{[i+1]}|)$ with probability at least $1-1/n^{d+1}$. Thus the new SGS will consist of $O(d \log |G| \log n)$ group elements and the sum of depths of the Schreier trees is $O(d \log n \log |G|)$. The time requirement of the algorithm is as stated, since we sift $O(d \log n \log |G|)$ random elements at an $O(nd \log n \log |G|)$ cost each. □

Although it is easy to give examples for which constructing an SGS from scratch is faster than the construction via a sequence of base point transpositions, in practice it is hard to recognize in advance the instances when the construction from scratch will be faster. The default method in GAP is via a sequence of base point transpositions combined with a conjugation, as described in this section.

5.5. Blocks of Imprimitivity

Let $G \leq \mathrm{Sym}(\Omega)$ be transitive, and let $n = |\Omega|$. Recall that $\Delta \subseteq \Omega$ is a *block of imprimitivity* for G if for all $g \in G$, $\Delta^g = \Delta$ or $\Delta^g \cap \Delta = \emptyset$. It is easy to see that G-images of a block of imprimitivity partition Ω. A transitive group $G \leq \mathrm{Sym}(\Omega)$ is called *primitive* if the only blocks of imprimitivity are $\emptyset$, Ω, and the one-element subsets of Ω. A characterization of primitive groups is that G is primitive if and only if G_ω is a maximal subgroup of G for some (and then, for all) $\omega \in \Omega$. More generally, the subgroups of G containing G_ω are in one-to-one correspondence with the blocks of imprimitivity containing ω. Namely, if $G_\omega \leq H \leq G$ then ω^H is a block of imprimitivity for G; conversely, if Δ is a block of imprimitivity for G with $\omega \in \Delta$ then $H := \{h \in G \mid \omega^h \in \Delta\}$ is a subgroup satisfying $G_\omega \leq H \leq G$.

Primitivity is a very strong structural property and, from the very beginning of group theory, primitive groups have played a central role in the study of permutation groups. Also, algorithms for finding special subgroups often use the action on blocks of imprimitivity of the input group to reduce the problem to smaller groups. Hence finding blocks of imprimitivity efficiently is crucial for many investigations.

Finding blocks of imprimitivity is one of the few tasks that can be accomplished without computing a strong generating set. The reason for this may be that computing blocks can be reduced to an orbit computation, albeit on a set of size $\Theta(n^2)$ (cf. Exercise 5.6). The first algorithm for computing blocks of imprimitivity is given in [Atkinson, 1975], where a variant of this orbit computation is performed without the quadratic increase of the permutation domain. Atkinson's algorithm runs in quadratic time for any transitive input group.

5.5.1. Blocks in Nearly Linear Time

The first nearly linear-time algorithm for computing blocks is given in [Beals, 1993a]. Here we reproduce a method from [Schönert and Seress, 1994], copyright © 1994 Association for Computing Machinery, Inc. Reprinted by Permission. This method has the same worst-case estimate as Beals's algorithm but is easier to implement and runs faster in practice. The algorithm uses some ideas and a subroutine of Beals.

Suppose that $G \le \mathrm{Sym}(\Omega)$ is transitive, and let $\alpha \in \Omega$ be fixed. For $\beta \in \Omega$, let $\bar{\beta}$ denote the smallest block of imprimitivity containing $\{\alpha, \beta\}$. Two elements $\beta, \gamma \in \Omega$ are called *equivalent*, in notation $\beta \sim \gamma$, if $\bar{\beta} = \bar{\gamma}$. Clearly, $\sim$ is an equivalence relation on Ω. We also introduce a binary relation $\preceq$ on $\Omega : \beta \preceq \gamma$ if and only if $\bar{\beta} \subseteq \bar{\gamma}$. Then $\preceq$ is a partial order (we allow that $\beta \preceq \gamma$ and $\gamma \preceq \beta$ for distinct β, γ; this happens exactly when $\beta \sim \gamma$). Each block containing α is a union of equivalence classes. A block is minimal if it is the union of $\{\alpha\}$ and exactly one other equivalence class. Our goal is to find a minimal block.

Let R be a transversal for G mod G_α. For $\beta \in \Omega$, let $r_\beta \in R$ be the unique element of R satisfying $\alpha^{r_\beta} = \beta$. Because the subgroups of G containing G_α are in one-to-one correspondence with the blocks of imprimitivity containing α, it is easy to see that $\bar{\beta}$ is the orbit of α in the group $\langle G_\alpha, r_\beta \rangle$, points in the orbits of G_α are equivalent, and $\gamma \in \bar{\beta}$ if and only if $r_\gamma \in \langle G_\alpha, r_\beta \rangle$.

Theorem 5.5.1. *Suppose that a set S of generators for some transitive $G \le \mathrm{Sym}(\Omega)$ is given and $|\Omega| = n$. Then a minimal block of imprimitivity can be computed in $O(n \log^3 |G| + n|S| \log |G|)$ time by a deterministic algorithm.*

The algorithm runs much faster in the case when the orbits of G_α and the coset representatives r_β are already known. For example, this situation occurs when a strong generating set for G was already computed.

Theorem 5.5.2. *Suppose that the orbits of G_α are known and R is coded in a Schreier tree of depth l. Then a minimal block can be computed in $O(nl \log n)$ time by a deterministic algorithm.*

We use Beals's idea to compute $\bar{\beta}$ as the orbit of α in $\langle G_\alpha, r_\beta \rangle$ rather than the component containing α in the graph $\{\alpha, \beta\}^G$, as in [Atkinson, 1975] (cf. Exercise 5.6). Also, as in [Beals, 1993a], the algorithm consists of two phases: Construction and checking. In the case when the orbits of G_α are known, the construction phase always returns a minimal block. In the case when only a refinement of the partition of Ω into orbits of G_α is known, the construction phase may return an incorrect answer. In this case, the checking phase returns an element of G_α that collapses some sets of the known partition, and we return to the construction phase. In fact, we use a small modification of Beals's Verify_Block routine for the checking phase (cf. Lemma 5.5.5).

This is an algorithm for which the claimed running times critically depend on the data structures used; in fact, the data structures we introduce are largely responsible for the nearly linear running time. Therefore, contrary to our general philosophy, we shall give a detailed description of the data structures and a pseudocode for the algorithm.

We start with our usual high-level description. The basic idea is that, while computing $\bar{\beta}$ for some $\beta \in \Omega$, we immediately abandon this computation as we encounter some $\gamma \preceq \beta$ for which we cannot exclude that $\bar{\gamma}$ is a proper subset of $\bar{\beta}$, and then we start to compute $\bar{\gamma}$. Let H be a subgroup of G_α for which generators are already known. In the construction phase, we maintain a list $\mathcal{L} = (\Lambda_1, \Lambda_2, \ldots, \Lambda_k)$ of disjoint subsets of $\Omega \backslash \{\alpha\}$ with the property that elements in Λ_i are already known to be equivalent, and $\Lambda_1 \succeq \Lambda_2 \succeq \cdots \succeq \Lambda_k$. Also, each set Λ_i is a union of H-orbits. Moreover, for each Λ_i, we store a representative $\lambda_i \in \Lambda_i$. The list $\mathcal{L}$ is initialized as $\mathcal{L} = (\Lambda_1)$, where Λ_1 is an orbit of H different from $\{\alpha\}$.

We always work in the last set Λ_k in $\mathcal{L}$ and try to compute $\bar{\lambda}_k$. To this end, we compute the images $\gamma^{r_{\lambda_k}}$ for all $\gamma \in \Lambda_k$. If $\mu := \gamma^{r_{\lambda_k}}$ is not in $\{1\} \cup \Lambda_1 \cup \cdots \cup \Lambda_k$ then we define Λ_{k+1} as the H-orbit of μ, abandon the computation of $\bar{\lambda}_k$, and start to compute $\bar{\lambda}_{k+1}$ for a representative $\lambda_{k+1} \in \Lambda_{k+1}$. If $\mu = \gamma^{r_{\lambda_k}} \in \Lambda_i$ for some $i < k$ then we see that $\Lambda_i \subseteq \bar{\lambda}_k$. Hence we redefine Λ_i as the union of the previous $\Lambda_i, \Lambda_{i+1}, \ldots, \Lambda_k$, and we start the procedure in Λ_i. Finally, if all images $\gamma^{r_{\lambda_k}}$ are in $\{\alpha\} \cup \Lambda_k$ then we stop the construction phase.

If the orbit structure of H is the same as the orbit structure of G_α then $\{\alpha\} \cup \Lambda_k$ is a minimal block (cf. Lemma 5.5.3). In the general case, we call Beals's Verify_Block routine, which checks that $\{\alpha\} \cup \Lambda_k$ is a minimal block or not; if not, it returns a permutation $g \in G_\alpha \backslash H$ with the property that the number of orbits of $\langle H, g \rangle$ is strictly less the number of orbits of H. We update $\mathcal{L}$ to achieve that each set Λ_i is the union of orbits of $\langle H, g \rangle$, and then we return to the construction phase.

The pseudocode for the algorithm is as follows:

MinimalBlock$[G]$

Input: Generators for $G \leq \mathrm{Sym}(\Omega)$ and for $H \leq G_\alpha$; H may be trivial.
Output: A minimal block of imprimitivity properly containing $\{\alpha\}$.
(Note that G is primitive if and only if the output is Ω.)

Step 1: Initialize: Compute coset representatives r_β for G mod G_α as words of length at most $2\log|G|$ over a set $T \cup T^{-1} \subseteq G$, $|T| \leq \log|G|$; $\Lambda_1 := \omega^H$ for some arbitrary $\omega \in \Omega\backslash\{\alpha\}$; $\lambda_1 := \omega$; $\mathcal{L} := (\Lambda_1)$.

Step 2: Λ :=last element of the list $\mathcal{L} = (\Lambda_1, \ldots, \Lambda_k)$; λ :=representative of Λ.

Step 3: for all $\gamma \in \Lambda$ **do**

(i) Compute $\mu := \gamma^{r_\lambda}$.

(ii) **if** $\mu \notin \{1\} \cup \Lambda_1 \cup \cdots \cup \Lambda_k$
then add a new set Λ_{k+1} to $\mathcal{L}$: let $\Lambda_{k+1} := H$-orbit of μ; λ_{k+1} :=first element of Λ_{k+1}; **goto Step 2**.
elseif $\mu \in \Lambda_i$, $i < k$
then $\Lambda_i := \Lambda_i \cup \cdots \cup \Lambda_k$; λ_i :=representative of largest (by cardinality) of previous $\Lambda_i, \Lambda_{i+1}, \ldots, \Lambda_k$; remove $\Lambda_{i+1}, \ldots, \Lambda_k$; **goto Step 2**.

Step 4: Verify that $\{\alpha\} \cup \Lambda$ is a block. If not, compute $g \in G_\alpha\backslash H$. If yes, **output** $\{\alpha\} \cup \Lambda$.

Step 5: $H := \langle H, g\rangle$; update $\mathcal{L}$ by possibly merging some of the Λ_i and adding points from $\Omega\backslash(\{\alpha\} \cup \Lambda_1 \cup \cdots \cup \Lambda_k)$ to achieve that $\Lambda_1 \succeq \cdots \succeq \Lambda_{k'}$ and each Λ_i is the union of H-orbits; **goto Step 2**.

We postpone details of Steps 4 and 5 until later.

It is clear from the description already given that, at any moment during the execution of the algorithm, each Λ_i is the union of H-orbits. Also, using the remarks preceding Theorem 5.5.1, it is easy to see the following facts:

Lemma 5.5.3. *(a) At any moment during the execution of the algorithm, each* Λ_i *is a subset of a* $\sim$*-equivalence class and* $\Lambda_1 \succeq \cdots \succeq \Lambda_k$.

(b) Suppose that Step 4 of the algorithm is entered and, at that moment, Λ *is the last element of* $\mathcal{L}$ *and* λ *is the representative of* Λ*. Then* $\alpha^{\langle H, r_\lambda\rangle} \subseteq \{\alpha\} \cup \Lambda \subseteq \alpha^{\langle G_\alpha, r_\lambda\rangle}$*; in particular, if* H *and* G_α *have the same orbits then* $\{\alpha\} \cup \Lambda$ *is a minimal block of imprimitivity.*

Proof. (a) We prove the two statements in (a) simultaneously, by induction on the number of executions of the loop in Steps 2 and 3. After 0 executions of this loop (i.e., after the initialization in Step 1), both statements are obviously true. Suppose that the statements hold for the list $\mathcal{L} = (\Lambda_1, \ldots, \Lambda_k)$; in particular, $\Lambda_k \subseteq \bar{\lambda}_k = \alpha^{\langle G_\alpha, r_{\lambda_k}\rangle}$ for the representative λ_k of Λ_k. If a new set Λ_{k+1} is added to $\mathcal{L}$ then $\Lambda_{k+1} \subseteq \alpha^{\langle G_\alpha, r_{\lambda_k}\rangle}$, so $\Lambda_{k+1} \preceq \Lambda_k$. Also, since Λ_{k+1} is an orbit of $H \le G_\alpha$, all elements of Λ_{k+1} are equivalent. Similarly, if a final segment $(\Lambda_i, \Lambda_{i+1}, \ldots, \Lambda_k)$ of $\mathcal{L}$ is merged then $\Lambda_i \subseteq \alpha^{\langle G_\alpha, r_{\lambda_k}\rangle}$, so $\lambda_i \preceq \lambda_k$ for the representative λ_i of Λ_i. But $\lambda_i \succeq \nu \succeq \lambda_k$ for all $\nu \in \Lambda_i \cup \cdots \cup \Lambda_k$; therefore all such ν are equivalent.

(b) $\{\alpha\} \cup \Lambda$ is the union of H-orbits and closed for the action of r_λ; hence it contains $\alpha^{\langle H, r_\lambda\rangle}$. Also, since the elements of Λ are equivalent, $\Lambda \subseteq \bar{\lambda} = \alpha^{\langle G_\alpha, r_\lambda\rangle}$. For the last assertion, we note that $\Delta = \alpha^{\langle H, r_\lambda\rangle}$ depends only on the partition of Ω into orbits of H, and not on H itself since Δ is the set containing α in the join of the two partitions defined by the orbits of H and by the orbits of $\langle r_\lambda\rangle$ in the partition lattice of Ω. □

Next we describe the data structures used in the algorithm. We also give some further details of Steps 4 and 5 and prove the claimed bounds on the running time.

The coset representatives r_β are stored in the usual Schreier tree data structure.

The sets $\Lambda_1, \ldots, \Lambda_k$ are stored as linked lists. We maintain four lists of length k, named *firstL, lastL, lengthL,* and *testL*, and two lists of length n, named *nextL* and *reprL*. *firstL*$[i]$ and *lastL*$[i]$ contain the first and last elements of Λ_i, respectively; *lengthL*$[i]$ is the cardinality of Λ_i. *reprL*$[\nu]$, if defined, contains the index i such that $\nu \in \Lambda_i$. The list *nextL* contains the links: *nextL*$[\nu]$, if defined, is the successor of ν in the set Λ_i, for $\nu \in L_i$ (note that the set Λ_i is stored as a list in the computer, so it makes sense to talk about the successor of ν). The representative λ_i of Λ_i will always be the first element of Λ_i. *testL*$[i]$ gives the latest element ν of Λ_i for which $\nu^{r_{\lambda_i}}$ was already computed.

The orbits of H are also stored in linked lists. We define *firstH, lastH, nextH,* and *reprH* analogously to the appropriate items for $\mathcal{L}$.

The implementation of Step 3 is quite straightforward; therefore we remark only on one subcase. Suppose $\mu = \gamma^{r_\lambda}$ was computed for some $\gamma \in \Lambda$ and, using the list *reprL*, we found that $\mu \in \Lambda_i$ for some $i < k$. The smallest index j such that $|\Lambda_j| \ge |\Lambda_{j'}|$ for all $j' \in [i, k]$ can be determined from the list *lengthL*. Then we link the sets $\Lambda_i, \ldots, \Lambda_k$ such that the elements of Λ_j come first; in this way, we can maintain the property that the coset representative $r_{\lambda'}$ belonging to the first element λ' of the new block was already applied to an initial segment of this new block, and the last element of this initial segment is given as *testL*$[j]$.

We use the same "biggest comes first" linking strategy in Step 5. Suppose that Step 4 returned some $g \in G_\alpha \backslash H$ such that g collapses some orbits of H;

our goal is to make the necessary merges of the sets Λ_i and orbits of H to maintain the properties described in Lemma 5.5.3(a).

Step 5:

(i) Order the entries of *lengthL* in decreasing order:
$|\Lambda_{j_1}| \geq |\Lambda_{j_2}| \geq \cdots \geq |\Lambda_{j_k}|$.

(ii) **for** $i = 1$ to k **do**
if *firstL*$[j_i]$ is bound (i.e., defined)
then for all $\nu \in \Lambda_{j_i}$ **do**
$\mu := \nu^g$;
if $\mu \notin \{\alpha\} \cup \Lambda_1 \cup \cdots \cup \Lambda_k$
then link the H-orbit of μ to Λ_{j_i}
elseif $\mu \in \Lambda_l$, $l \neq j_i$
then link all Λ_s with *firstL*$[s]$ bound and s between l and j_i to Λ_{j_i}; unbind *firstL*$[s]$.

(iii) Left-justify bound entries of *firstL* (i.e., write the entries of *firstL* that are still defined in a sequence); update *lastL, reprL, testL, lengthL, nextL.*

(iv) Compute orbits of (new) H; update *firstH, lastH, reprH, nextH.*

At the end of Step 5, the sets in $\mathcal{L}$ are closed for the action of g.

The following lemma plays a crucial role in estimating the running time.

Lemma 5.5.4. *For each $\nu \in \Omega$, the representative λ_i of the set $\Lambda_i \in \mathcal{L}$ containing ν changes at most* $\log n$ *times during the algorithm.*

Proof. The representative may change only when Λ_i is merged into a larger set in Step 3 or Step 5. At these occasions, since the representative of the new set Λ is defined as the representative of the *largest* original set Λ_j that was merged, $|\Lambda| \geq 2|\Lambda_i|$ for all sets Λ_i whose representative was changed. Clearly, such doubling of size can occur only at most $\log n$ times. □

Step 4 is handled by the following lemma. Recall that the input is a set $\Lambda \subseteq \Omega\backslash\{\alpha\}$ such that all elements of Λ are known to be equivalent and a representative $\lambda \in \Lambda$. The set $\{\alpha\} \cup \Lambda$ is closed for the action of r_λ.

Lemma 5.5.5. *(a) In $O(nl + n|S|)$ time, it can be decided whether $\{\alpha\} \cup \Lambda$ is a block of imprimitivity.*

(b) If $\{\alpha\} \cup \Lambda$ is not a block of imprimitivity then some $g \in G_\alpha\backslash H$ can be constructed in $O(n \log^2 |G| + n|S_1|)$ time, where S_1 is the known generating set of H.

Proof. (a) Let $\Delta := \{\alpha\} \cup \Lambda$. We try to partition Ω into disjoint G-images of Δ. Suppose that the disjoint sets $\Delta = \Delta_1, \Delta_2, \ldots, \Delta_{j-1}$ are already constructed; we pick $\beta_j \in \Omega \backslash (\Delta_1 \cup \cdots \cup \Delta_{j-1})$ and compute $\Delta^{r_{\beta_j}}$. If $\Delta^{r_{\beta_j}} \subseteq \Omega \backslash (\Delta_1 \cup \cdots \cup \Delta_{j-1})$ then we define $\Delta_j := \Delta^{r_{\beta_j}}$; otherwise, we have $\emptyset \subsetneqq \Delta^{r_{\beta_i}} \cap \Delta^{r_{\beta_j}} \subsetneqq \Delta^{r_{\beta_j}}$ for some $i < j$, which proves that Δ is not a block. Since the r_{β_j} are stored as words of length at most l, this part of the algorithm runs in $O(nl)$ time.

If the partition of Ω into G-images of Δ succeeds then we check that each generator of G respects this partition. This can be done in $O(n|S|)$ time.

(b) If Δ is not a block then we claim that $\Gamma := \alpha^{\langle H, r_\lambda \rangle}$ is not a block either. By Lemma 5.5.3, $\Gamma \subseteq \Delta$. If $\Gamma = \Delta$ then obviously Γ is not a block. If $\Gamma \subsetneqq \Delta$ then $\lambda \in \Gamma \backslash \{\alpha\} \subsetneqq \Delta \backslash \{\alpha\} \subseteq \bar{\lambda}$, so Γ is not a block.

We repeat the algorithm described in part (a) with Γ playing the role of Δ. This means that we try to partition Ω into G-images of Γ and, if the partition succeeds, compute the action of the generators of G on the sets in the partition. Since Γ is not a block, we shall encounter $h_1, h_2 \in G$ such that $\emptyset \neq \Gamma^{h_1} \cap \Gamma^{h_2} \subsetneqq \Gamma^{h_1}$. (If h_1, h_2 are found during the partition phase then $h_1 = r_{\beta_i}, h_2 = r_{\beta_j}$ for some coset representatives; if they are found at the checking of the action of generators then $h_1 = r_{\beta_i} s, h_2 = r_{\beta_j}$ for some $s \in S$.) Defining $g_1 := h_1 h_2^{-1}$, we have $\emptyset \neq \Gamma \cap \Gamma^{g_1} \subsetneqq \Gamma$. We multiply out g_1 as a permutation. As in part (a), the time requirement so far is $O(nl + n|S|)$.

After that, we use the algorithm in the proof of Lemma 4.4.2 to build a Schreier tree (S^*, T) of depth at most $2 \log |\langle H, r_\lambda \rangle|$, coding a transversal for $\langle H, r_\lambda \rangle$ mod $\langle H, r_\lambda \rangle_\alpha$. This Schreier tree computation takes $O(n \log^2 |G| + |S_1| n)$ time.

Let $\beta, \gamma \in \Gamma$ such that $\beta^{g_1} = \gamma$, and let w_β, w_γ be the corresponding coset representatives from (S^*, T). In particular, $\Gamma^{w_\beta} = \Gamma^{w_\gamma} = \Gamma$. Then $g := w_\beta g_1 w_\gamma^{-1} \in G_\alpha$ and $\Gamma^g \neq \Gamma$. Thus $g \notin H$ and $\alpha^{\langle H, g \rangle} \supsetneqq \alpha^H$; otherwise we would have $\alpha^{\langle H, g, r_\lambda \rangle} = \alpha^{\langle H, r_\lambda \rangle} = \Gamma$, a contradiction. □

Lemma 5.5.6.

(a) In $O(n \log^2 |G| + n|S|)$ time, coset representatives r_β as words of length l for some $l \leq 2 \log |G|$ can be computed.
(b) The total amount of time spent in Step 3 is $O(nl \log n)$.
(c) The time spent in Step 4 is $O(n \log^3 |G| + n|S| \log |G|)$.
(d) The time spent in Step 5 is $O(n \log n \log |G|)$.

Proof. (a) This is proven by Lemma 4.4.2. Recall that, although a Schreier tree codes inverses of coset representatives, the set S^* of labels in the Schreier tree (S^*, T) constructed in Lemma 4.4.2 is closed for inverses. Hence, after (S^*, T) is constructed, the coset representatives themselves can also be written as words in S^*.

(b) By (a), computing an image γ^{r_λ} takes $O(l)$ time. By Lemma 5.5.4, for each fixed $\gamma \in \Omega$, γ^{r_λ} for some λ is computed at most $\log n$ times; hence the total cost of these computations is $O(nl \log n)$. Updating *reprL* and the linking of lists cost $O(n \log n)$.

(c) Step 4 is entered at most $\log |G_\alpha|$ times, since H increases between consecutive calls.

(d) Step 5 is entered always after a call of Step 4, that is, at most $\log |G_\alpha|$ times. At one call, ordering *lengthL* costs $O(n \log n)$ whereas the linking and updating of lists runs in $O(n)$ time. □

Lemma 5.5.6 immediately implies Theorem 5.5.1. Theorem 5.5.2 is obtained by combining Lemma 5.5.6(b) and Lemma 5.5.3(b). We also remark that within the time bounds claimed in Theorems 5.5.1 and 5.5.2, *all* minimal blocks containing α can be constructed.

Our running time estimate is overly pessimistic. It is dominated by the applications of Lemma 4.4.2, which, in practice, are much faster than our worst-case estimate. Also, if the orbit structure of G_α is not known, a few (five in the GAP implementation) random Schreier generators may be computed to approximate G_α, and this seems to be enough: The construction phase may return a correct block of imprimitivity even in the case when the orbit structure of H is not the same as the orbit structure of G_α, and in practice it is extremely rare that the output of the construction phase is incorrect.

5.5.2. The Smallest Block Containing a Given Subset

In a number of applications, we need to construct the block system Σ with the smallest possible block sizes for some transitive $G = \langle S \rangle \leq \mathrm{Sym}(\Omega)$, $\Omega = [1, n]$, satisfying the property that points of a given subset $\Delta \subseteq \Omega$ belong to the same block. The algorithm in Section 5.5.1 can be modified to do that, but we leave the details as an exercise (cf. Exercise 5.8). In this section we describe a faster method from [Atkinson et al., 1984].

Computation of the Smallest Block Containing Δ

During the algorithm, we maintain a partition Π of Ω. Initially, Π consists of Δ and $|\Omega| - |\Delta|$ sets of size 1, the latter ones containing the elements of $\Omega \backslash \Delta$. Later we merge sets of Π, but Π always satisfies the property that if two points of Ω belong to the same set in Π then they must belong to the same block in Σ.

If $\alpha, \beta \in \Omega$ belong to the same set A in Π then, for all $s \in S$, the points α^s, β^s must belong to the same set. Hence if α^s and β^s belong to different sets $A_1, A_2 \in \Pi$ for some $\alpha, \beta \in \Omega$ and $s \in S$ then we delete A_1 and A_2 from Π, and add the set $A_1 \cup A_2$ to Π. The algorithm terminates when for all $s \in S$ and $A \in \Pi$ there exists $B \in \Pi$ such that $A^s = B$. At termination, Π consists of the blocks of the desired block system Σ.

The efficiency of the algorithm depends on how quickly we can decide which set of Π a point α^s belongs to, and then how quickly we can perform the necessary merges of sets in Π.

We store the sets in Π in a *Union-Find data structure*. Each set A is stored in a rooted tree T_A. Each vertex of T_A represents an element in A, and the root represents the entire set as well. A vertex v of T_A consists of three cells. One of them contains the appropriate element of A, and the second one contains a pointer to the parent of v. The pointer of the root points to itself. The content of the third cell is considered only if v is the root of T_A, and in this case the third cell contains $|A|$. In an implementation, the entire data structure for Π can be stored in two lists of length n. For $\alpha \in \Omega$, let $v(\alpha)$ denote the tree vertex representing α. In one of the lists, at position α we store the element of Ω represented by the second cell of $v(\alpha)$, and in the other one at position α we store $|A|$ if $v(\alpha)$ happens to be the root of T_A for some set A.

We perform two operations on the data structure. One of them is **Find**(α), which finds the set A to which $\alpha \in \Omega$ belongs. This amounts to following the path (determined by the second cells of vertices) from the vertex $v(\alpha)$ representing α until we reach the root of T_A. In addition, after each **Find** operation, we also perform a *collapse* of the appropriate tree: We go along the path from α to the root again and change the parent of each vertex to the (already known) root. The cost of **Find**(α), with or without applying the collapse rule, is $\Theta(l)$, where l is the length of the path from $v(\alpha)$ to the root in T_A.

The second operation is **Union**(A, B), which replaces the sets $A, B \in \Pi$ by their union. We make one of the roots of T_A, T_B a child of the other, and we update its second cell as the pointer to the new root. We also update the third cell of the new root. It is beneficial to use the *weighted union rule*: If $|A| \geq |B|$ then the new root is the root of T_A, and otherwise the new root is the root of T_B. We consider the cost of **Union**(A, B) as constant, regardless of whether we apply the weighted union rule (i.e., we ignore the cost of the arithmetic operations necessary for updating the third cell).

Theorem 5.5.7. *Given a transitive* $G = \langle S \rangle \leq \mathrm{Sym}(\Omega)$, $\Omega = [1, n]$, *and* $\Delta \subseteq \Omega$, *the smallest block of imprimitivity of* G *containing* Δ *can be computed using at most* $2|S|n$ **Find** *operations and less than* n **Union** *operations.*

Proof. As indicated in the preceding paragraphs, we initialize a partition Π of Ω to consist of Δ and $|\Omega| - |\Delta|$ sets of size 1. We pick an arbitrary element δ

of Δ as root, and we store Δ in a tree T_Δ of depth 1. The other sets $A = \{\alpha\}$ are represented by a tree T_A with one vertex. We also maintain a list L of pairs of points from Ω. This list contains the pairs $\{\alpha, \beta\} \subseteq \Omega$ for which we want to check whether α^s and β^s belong to the same set of Π, for $s \in S$. We initialize L to contain all pairs $\{\delta, \beta\}$, for $\beta \in \Delta \backslash \{\delta\}$.

We process the list L as follows: If $\{\alpha, \beta\}$ is the next element of L to process then for each $s \in S$ we compute α^s and β^s, and we use **Find**(α^s) and **Find**(β^s) to determine which sets A and B in Π these points belong to. If $A \neq B$ then we perform **Union**(A, B) and add the pair $\{\rho_A, \rho_B\}$ to L, where ρ_A and ρ_B are the elements of Ω represented by the roots of T_A and T_B, respectively. The algorithm terminates when all pairs in L are processed.

We claim that at termination Π consists of the sets in a block system. Since it is clear from the definition of blocks that, at any moment during the execution of the algorithm, points belonging to the same set A in Π must belong to the same block in any block system where Δ is a subset of a block, our claim implies that, at termination, Π consists of blocks of minimal possible size.

For $\alpha \in \Omega$, let $A_1 \subsetneqq A_2 \subsetneqq \cdots \subsetneqq A_m$ be the sets in Π containing α during the algorithm. At termination, the set A_m is in Π. Moreover, let $(\rho_1, \ldots, \rho_m)$ be the sequence of elements of Ω that are represented by the roots of $T_{A_1}, \ldots, T_{A_m}$. For $i \in [1, m-1]$, if $\rho_i \neq \rho_{i+1}$ then the pair $\{\rho_i, \rho_{i+1}\}$ occurs in L, since A_{i+1} is obtained as the union of A_i with some set in Π. Moreover, if $\rho_1 \neq \alpha$ (i.e., $\alpha \in \Delta \backslash \{\delta\}$) then the pair $\{\alpha, \rho_1\}$ is in L. Therefore, at termination α^s belongs to the same set of Π as ρ_m^s, for all $s \in S$. Since any generator of G maps each element of A_m to the same set as it maps ρ_m, applying this argument for all sets of Π at termination gives that G permutes the sets in Π.

Because initially Π consists of $|\Omega| - |\Delta| + 1$ sets, we obviously perform at most $|\Omega| - |\Delta| \leq n - 1$ **Union** operations. Initially $|L| = |\Delta| - 1$, and we add at most $|\Omega| - |\Delta|$ pairs to L. Hence $|L| \leq n - 1$ at termination. For each pair in L, we perform $2|S|$ **Find** operations. □

Let $t(m, n)$ denote the worst-case time requirement of at most m **Find** operations and at most $n-1$ **Union** operations. If we do not use the collapse rule then it is fairly easy to estimate $t(m, n)$, regardless of whether the weighted union rule is used or not (cf. Exercise 5.9). However, if both the collapse and weighted union rules are used then the analysis becomes much harder. The asymptotic behavior of $t(m, n)$ was determined in [Tarjan, 1975]. To state the result, we have to introduce the Ackerman function.

Let $A : \mathbb{N} \times \mathbb{N} \to \mathbb{N}$ be defined recursively as follows: For all $i, x \in \mathbb{N}$, let $A(0, x) := 2x$, $A(i, 0) := 0$, and $A(i, 1) := 2$. Moreover, for $i \geq 1$ and $x \geq 2$, let

$$A(i, x) := A(i-1, A(i, x-1)). \tag{5.3}$$

Using (5.3), it is easy to see by induction that $A(1, x) = 2^x$ and $A(i, 2) = 4$ for all $x \geq 1$ and $i \geq 0$. Also, $A(2, x+1) = A(1, A(2, x)) = 2^{A(2,x)}$, and so

$$A(2, x) = 2^{2^{\cdot^{\cdot^{2}}}} \quad \text{(a tower of } x \text{ twos).}$$

The function $A(2, x)$ already grows very fast, but $A(3, x)$ is mind-boggling. We have $A(3, 0) = 0$, $A(3, 1) = 2$, $A(3, 2) = 4$, $A(3, 3) = A(2, A(3, 2)) = 2^{2^{2^2}} = 65536$, and

$$A(3, 4) = A(2, A(3, 3)) = 2^{2^{\cdot^{\cdot^{2}}}} \quad \text{(a tower of 65536 twos).}$$

For $i \geq 4$, the functions $A(i, x)$ are growing even faster. We just compute one more value: $A(4, 3) = A(3, A(4, 2)) = A(3, 4)$.

For fixed $m \geq 1$, we define the inverse of $A(m, n)$ as $A^{-1}(m, n) := \min\{x \mid A(m, x) > \log n\}$. If $m \geq 3$ then, for all practical purposes and beyond, $A^{-1}(m, n) \leq 4$. The main result of [Tarjan, 1975] is the following theorem.

Theorem 5.5.8. *Let $t(m, n)$ denote the worst-case time requirement of at most m* **Find** *operations and at most $n - 1$* **Union** *operations, and suppose that $m \geq n$. If both the collapse rule and the weighted union rule are used then $t(m, n) \in \Theta(mA^{-1}(m, n))$.*

Corollary 5.5.9. *Given a transitive $G = \langle S \rangle \leq \mathrm{Sym}(\Omega)$, with $|\Omega| = n$, and $\Delta \subseteq \Omega$, the smallest block of imprimitivity of G containing Δ can be computed in $O(|S|nA^{-1}(2|S|n, n))$ time by a deterministic algorithm.*

The method described in this section can be used to test the primitivity of G, by computing the smallest block containing $\{\alpha, \beta\}$ for some fixed $\alpha \in \Omega$ and all $\beta \in \Omega \backslash \{\alpha\}$. However, this multiplies the running time indicated in Corollary 5.5.9 by a factor n. There is a possible shortcut. If we already know that the smallest block containing $\{\alpha, \beta_1\}$ is Ω and at a subsequent computation we deduce that the smallest block containing $\{\alpha, \beta_2\}$ must contain β_1, then we can terminate that computation. This observation usually helps in practice, but the worst-case running time is quadratic in n.

5.5.3. Structure Forests

In many problems dealing with permutation groups, both in theoretical and computational settings, we can use the orbit structure to reduce the original problem to smaller groups. A further reduction is possible by using the imprimitivity structure of transitive groups. This way, the original problem is reduced to primitive groups, and the strong structural properties of primitive groups can be utilized. In computations, sometimes it is convenient to organize this reduction process by using a *structure forest*. A structure forest $\mathcal{F}$ for a permutation group $G \leq \mathrm{Sym}(\Omega)$ is a forest of rooted trees where each vertex is a subset of Ω, and the leaves are the one-element subsets of Ω. The elements of G act as automorphisms of $\mathcal{F}$, fixing the roots, such that for each nonleaf vertex v of the forest, G_v acts primitively on the children of v. Thus, in particular, there is exactly one tree $T_{\mathcal{O}}$ per orbit $\mathcal{O}$ in Ω. In $T_{\mathcal{O}}$, the vertices are blocks of imprimitivity in the transitive G-action on $\mathcal{O}$ such that each level of the tree defines a partition of $\mathcal{O}$. It is not possible to insert intermediate levels in that tree, with nontrivial branching, and remain consistent with the G-action on the tree.

As a simple example, consider the cyclic group $G = \langle(1, 2, 3, 4)(5, 6)\rangle \leq \mathrm{Sym}([1, 6])$. The structure forest consists of two trees. In the first one, the root is the orbit $\{1, 2, 3, 4\}$, it has two children $\{1, 3\}$ and $\{2, 4\}$, and the tree has the leaves $\{1\}$, $\{2\}$, $\{3\}$, and $\{4\}$. In the second tree, the root is the orbit $\{5, 6\}$ and it has two children $\{5\}$ and $\{6\}$. The edges of the forest are defined by set containment.

In general, a group $G \leq \mathrm{Sym}(\Omega)$ may have many nonisomorphic structure forests. We may construct one by taking a maximal block system on each orbit of G, then taking maximal block systems of the action on these blocks, etc. Therefore, a structure forest for G can be computed in nearly linear time by a deterministic algorithm.

Structure forests play an important role in the current fastest deterministic SGS construction for arbitrary input groups (cf. [Babai et al., 1997b]) and in the parallel (NC) handling of permutation groups (cf. [Babai et al., 1987]).

Exercises

5.1. Suppose that a nonredundant base B is given for some $G \leq \mathrm{Sym}(\Omega)$. Prove that for $H \leq G$ and $T \subseteq G$, the subgroup $\langle H \cup T\rangle$ and normal closure $\langle H^G\rangle$ can be computed in nearly linear time by a deterministic algorithm.

5.2. Let S be an SGS for G relative to the base $(\beta_1, \ldots, \beta_m)$. Let $S_i = S \cap G_{(\beta_1,\ldots,\beta_{i-1})}$ and let R_i be the transversal computed from S_i for $G_{(\beta_1,\ldots,\beta_{i-1})}$

mod $G_{(\beta_1,\ldots,\beta_i)}$. Consider the elements of R_i as words in S_i. For all $j \in [1, m]$, $r \in R_j$, and $g \in S_j$, let

$$rg = r_m r_{m-1} \cdots r_1,\ r_i \in R_i \tag{5.4}$$

be the factorization of rg as a product of coset representatives.

Introduce a symbol x_g for each $g \in S$. Prove that the set of words obtained by rewriting all equations of the form (5.4) in terms of the x_g gives a presentation for G. *Hint:* Rewrite any word in the x_g representing the identity in the spirit of the proof of Lemma 4.2.1.

5.3. Give an example of some transitive $G \leq S_n$ with minimal base size b such that there exists $g \in G$ with $|\mathrm{supp}(g)| = n/b$. *Hint:* Take a wreath product of cyclic groups.

5.4. (Cauchy) Suppose that some group $G \leq \mathrm{Sym}(\Omega)$ has t orbits. For $g \in G$, define $\mathrm{fix}(g) := \Omega \backslash \mathrm{supp}(g)$. Prove that

$$\sum_{g \in G} |\mathrm{fix}(g)| = t|G|.$$

Hint: Count the number of pairs (α, g) with $\alpha \in \Omega, g \in G, \alpha \in \mathrm{fix}(g)$ two different ways.

For the history of this lemma, see [Neumann, 1979].

5.5. [Sims, 1971a] Let $(\beta_1, \ldots, \beta_m)$ be a base for G with the corresponding point stabilizer chain $G = G^{[1]} \geq \cdots \geq G^{[m+1]} = 1$, fundamental orbits $\Delta_1, \ldots, \Delta_m$, and transversals $R_1, \ldots, R_m$. Let $\bar{\Delta}_1, \ldots, \bar{\Delta}_m$ be the fundamental orbits of G relative to the base $\bar{B} = (\beta_1, \ldots, \beta_{i-1}, \beta_{i+1}, \beta_i, \beta_{i+2}, \ldots, \beta_m)$. Prove that $\bar{\Delta}_{i+1} \subseteq \Delta_i$. Moreover, for $\delta \in \Delta_i$, let $g \in R_i$ be such that $\beta_i^g = \delta$. Prove that $\delta \in \bar{\Delta}_{i+1}$ if and only if $\beta_{i+1}^{g^{-1}} \in \Delta_{i+1}$. Based on this observation, design an algorithm for computing an SGS relative to $\bar{B}$.

5.6. Let $G \leq \mathrm{Sym}(\Omega)$ be transitive, and let $\alpha, \beta \in \Omega$. Prove that the smallest block of imprimitivity containing α and β can be obtained as the component containing α in the graph with vertex set Ω and edge set $\{\alpha, \beta\}^G$ (i.e., the edges are the G-images of the unordered pair $\{\alpha, \beta\}$) or as the strongly connected component containing α in the directed graph with edge set $(\alpha, \beta)^G$ (i.e., the edges are the G-images of the ordered pair (α, β)).

5.7. Modify the algorithm described in Section 5.5.1 to find *all* minimal blocks containing α within the time bounds given in Theorems 5.5.1 and 5.5.2.

5.8. Let $G \leq \mathrm{Sym}(\Omega)$ be transitive. Modify the algorithm in Section 5.5.1 to compute a minimal block containing a given subset $\Delta \subseteq \Omega$ by a nearly linear-time deterministic algorithm.

5.9. Let $t(m, n)$ denote the worst-case time requirement of at most m **Find** operations and at most $n - 1$ **Union** operations, and suppose that $m \geq n$. Prove that

(i) if neither the collapse rule nor the weighted union rule is used then $t(m, n) \in \Theta(mn)$; and

(ii) if only the weighted union rule is used then $t(m, n) \in \Theta(m \log n)$.

5.10. Let $G \leq \mathrm{Sym}(\Omega)$. Prove that any structure forest of G has less than $2|\Omega|$ vertices.

5.11. Give examples of groups $G \leq \mathrm{Sym}(\Omega)$ with exponentially many (as a function of $|\Omega|$) structure forests.

The following four exercises lead to the regularity test from [Acciaro and Atkinson, 1992].

5.12. Let $G = \langle S \rangle \leq \mathrm{Sym}(\Omega)$ be transitive, and let $\alpha \in \Omega$. Prove that G acts regularly on Ω if and only if $G_\alpha = G_{\alpha^g}$ for all $g \in S$.

5.13. Let $G = \langle S \rangle \leq \mathrm{Sym}(\Omega)$ be transitive, and let $\alpha, \omega \in \Omega$. Construct a list $(T[\beta] \mid \beta \in \Omega)$ in the following way: Initialize $T[\alpha] := \omega$. Build a breadth-first-search tree L that computes the orbit α^G by the algorithm described in Section 2.1.1. Each time an edge $\overrightarrow{(\beta, \gamma)}$ is added to L, because $\gamma = \beta^s$ for some already constructed vertex β of L and some $s \in S$, we also define $T[\gamma] := T[\beta]^s$.

Prove that $G_\alpha = G_\omega$ if and only if $T[\beta^g] = T[\beta]^g$ for all $\beta \in \Omega$ and for all $g \in S$.

5.14. Based on Exercises 5.12 and 5.13, design a deterministic algorithm that tests whether a group $G = \langle S \rangle \leq \mathrm{Sym}(\Omega)$, with $|\Omega| = n$, acts regularly on Ω, in $O(|S|^2 n)$ time.

5.15. Let $G = \langle S \rangle \leq \mathrm{Sym}(\Omega)$, with $|\Omega| = n$, be transitive. Suppose that for some $\alpha, \beta \in \Omega$ we have $G_\alpha = G_\beta$, and let Δ be the smallest block of imprimitivity for G that contains α and β. Prove that $G_\alpha = G_\delta$ for each $\delta \in \Delta$.

Based on this observation, modify the algorithm in Exercise 5.14 to run in $O(|S| n \log n)$ time.

5.16. Let $G = \langle S \rangle \leq \mathrm{Sym}(\Omega)$, with $|\Omega| = n$, be transitive. Based on Lemma 4.4.2, design a deterministic algorithm that in $O((|S| + \log n) n \log n)$ time either decides that G is not regular, or outputs at most $\log n$ generators for G.

6

A Library of Nearly Linear-Time Algorithms

In this chapter, we develop a nearly linear-time library for constructing certain important subgroups of a given group. All algorithms are of the Monte Carlo type, since they are based on results of Section 4.5. However, if a base and SGS are known for the input group, then all algorithms in this chapter are of Las Vegas type (and in most cases there are even deterministic versions).

A current research project of great theoretical and practical interest is the upgrading of Monte Carlo permutation group algorithms to Las Vegas type. The claim we made in the previous paragraph implies that it is enough to upgrade the SGS constructions; we shall present a result in this direction in Section 8.3. To prove that all algorithms in this chapter are of Las Vegas type or deterministic, we always suppose that the input is an SGS S for some $G \leq \mathrm{Sym}(\Omega)$ relative to some base B, S satisfies $S = S^{-1}$, and transversals corresponding to the point stabilizer chain defined by B are coded in shallow Schreier trees. Throughout this chapter, *shallow Schreier tree* means a Schreier tree of depth at most $2\log|G|$. We remind the reader that, by Lemma 4.4.2, given an arbitrary SGS for G, a new SGS S satisfying $S = S^{-1}$ and defining a shallow Schreier tree data structure can be computed in nearly linear time by a deterministic algorithm. Therefore, without further mention, we also suppose that the algorithms output shallow Schreier tree data structures for the constructed subgroups.

In the descriptions, we do not compute the exponent of $\log|G|$ in the running time estimates; rather, for clarity, we concentrate only on making clear the nearly linear running time. For example, several $\log|G|$ factors can be eliminated from the running times by applying Theorem 4.4.6 at the construction of Schreier trees. The algorithms we present are practical enough for implementation and, in fact, most of them are already available in the GAP system.

For most algorithms, we have deterministic and Las Vegas nearly linear-time versions, depending on which variants of the low-level algorithms from Chapter 5 are used as subroutines. Whenever possible, we shall describe *deterministic* nearly linear-time algorithms. In implementations, usually the faster Las Vegas versions are used.

In the entire chapter, there is only one algorithm where the worst-case running time is not nearly linear: The computation of $C_{\mathrm{Sym}(\Omega)}(G)$ in Section 6.1.2. We included that algorithm here because its method is similar to the accompanying nearly linear-time algorithms.

6.1. A Special Case of Group Intersection and Applications

The common theme in this section is the application of a subroutine for a special case of the group intersection problem. As mentioned in Section 3.3, no polynomial-time algorithm is known for computing the intersection of two arbitrary permutation groups. However, if one of the groups normalizes the other then their intersection is computable in polynomial time. The method has been used for a long time in linear algebra to obtain a basis for the intersection of subspaces. It was generalized for both the theoretical (cf. [Rose, 1965]) and computational (cf. [Laue et al., 1984]) study of solvable groups. In the permutation group setting, it was introduced in [Cooperman et al., 1989] and our discussion in Sections 6.1.1 and 6.1.2 follows this paper.

6.1.1. Intersection with a Normal Closure

Given $G, H \leq \mathrm{Sym}(\Omega)$, we introduce subgroups of $\mathrm{Sym}(\Omega_1)\times\mathrm{Sym}(\Omega_2)$, where Ω_1, Ω_2 are disjoint copies of the set Ω. We write elements of $\mathrm{Sym}(\Omega_1) \times \mathrm{Sym}(\Omega_2)$ as ordered pairs (a, b), where $a \in \mathrm{Sym}(\Omega_1)$ and $b \in \mathrm{Sym}(\Omega_2)$. The subgroups we consider are $D = \mathrm{Diag}(G \times G) := \{(g, g) \mid g \in G\}$, $1 \times H := \{(1, h) \mid h \in H\}$, and $K := \langle D, 1 \times H\rangle$.

An arbitrary element $(x_1, x_2) \in K$ can be written in the form

$$\begin{aligned}(x_1, x_2) &= (g_1, g_1)(1, h_1)(g_2, g_2)(1, h_2)\cdots(g_k, g_k)(1, h_k)(g_{k+1}, g_{k+1})\\ &= (g_1 g_2 \cdots g_{k+1}, g_1 h_1 g_2 h_2 \cdots g_k h_k g_{k+1})\\ &= \left(g_1 g_2 \cdots g_{k+1}, h_1^{g_1^{-1}} h_2^{(g_1 g_2)^{-1}} \cdots h_k^{(g_1 g_2 \cdots g_k)^{-1}} g_1 g_2 \cdots g_k g_{k+1}\right), \qquad (6.1)\end{aligned}$$

for some $g_1, \ldots, g_{k+1} \in G$ and $h_1, \ldots, h_k \in H$.

Lemma 6.1.1. $K_{(\Omega_1)} = \{(x_1, x_2) \mid x_1 = 1\} \cong \langle H^G\rangle$.

Proof. If $(x_1, x_2) \in K_{(\Omega_1)}$ then from (6.1) we obtain $x_1 = g_1 g_2 \cdots g_{k+1} = 1$ and $x_2 \in \langle H^G\rangle = \langle h^g \mid h \in H, g \in G\rangle$. Conversely, if $x_2 = h_1^{\tilde{g}_1} h_2^{\tilde{g}_2} \cdots h_k^{\tilde{g}_k} \in \langle H^G\rangle$ then successively define $g_1, g_2, \ldots, g_{k+1} \in G$ such that $(g_1 g_2 \cdots g_i)^{-1} = \tilde{g}_i$ for

$1 \leq i \leq k$ and $g_{k+1} = (g_1 g_2 \cdots g_k)^{-1}$. Then

$$(1, x_2) = (g_1, g_1)(1, h_1)(g_2, g_2)(1, h_2) \cdots (g_k, g_k)(1, h_k)(g_{k+1}, g_{k+1}),$$

so $(1, x_2) \in K_{(\Omega_1)}$. □

Lemma 6.1.2. $K_{(\Omega_2)} = \{(x_1, x_2) \mid x_2 = 1\} \cong \langle H^G \rangle \cap G$.

Proof. If $(x_1, x_2) \in K_{(\Omega_2)}$ then from (6.1) we obtain $g_1 g_2 \cdots g_{k+1} \in \langle H^G \rangle$ and, obviously, $g_1 g_2 \cdots g_{k+1} \in G$. Therefore, $x_1 \in \langle H^G \rangle \cap G$ and $x_2 = 1$. Conversely, if $x_1 \in \langle H^G \rangle \cap G$ then, by Lemma 6.1.1, $(1, x_1^{-1}) \in K$. Also, $(x_1, x_1) \in D \leq K$, so $(1, x_1^{-1})(x_1, x_1) = (x_1, 1) \in K$. Since $(x_1, 1) \in K$ and it stabilizes Ω_2 pointwise, we have $(x_1, 1) \in K_{(\Omega_2)}$. □

Corollary 6.1.3. *Suppose that strong generating sets defining shallow Schreier tree data structures are given for G, $H \leq \mathrm{Sym}(\Omega)$. If G normalizes H then the intersection of G and H can be computed by a deterministic nearly linear-time algorithm.*

Proof. We use the notation just introduced. Observe that the action of K on Ω_1 is permutation isomorphic to the action of G on Ω, so Lemma 6.1.1 implies that $|K| = |G| \cdot |\langle H^G \rangle|$. In particular, if G normalizes H then $|K| = |G||H|$ and it follows from Lemma 6.1.1 that $B_1 \cup B_2$ is a base for K, where B_1 is a copy of the base of G in Ω_1 and B_2 is a copy of the base of H in Ω_2. Therefore, by Lemma 5.2.3, an SGS for K is computable by a deterministic algorithm with running time estimate of the form $O(n(\log|G| + \log|H|)^c)$. If we order $B_1 \cup B_2$ such that the elements of B_2 precede the elements of B_1 then $(G \cap H) \times 1$ occurs as a member of the point stabilizer subgroup chain of K. □

We remark that Lemma 6.1.1 also implies that the computation of normal closures can also be reduced to point stabilizer computations. However, implementations of the method presented in Section 5.1.4 are faster in practice.

It is noted in [Cooperman et al., 1989] that if G normalizes H and strong generating sets for H and G are already known then an SGS for K (relative to a base that first fixes the points of Ω_1) can be constructed directly, without invoking the Schreier–Sims algorithm. After that, generators for $G \cap H$ may be obtained by a base change.

6.1.2. Centralizer in the Symmetric Group

First we handle the transitive case. Let $G \le \mathrm{Sym}(\Omega)$ be transitive and, for some fixed $\alpha \in \Omega$, let A be the set of fixed points of G_α:

$$A := \mathrm{fix}(G_\alpha) = \{\beta \in \Omega \mid (\forall g \in G_\alpha)(\beta^g = \beta)\}.$$

Observe that $G_\beta = G_\alpha$ for all $\beta \in A$ since $G_\alpha \le G_\beta$ by the definition of A and the transitivity of G implies that $|G_\alpha| = |G_\beta|$ (because any two point stabilizer groups are conjugate).

We denote $C_{\mathrm{Sym}(\Omega)}(G)$ by C. Then the following lemma holds.

Lemma 6.1.4. *C is semiregular and A is an orbit of C.*

Proof. Suppose that $c \in C$ and $\beta^c = \beta$ for some $\beta \in \Omega$. Let $\delta \in \Omega$ be arbitrary. Since G is transitive, there exists $g \in G$ such that $\beta^g = \delta$. Then

$$\delta^c = \beta^{gc} = \beta^{cg} = \beta^g = \delta.$$

This proves that C is semiregular. Next, suppose that $c \in C$ and $\alpha^c = \beta$. Let $h \in G_\alpha$ be arbitrary. Then

$$\beta^h = \alpha^{ch} = \alpha^{hc} = \alpha^c = \beta,$$

which shows that $\alpha^C \subseteq A$.

Conversely, suppose that $\beta \in A$; we have to construct $c \in C$ such that $\alpha^c = \beta$. Given an arbitrary $\gamma \in \Omega$, γ^c can be determined as follows: Take $g \in G$ such that $\alpha^g = \gamma$ and define

$$\gamma^c := \beta^g. \tag{6.2}$$

This definition is independent of which representative of the coset of G_α carrying α to γ is chosen since if $\alpha^{g_1} = \alpha^{g_2}$ then $g_1 g_2^{-1} \in G_\alpha$ and so $\beta^{g_1} = \beta^{g_2}$. We have to show that the function $c : \Omega \to \Omega$ defined by (6.2) is a permutation and that it centralizes G. For all $g \in G$, $\alpha^{gc} = \beta^g = \alpha^{cg}$ since any $g \in G$ can be used in (6.2) when we define γ^c for $\gamma := \alpha^g$. Next we show that $\delta^{gc} = \delta^{cg}$ holds for all $g \in G$ and $\delta \in \Omega$. Fix g, δ and take $h \in G$ such that $\alpha^h = \delta$. Then

$$\delta^{gc} = \alpha^{(hg)c} = \alpha^{c(hg)} = \alpha^{(ch)g} = \alpha^{(hc)g} = \delta^{cg}.$$

Finally, we show that c is a permutation. Suppose that $\gamma^c = \delta^c$ for some $\gamma, \delta \in \Omega$ and choose $g, h \in G$ such that $\alpha^g = \gamma$ and $\alpha^h = \delta$. Then

$$\beta^g = \alpha^{cg} = \alpha^{gc} = \gamma^c = \delta^c = \alpha^{hc} = \alpha^{ch} = \beta^h,$$

so $gh^{-1} \in G_\beta$. However, $G_\beta = G_\alpha$ and so $\gamma = \alpha^g = \alpha^h = \delta$. □

Corollary 6.1.5. *If $G \leq \mathrm{Sym}(\Omega)$ is transitive and abelian then G is regular and $C_{\mathrm{Sym}(\Omega)}(G) = G$.*

Theorem 6.1.6. *Suppose that $G \leq \mathrm{Sym}(\Omega)$ is transitive. If a nonredundant base B and an SGS S relative to B are given for G then $C_{\mathrm{Sym}(\Omega)}(G)$ can be computed by a nearly linear-time deterministic algorithm.*

Proof. Let $n := |\Omega|$ and $B = (\beta_1, \ldots, \beta_m)$. The construction of $C = C_{\mathrm{Sym}(\Omega)}(G)$ is straightforward from the proof of Lemma 6.1.4. We use the notation of that lemma. We can choose $\alpha = \beta_1$, and A is obtained as the set of points fixed by all elements of $S \cap G_\alpha$. Let R be the transversal, coded in a Schreier tree, for G mod G_α. For $\beta \in A$, an element $c \in C$ carrying α to β can be constructed using (6.2). Note that the elements of R may be used as the permutations g required in (6.2). Moreover, we do not need to multiply out g explicitly, since we are interested only in the image of one point of Ω. It is enough to consider g as a word and follow the image of β. Note that since $S = S^{-1}$ by our convention in this chapter, we can switch easily to the word representing g from the word representing g^{-1}, which can actually be obtained from the Schreier tree (cf. Section 4.1). Hence, because the tree coding R is shallow, an element of C can be computed in nearly linear time by a deterministic algorithm. Finally, observe that at most $\log |C| \leq \log n$ elements of C have to be constructed via (6.2) since we need only generators for C. If $c_1, \ldots, c_k \in C$ are already known, we choose $\beta_{k+1} \in A \backslash \alpha^{\langle c_1, \ldots, c_k \rangle}$ and construct $c_{k+1} \in C$ such that $\alpha^{c_{k+1}} = \beta_{k+1}$. □

Next, we treat the case of intransitive $G \leq \mathrm{Sym}(\Omega)$. We start with a simple but important lemma about groups normalized by certain elements of $\mathrm{Sym}(\Omega)$. This lemma is useful in other situations as well and it is one of the most frequently quoted results in this book (cf. Exercise 6.1, Theorem 6.3.1, Lemmas 6.1.10, 6.1.11, 7.1.1, and Sections 7.3.2 and 7.4.1).

Lemma 6.1.7. *Suppose that $H \leq \mathrm{Sym}(\Omega)$, $y \in \mathrm{Sym}(\Omega)$, and y normalizes H. Then y permutes the orbits of H.*

Proof. Let $\beta, \gamma \in \Omega$ be in the same H-orbit. Then there exists $h \in H$ such that $\beta^h = \gamma$ and so $\gamma^y = \beta^{yy^{-1}hy} = \beta^{yh^*}$ for $h^* = y^{-1}hy \in H$. This means that if two points $\beta, \gamma \in \Omega$ are in the same H-orbit then their images β^y, γ^y are in the same H-orbit again. □

Since $C = C_{\mathrm{Sym}(\Omega)}(G)$ normalizes G, Lemma 6.1.7 implies that C permutes the orbits of G. We say that two orbits Δ_1, Δ_2 of G are *equivalent*, denoted $\Delta_1 \equiv \Delta_2$, if there is a bijection $\varphi : \Delta_1 \to \Delta_2$ such that for all $g \in G$ and $\delta \in \Delta_1$,

$$\varphi(\delta^g) = \varphi(\delta)^g. \tag{6.3}$$

It is clear that $\equiv$ is an equivalence relation. Also, it is clear that if $c \in C$ maps Δ_1 to Δ_2 then the restriction $c|_{\Delta_1}$ is a bijection satisfying (6.3). Conversely, if φ is a bijection satisfying (6.3) then $\varphi \cup \varphi^{-1}$ is an involution in C that fixes the set $\Omega \backslash (\Delta_1 \cup \Delta_2)$ pointwise. Therefore, for any two orbits Δ_1, Δ_2 of G, $\Delta_1 \equiv \Delta_2$ if and only if some $c \in C$ maps Δ_1 to Δ_2. Moreover, if $\Delta_1 \equiv \Delta_2$ then $C_{\mathrm{Sym}(\Delta_1)}(G|_{\Delta_1}) \cong C_{\mathrm{Sym}(\Delta_2)}(G|_{\Delta_2})$. From these, we obtain the following lemma.

Lemma 6.1.8. *Let $\mathcal{C}_1, \ldots, \mathcal{C}_t$ be the $\equiv$-equivalence classes of orbits of G, $|\mathcal{C}_i| = k_i$, and for each i choose a representative $\Gamma_i \in \mathcal{C}_i$. Then*

$$C_{\mathrm{Sym}(\Omega)}(G) \cong \prod_{i=1}^{t} C_{\mathrm{Sym}(\Gamma_i)}(G|_{\Gamma_i}) \wr S_{k_i}.$$

Hence, to obtain $C_{\mathrm{Sym}(\Omega)}(G)$, we have to determine which orbits of G are equivalent and we have to construct bijections satisfying (6.3) between a representative of an equivalence class and the other members of the class.

Lemma 6.1.9. *Let Δ_1, Δ_2 be orbits of G of the same size, let $\alpha \in \Delta_1$, and let A be the set of fixed points of G_α. Then $\Delta_1 \equiv \Delta_2$ if and only if $A \cap \Delta_2 \neq \emptyset$.*

Proof. If $\varphi : \Delta_1 \to \Delta_2$ witnesses $\Delta_1 \equiv \Delta_2$ then $\varphi(\alpha) \in A \cap \Delta_2$. To see this, it is enough to observe that (6.3) implies that, for all $g \in G_\alpha$, $\varphi(\alpha)^g = \varphi(\alpha^g) = \varphi(\alpha)$.

Conversely, let $\beta \in A \cap \Delta_2$. Then we can define $\varphi : \Delta_1 \to \Delta_2$ as follows: For $\gamma \in \Delta_1$, take any $g \in G$ such that $\alpha^g = \gamma$ and let $\varphi(\gamma) := \beta^g$. Similarly to the proof of Lemma 6.1.4, it can be shown that φ is a bijection satisfying (6.3). Details are left to the reader. □

If α is the first base point of G then, using the set of fixed points of G_α and a shallow Schreier tree coding the transversal for G mod G_α, we can compute the orbits equivalent to the orbit of α and bijections between them by a nearly linear-time deterministic algorithm. Equivalence of the orbit containing some $\gamma \in \Omega$ with the other orbits can be determined by first applying a base change algorithm to obtain a base starting with γ. However, if G has numerous inequivalent orbits (but of the same size, so equivalence must be tested) then this computation of $C_{\mathrm{Sym}(\Omega)}(G)$ may require $\Omega(n^2/2^{\sqrt{\log n}})$ time even for small-base inputs (cf. Exercise 6.3). It is an open problem whether there is a nearly linear-time algorithm that computes $C_{\mathrm{Sym}(\Omega)}(G)$ for all $G \leq \mathrm{Sym}(\Omega)$.

6.1.3. The Center

Since $C_{\mathrm{Sym}(\Omega)}(G)$ is normalized by G, it is possible to compute $Z(G) = C_{\mathrm{Sym}(\Omega)}(G) \cap G$ by combining the methods of Sections 6.1.1 and 6.1.2. However, computing $C_{\mathrm{Sym}(\Omega)}(G)$ may not be possible by a nearly linear-time algorithm. Moreover, even if $C_{\mathrm{Sym}(\Omega)}(G)$ can be computed, it is possible that $|C_{\mathrm{Sym}(\Omega)}(G)|$ is much larger than $|G|$ and so the intersection algorithm does not run in time $O(n \log^c |G|)$. Hence we need a small additional trick.

Our treatment follows that of [Beals and Seress, 1992]. Let B be a nonredundant base for G, and let $\{\Delta_1, \Delta_2, \ldots, \Delta_k\}$ be the set of orbits of G that have nonempty intersections with B. Note that $k \leq |B| \leq \log |G|$. Let $\Delta := \bigcup_{i=1}^{k} \Delta_i$, and let G^* be the restriction of G to Δ. Since Δ contains a base for G, $|G^*| = |G|$ and, by Lemma 5.2.1, each element of G^* is extendible uniquely to an element of G by a deterministic nearly linear-time algorithm. Hence it is enough to construct generators for $Z(G^*)$.

Note that G^* has at most $\log |G| = \log |G^*|$ orbits, so $C_{\mathrm{Sym}(\Delta)}(G^*)$ can be obtained by a deterministic nearly linear-time algorithm. A further speedup is possible by observing that it is enough to construct generators for some $K \leq \mathrm{Sym}(\Delta)$ satisfying $Z(G^*) \leq K \leq C_{\mathrm{Sym}(\Delta)}(G^*)$, since $K \cap G^* = Z(G^*)$ for any such K. Clearly,

$$K = \prod_{i=1}^{k} C_{\mathrm{Sym}(\Delta_i)}(G^*|_{\Delta_i})$$

is appropriate, since elements of $Z(G^*)$ cannot move the orbits of G^*.

6.1.4. Centralizer of a Normal Subgroup

If $N \lhd G \leq \mathrm{Sym}(\Omega)$ then $C_{\mathrm{Sym}(\Omega)}(N)$ is normalized by G (cf. Exercise 6.4) and so $C_G(N) = C_{\mathrm{Sym}(\Omega)}(N) \cap G$ can be constructed by the methods of Sections 6.1.1 and 6.1.2. As in the case of $Z(G)$, the difficulty is that $C_{\mathrm{Sym}(\Omega)}(N)$ may not be computable in nearly linear time and/or it may have much larger order than G.

A deterministic nearly linear-time algorithm is described in [Luks and Seress, 1997]. We start with a trick similar to the one applied for computing the center. As in Section 6.1.3, let B be a nonredundant base for G, let $\{\Delta_1, \Delta_2, \ldots, \Delta_k\}$ be the set of orbits of G that have nonempty intersections with B, let $\Delta := \bigcup_{i=1}^{k} \Delta_i$, and let G^*, N^* be the restrictions of G, N respectively, to Δ. It is enough to compute $C_{G^*}(N^*)$. Moreover,

$$K = \prod_{i=1}^{k} C_{G^*|_{\Delta_i}}(N^*|_{\Delta_i})$$

contains $C_{G^*}(N^*)$, K is normalized by G^*, and $|K| \le |G|^{\log|G|}$, so $\log|K| \le \log^2|G|$. Hence $C_{G^*}(N^*)$ can be obtained as $C_{G^*}(N^*) = K \cap G^*$, provided that the groups $C_{G^*|_{\Delta_i}}(N^*|_{\Delta_i})$ are available, and so we reduced the computation of $C_G(N)$ to the computation of the groups $C_{G^*|_{\Delta_i}}(N^*|_{\Delta_i})$. Therefore it is enough to find a nearly linear-time solution to the following problem:

$$\text{Given a transitive } G \le \mathrm{Sym}(\Omega) \text{ and } N \lhd G, \text{ compute } C_G(N) \text{ by a deterministic nearly linear-time algorithm.} \tag{6.4}$$

Note that despite the fact that G is transitive in (6.4), N may have numerous orbits and so the difficulties with the nearly linear-time construction of $C_{\mathrm{Sym}(\Omega)}(N)$ seem to persist. Although in this special case it is possible to construct generators for $C_{\mathrm{Sym}(\Omega)}(N)$ in nearly linear time, $C_{\mathrm{Sym}(\Omega)}(N)$ may be too large. Therefore we still cannot apply the intersection method of Section 6.1.1. Instead, we proceed by constructing a block homomorphism whose kernel $\tilde{G}$ contains $C_G(N)$ and a second homomorphism from $\tilde{G}$ whose kernel is exactly $C_G(N)$.

Lemma 6.1.10. *Let $G \le \mathrm{Sym}(\Omega)$ be transitive and let $N \lhd G$.*

(i) *For some $\alpha \in \Omega$, let $A = \mathrm{fix}(N_\alpha)$ be the set of fixed points of N_α. Then A is a block of imprimitivity for G.*
(ii) *Let $\mathcal{B}$ be the block system consisting of the G-images of A and let $\tilde{G}$ be the kernel of the G-action on $\mathcal{B}$. Then $C_G(N) \le \tilde{G}$.*

Proof. (i) By Lemmas 6.1.4, 6.1.8, and 6.1.9, A is an orbit of $C_{\mathrm{Sym}(\Omega)}(N)$. So, by Exercise 6.4 and Lemma 6.1.7, A is a block of imprimitivity for G.

(ii) The blocks in $\mathcal{B}$ are the orbits of $C_{\mathrm{Sym}(\Omega)}(N)$; therefore, in particular, the elements of $C_G(N)$ cannot move these blocks. Thus $C_G(N) \le \tilde{G}$. □

The tricky part of the computation of $C_G(N)$ is the definition of a homomorphism $\varphi : \tilde{G} \to \mathrm{Sym}(\Omega)$ with $\ker(\varphi) = C_G(N)$. We shall define φ in Lemma 6.1.12, but first we need some preparatory steps.

Lemma 6.1.11. *Let $\Delta_1, \Delta_2, \ldots, \Delta_k$ be the orbits of N that have nonempty intersection with A and let $\Delta := \bigcup_{i=1}^k \Delta_i$. Then Δ is a block of imprimitivity for G.*

Proof. Observe that Δ is the union of the N-images of A and so $\{\Delta_1, \Delta_2, \ldots, \Delta_k\}$ is an orbit of N in the action on $\mathcal{B}$. By Lemma 6.1.7, $\{\Delta_1, \Delta_2, \ldots, \Delta_k\}$ is a block for G in the action on $\mathcal{B}$ and so $\bigcup_{i=1}^k \Delta_i$ is a block for the G-action on Ω. □

Next, choose $\alpha_i \in A \cap \Delta_i$ for $1 \le i \le k$. Since Δ is a block of imprimitivity for G, we can take G-images $\Delta^{g_1}, \ldots, \Delta^{g_m}$ of Δ that partition Ω and so the images $\Gamma := \{\alpha_i^{g_j} \mid 1 \le i \le k, 1 \le j \le m\}$ define a system of representatives of all N-orbits. Recall again that we have an equivalence relation on the orbits of N: Two orbits are equivalent if and only if there is an element of $C_{\mathrm{Sym}(\Omega)}(N)$ mapping one to the other. The orbits $\Delta_1, \Delta_2, \ldots, \Delta_k$ form one of the equivalence classes. If two N-orbits are equivalent then their representatives in Γ are in the same orbit of $C_{\mathrm{Sym}(\Omega)}(N)$.

Lemma 6.1.12. *Let $G \le \mathrm{Sym}(\Omega)$ be transitive, let $N \lhd G$, and define $\mathcal{B}, \tilde{G}, \Gamma$ as in Lemma 6.1.10 and in the previous paragraph.*

(i) *For each $g \in \tilde{G}$, there exists a unique $c_g \in C_{\mathrm{Sym}(\Omega)}(N)$ such that gc_g^{-1} fixes Γ pointwise.*
(ii) *The map $\varphi : \tilde{G} \to \mathrm{Sym}(\Omega)$ defined by $\varphi : g \mapsto gc_g^{-1}$ is a group homomorphism.*
(iii) *$C_G(N)$ is the kernel of the homomorphism φ.*

Proof. (i) Let $g \in \tilde{G}$. Recall that, by Lemma 6.1.8,

$$C_{\mathrm{Sym}(\Omega)}(N) \cong \prod_{j=1}^{m} \left(C_{\mathrm{Sym}(\Delta_1^{g_j})}(N|_{\Delta_1^{g_j}}) \wr S_k \right). \tag{6.5}$$

Since $g \in \tilde{G}$, it cannot move the blocks in $\mathcal{B}$. In particular, g fixes Δ^{g_j} for $1 \le j \le m$. Let $j \in [1, m]$ be arbitrary. There are k N-orbits in Δ^{g_j} and g permutes them; this permutation π_j, regarded as an element of S_k in the wreath product in the jth term of (6.5), must also be induced by any $c_g \in C_{\mathrm{Sym}(\Omega)}(N)$ satisfying that gc_g^{-1} fixes $\alpha_i^{g_j}$ for $i = 1, 2, \ldots, k$.

Now, for each $i \in [1, k]$, consider $(\alpha_i^{g_j})^g$. Since g fixes the G-images of A and $C_{\mathrm{Sym}(\Delta_i^{g_j})}(N|_{\Delta_i^{g_j}})$ acts transitively on $A^{g_j} \cap \Delta_i^{g_j}$, there exist $c_{ij} \in C_{\mathrm{Sym}(\Delta_i^{g_j})}(N|_{\Delta_i^{g_j}})$ such that $(\alpha_i^{g_j})^g = (\alpha_i^{g_j})^{c_{ij}}$. Moreover, since $C_{\mathrm{Sym}(\Delta_i^{g_j})}(N|_{\Delta_i^{g_j}})$ is semiregular, there is a unique c_{ij} with this property. Then for

$$c_g := \prod_{j=1}^{m} (c_{1j}, \ldots, c_{kj}; \pi_j) \in C_{\mathrm{Sym}(\Omega)}(N),$$

gc_g^{-1} fixes Γ pointwise and it is clear from the discussion above that there is only one element in $C_{\mathrm{Sym}(\Omega)}(N)$ with this property.

(ii) Observe that for $g, h \in \tilde{G}$, we have $gc_g^{-1}hc_h^{-1} = gh(c_g^{-1})^h c_h^{-1}$. Here $(c_g^{-1})^h c_h^{-1} \in C_{\mathrm{Sym}(\Omega)}(N)$ and $gh(c_g^{-1})^h c_h^{-1}$ fixes Γ pointwise, so the uniqueness of c_{gh} implies that $c_{gh}^{-1} = (c_g^{-1})^h c_h^{-1}$ and $\varphi(g)\varphi(h) = \varphi(gh)$.

(iii) $C_G(N)$ is the kernel of the homomorphism φ since $\varphi(g) = 1$ if and only if $g = c_g$, that is, $g \in \tilde{G} \cap C_{\mathrm{Sym}(\Omega)}(N) = C_G(N)$. □

In summary, the algorithm for solving problem (6.4) is the following:

Centralizer of Normal Subgroup in Transitive Group[G, N]

Input: An SGS for a transitive $G \le \mathrm{Sym}(\Omega)$ relative to a nonredundant base B; generators for $N \lhd G$.

Outout: An SGS for $C_G(N)$.

(*Note that N is not necessarily transitive.*)

Step 1: Compute an SGS relative to B for N.

Step 2: Let α be the first point of B. Compute $A := \mathrm{fix}(N_\alpha)$.

Step 3: Compute the block system $\mathcal{B}$ consisting of the G-images of A and $\tilde{G}$, the kernel of the G-action on $\mathcal{B}$.

Step 4: Compute those N-orbits $\Delta_1, \ldots, \Delta_k$ that have nonempty intersection with A, and compute $\Delta := \bigcup_{i=1}^k \Delta_i$. For $1 \le i \le k$, pick $\alpha_i \in A \cap \Delta_i$.

Step 5: Compute G-images $\Delta^{g_1}, \ldots, \Delta^{g_m}$ of Δ that partition Ω. Compute $\Gamma := \{\alpha_i^{g_j} \mid 1 \le i \le k, 1 \le j \le m\}$.

Step 6: For each generator g of $\tilde{G}$, compute $c_g \in C_{\mathrm{Sym}(\Omega)}(N)$ such that gc_g^{-1} fixes Γ pointwise.

Step 7: Compute and output the kernel of the homomorphism $\varphi : \tilde{G} \to \mathrm{Sym}(\Omega)$ defined by $\varphi : g \mapsto gc_g^{-1}$.

Theorem 6.1.13. *The algorithm* **Centralizer of Normal Subgroup in Transitive Group**(G, N) *is deterministic and it runs in nearly linear time.*

Proof. Appealing to the results of Chapter 5 about computing an SGS relative to a given base (Theorem 5.2.3) and computing kernels of homomorphisms (Section 5.1.2), we clearly see that Steps 1–5 and 7 can be performed by nearly linear-time deterministic algorithms. For Step 6, observe that for fixed i, j, generators and a shallow Schreier tree data structure for $C_{\mathrm{Sym}(\Delta_i^{g_j})}(N|_{\Delta_i^{g_j}})$ can be computed in $O(|\Delta_i|\log^c|N|)$ time with an absolute constant c, as described in Section 6.1.2; in particular, c_{ij} can be obtained in $O(|\Delta_i|\log^c|N|)$ time. Hence, for each generator g of $\tilde{G}$, we can compute c_g in nearly linear time. □

6.1.5. Core of a Subnormal Subgroup

By repeated applications of Corollary 6.1.3, we can compute intersections with subnormal subgroups.

Lemma 6.1.14. *Suppose that $H \lhd\lhd M$ and $G \le M$. Given an SGS for M and generators for H and G, an SGS for $H \cap G$ can be computed in nearly linear time by a deterministic algorithm.*

Proof. Define $H_0 := M$ and $H_i := \langle H^{H_{i-1}}\rangle$ for $i > 0$. The series of subgroups $H_0 \rhd H_1 \rhd \cdots$ is computable in nearly linear time and, since $H \lhd\lhd M$, $H = H_k$ for some integer k (cf. Exercise 6.5). Note that the first such k is at most $\log|M|$, since the previous H_i define a strictly decreasing subgroup chain in M.

For all $i \ge 0$, $G \cap H_i$ normalizes H_{i+1}. Hence Corollary 6.1.3 implies that $(G \cap H_i) \cap H_{i+1} = G \cap H_{i+1}$ is successively computable for $i = 0, 1, \ldots, k-1$ by a nearly linear-time deterministic algorithm. □

Computation of the Core

We can apply Lemma 6.1.14 for computing $\mathrm{Core}_G(H) = \bigcap\{H^g \mid g \in G\}$ in the case when $H \lhd\lhd G$. Let $G = \langle S\rangle$. We compute a series of subgroups $H = C_0 \gneqq C_1 \gneqq \cdots$ and permutations $g_0 = 1, g_1, g_2, \ldots \in G$ such that, for all $i \ge 0$,

$$C_i = H \cap \bigcap_{j=1}^{i} H^{g_j}. \tag{6.6}$$

The process stops when $C_m = \mathrm{Core}_G(H)$. Since the C_i define a strictly decreasing subgroup chain and, by definition, $C_i \ge \mathrm{Core}_G(H)$ for all i, we reach $\mathrm{Core}_G(H)$ in $m \le \log|G|$ steps.

Suppose that C_i and $g_0, \ldots, g_i$ are already defined for some $i \ge 0$. Then we compute the conjugates of C_i by the generators of G. If $C_i^g = C_i$ for all

$g \in S$ then $C_i^h = C_i$ for all $h \in G$. This follows from induction on the length of the word as h is written as a product of generators. Also, (6.6) implies that $C_i \leq H^h$. Hence $C_i = \mathrm{Core}_G(H)$ and we are done.

Otherwise, there exists $g \in S$ such that $C_i^g \neq C_i$. Since $|C_i^g| = |C_i|$, this means that C_i^g is not a subgroup of H^{g_j} for some $j \in [0, i]$. Such j can be found by testing for each generator c of C_i whether c^g is in the group H^{g_j} (or, for easier implementation, testing whether $c^{gg_j^{-1}}$ is in H). After the appropriate index j is found, we define $g_{i+1} := g_j g^{-1}$, use the algorithm in the proof of Lemma 6.1.14 to compute $C_{i+1} := C_i \cap H^{g_j g^{-1}} = (C_i^{gg_j^{-1}} \cap H)^{g_j g^{-1}}$, and proceed with the recursion.

We remark that [Kantor and Luks, 1990] contains a polynomial-time algorithm for computing the core of arbitrary subgroups of G. However, that algorithm uses Sylow subgroup computations, which, at the moment, have nearly linear-time versions only for groups with no exceptional Lie-type composition factors (cf. [Morje, 1995]), and the algorithm is much more complicated than the elementary procedure described here.

6.2. Composition Series

The composition factors reveal substantial information about the structure of a group, and computing a composition series is the starting point in many algorithms (cf. Sections 7.3.1, 6.3.1, 8.3, 8.4, and 9.4). Therefore, significant effort has been devoted to the construction of a compostion series [Neumann, 1986; Luks, 1987; Kantor, 1991; Babai et al., 1993; Beals and Seress, 1992; Beals, 1993b; Cannon and Holt, 1997].

Given $G = \langle S \rangle \leq \mathrm{Sym}(\Omega)$, with $|\Omega| = n$, the goal is to find generators for the subgroups N_i in a composition series $G = N_1 \rhd N_2 \rhd \cdots \rhd N_l = 1$ and homomorphisms $\varphi_i : N_i \to S_n$ with $\ker(\varphi_i) = N_{i+1}$ for $i = 1, 2, \ldots, l-1$. We remark that it is not true in general that if $N \lhd G$ then G/N has a faithful permutation representation of degree at most n (cf. Exercise 6.6), but we shall prove that if N is a maximal normal subgroup of G then such a small-degree faithful permutation representation exists. Hence the homomorphisms φ_i exist as well.

The preliminary version of [Luks, 1987], circulated since 1981, was the first permutation group algorithm that utilized the classification of finite simple groups (CFSG). Although it turned out later that a compsosition series can be computed without appealing to this classification (cf. [Beals, 1993b]), the CFSG is a powerful and indispensible tool in many other algorithms. Also, implementations of the composition series algorithm in both GAP and MAGMA use consequences of the CFSG.

Our goal is to prove the following theorem.

Theorem 6.2.1. *Given $G = \langle S \rangle \leq \mathrm{Sym}(\Omega)$, with $|\Omega| = n$, and an SGS relative to a nonredundant base for G, a composition series for G can be computed by a nearly linear-time Las Vegas algorithm.*

The proof of Theorem 6.2.1 is spread out in Sections 6.2.1–6.2.4 and 6.2.6. In Sections 6.2.1–6.2.4 we describe a nearly linear-time Monte Carlo composition series construction. This construction is Monte Carlo, even if we have a nonredundant base and a strong generating set for the input group. The upgrade to a Las Vegas algorithm (supposing that a base and SGS are known) is given in Section 6.2.6. Moreover, in Section 8.3, we describe how to compute a nonredundant base, SGS, and composition series by a Las Vegas nearly linear-time algorithm, in groups satisfying some restrictions on their composition factors.

6.2.1. Reduction to the Primitive Case

Given $G = \langle S \rangle \leq \mathrm{Sym}(\Omega)$, with $|\Omega| = n$, and an SGS relative to a nonredundant base for G, the basic strategy for constructing a composition series of G is to find a normal subgroup $M \lhd G$ and a faithful permutation representation of degree at most n for G/M. Moreover, either $1 \neq M \neq G$ or $M = 1$ but the faithful permutation representation of G/M is of degree at most $n/2$. Applying this procedure recursively for G/M and M, we find a composition series of G in at most $\log |G| \log n$ iterations.

If G is not transitive then let Δ be the smallest nontrivial orbit of G and construct a transitive constituent homomorphism $\varphi : G \to \mathrm{Sym}(\Delta)$ and its kernel $M := \ker(\varphi)$. By Section 5.1.3, this construction can be done in nearly linear time.

If G is transitive then we use the nearly linear-time algorithm of Section 5.5 to check whether G is primitive. If G is imprimitive then this algorithm outputs a nontrivial block, and so we can construct a nontrivial block system Σ, the action $\varphi : G \to \mathrm{Sym}(\Sigma)$ on the blocks, and the kernel $M := \ker(\varphi)$ of this action. Using the results of Section 5.1.3 again, this construction can be done in nearly linear time.

If G is primitive then, by the following lemma from [Luks, 1987], it is enough to find *any* proper normal subgroup of G.

Lemma 6.2.2. *Let $G \leq Sym(\Omega)$, with $|\Omega| = n$, let $B = (\beta_1, \ldots, \beta_m)$ be a base for G, and let $G = G^{[1]} \geq G^{[2]} \geq \cdots \geq G^{[m+1]} = 1$ be the corresponding point stabilizer chain. Suppose that $N \lhd G$, and let i be the index such that*

$G = G^{[i]}N \gneq G^{[i+1]}N$. Then $|G : G^{[i+1]}N| \le n$ and the action of G on the cosets of $G^{[i+1]}N$ has N in its kernel. Moreover, given B and generators for N, this action can be constructed by a deterministic algorithm in nearly linear time.

Proof. We have $|G : G^{[i+1]}N| = |G^{[i]}N : G^{[i+1]}N| \le |G^{[i]} : G^{[i+1]}|$. Since $N \lhd G^{[i+1]}N$, N acts trivially on the G-cosets of $G^{[i+1]}N$.

To show the algorithmic feasibility of the construction of the action on the G-cosets of $G^{[i+1]}N$, first observe that, by Theorem 5.2.3, the index i can be found in nearly linear time. Let $H := G^{[i]} \cap G^{[i+1]}N$. Note that H can be constructed as the pointwise stabilizer of $\{\beta_1, \ldots, \beta_{i-1}\}$ in $G^{[i+1]}N$. Since $G^{[i]} \ge H \ge G^{[i+1]} = (G^{[i]})_{\beta_i}$, the set $\Delta := \beta_i^H$ is a block of imprimitivity for the action of $G^{[i]}$ on the cosets of $G^{[i+1]}$. Let Σ denote the block system consisting of the $G^{[i]}$-images of Δ. The homomorphism $\psi : G^{[i]} \to \mathrm{Sym}(\Sigma)$ defining the action of $G^{[i]}$ on the block system Σ can be obtained in nearly linear time.

We claim that the image of ψ is permutation isomorphic to the action of $G^{[i]}$ on the G-cosets of $G^{[i+1]}N$. Indeed, the map $\varphi : \Delta^g \mapsto G^{[i+1]}Ng$ is a bijection between the permutation domains. This map φ is well defined, since for $g, h \in G^{[i]}$,

$$\Delta^g = \Delta^h \iff gh^{-1} \in H = G^{[i]} \cap G^{[i+1]}N \iff gh^{-1} \in G^{[i+1]}N$$
$$\iff G^{[i+1]}Ng = G^{[i+1]}Nh.$$

Moreover, $(\varphi(\Delta^g))h = \varphi((\Delta^g)^h)$, so φ induces a permutation isomorphism.

Since we can construct the permutation action of $G^{[i]}$ on the cosets of $G^{[i+1]}N$ and N acts trivially on these cosets, the permutation action of a generating set of $G = \langle G^{[i]}, N\rangle$ is constructed in nearly linear time. □

Corollary 6.2.3. *Let $G \le \mathrm{Sym}(\Omega)$, with $|\Omega| = n$, and let N be a maximal normal subgroup of G. Then G/N has a faithful permutation representation of degree at most n.*

By the preceding discussion, finding a composition series is reduced to the following problem:

$$\begin{array}{c}\text{Given a primitive group } G \le \mathrm{Sym}(\Omega)\text{, with } |\Omega| = n, \text{ find generators}\\ \text{for a proper normal subgroup, or find a faithful representation}\\ \text{on at most } n/2 \text{ points, or prove that } G \text{ is simple.}\end{array} \tag{6.7}$$

The fact that the construction of the action of G on the cosets of $G^{[i+1]}N$ in Lemma 6.2.2 can be done in nearly linear time was observed in [Beals and Seress, 1992]. That paper also contains a strengthening of the method, when we do not have generators for N. We shall need this extension in Section 6.2.3.

Lemma 6.2.4. *Let $G \le Sym(\Omega)$, with $|\Omega| = n$, let $B = (\beta_1, \dots, \beta_m)$ be a base for G, and let $G = G^{[1]} \ge G^{[2]} \ge \cdots \ge G^{[m+1]} = 1$ be the corresponding point stabilizer chain. Suppose that $N \lhd G$, and suppose that the index i such that $G = G^{[i]}N \gneqq G^{[i+1]}N$ is known. Then, given an SGS for G relative to B and generators for a subgroup $K \ge G^{[i+1]}N$, the action of G on the cosets of K can be constructed in nearly linear time by a deterministic algorithm, even if we have no generators for N.*

Proof. The first part of the algorithm is similar to the one described in Lemma 6.2.2. Namely, we construct the permutation action of $G^{[i]}$ on the cosets of K. The only difference from Lemma 6.2.2 is that we compute $H := G^{[i]} \cap K$ (instead of the unknown $G^{[i]} \cap G^{[i+1]}N$). Then, as in the proof of Lemma 6.2.2, we compute $\Delta := \beta_i^H$ and the block system Σ consisting of the $G^{[i]}$-images of Δ in the fundamental orbit $\beta_i^{G^{[i]}}$. The same proof as in Lemma 6.2.2, just replacing each reference to $G^{[i+1]}N$ by K, shows that the $G^{[i]}$-action on the cosets of K is permutation isomorphic to the $G^{[i]}$-action on Σ. Let $\Delta_1 := \Delta, \Delta_2, \dots, \Delta_m$ denote the blocks in Σ (with $m = |G : K| \le n$).

The new part of the algorithm is that we also have to be able to construct the action of arbitrary elements of G on the cosets of K. To this end, we compute a shallow Schreier tree data structure for K relative to the base B, and for each $j \in [1, m]$ we choose a point σ_j from Δ_j. Using the ith tree in the shallow Schreier tree data structure of G, we write the coset representatives h_j carrying β_i to σ_j as words in the strong generators. The set of all h_j is a transversal system for G mod K. Note that because of the nearly linear time constraint, we cannot compute the permutations h_j explicitly.

Given some $g \in G$, for each $j \in [1, m]$ we want to determine to which coset of K the permutation $h_j g$ belongs. First, we compute g^{-1}. We claim that after that, for a fixed j, the coset of $h_j g$ can be found in $O(\log^3 |G|)$ time. Applying this procedure for the generators g of G, the action of G on the cosets of K is found in nearly linear time.

Recall our convention that strong generating sets are closed for taking inverses. Hence we can write a word w of length at most $2 \log |G| + 1$ whose product is $g^{-1}h_j^{-1}$. We sift w as a word (cf. Lemma 5.2.2) through the first

$i - 1$ levels of the Schreier tree data structure of K. Since any element of $G = G^{[i]}K$ can be written in the form $g_i k$ for some $k \in K$ and $g_i \in G^{[i]}$, this sifting is feasible. By Lemma 5.2.2, the time requirement is $O(\log^3 |G|)$ and the siftee $s := g^{-1}h_j^{-1}k \in G^{[i]}$, for some $k \in K$, is represented as a word of length $O(\log^2 |G|)$. So we can write a word (in terms of g and the SGS for G, K) representing s^{-1} in $O(\log^2 |G|)$ time. Since $ks^{-1} = h_j g$, the coset of $h_j g$ can be found by looking up to which block Δ_l the point $\beta_i^{s^{-1}}$ belongs. □

6.2.2. The O'Nan–Scott Theorem

Every abstract group has a faithful transitive permutation representation, but the structure of primitive permutation groups is quite restricted. Primitive groups can be divided into classes according to their socle and the stabilizer of a point of the permutation domain in the socle. This characterization is known as the O'Nan–Scott theorem (cf. [Scott, 1980]), but there are numerous versions. In fact, we shall also state two versions: The first one is more of a group theoretic nature, whereas the second one is a rearrangement of the cases and is suited better for our algorithmic purposes.

We need some notation. If T is a simple nonabelian group and $H = T_1 \times \cdots \times T_k$ is a group such that for $i = 1, 2, \ldots, k$ we have isomorphisms $\varphi_i : T \to T_i$ then we denote the subgroup $\{(\varphi_1(t), \ldots, \varphi_k(t)) \in H \mid t \in T\}$ by $\mathrm{Diag}(H)$ and call it a *diagonal subgroup* of H. Although $\mathrm{Diag}(H)$ depends on the isomorphisms φ_i, there will be no confusion from omitting these isomorphisms from the notation.

Theorem 6.2.5. *Let $G \leq \mathrm{Sym}(\Omega)$ be primitive and let $|\Omega| = n$. Then one of the following holds:*

(I) *G has a minimal normal subgroup N with $C_G(N) \neq 1$. Moreover,*
 (i) *if N is abelian then $n = p^d$ for some prime p, N is regular, and it is the only minimal normal subgroup of G;*
 (ii) *if N is nonabelian then there are exactly two minimal normal subgroups of G and both of them are regular.*

(II) *G has a unique minimal normal subgroup $N = \mathrm{Soc}(G) = T_1 \times \cdots \times T_r$, where each T_i is isomorphic to the same simple nonabelian group T and $G \lesssim \mathrm{Aut}(T) \wr S_r$. The group G permutes the set $\{T_1, \ldots, T_r\}$ by conjugation, and this permutation action is transitive. Moreover, let $\alpha \in \Omega$. Then one of the following three cases occurs:*
 (i) *$N_\alpha = 1$ and $n = |T|^r$;*

(ii) $N_\alpha = (T_1)_\alpha \times \cdots \times (T_r)_\alpha$ *with isomorphic subgroups* $(T_i)_\alpha$ *satisfying* $1 \neq (T_i)_\alpha \neq T_i$, *and* $n = |T_1 : (T_1)_\alpha|^r$; *or*

(iii) $r = kl$ *with* $k \geq 2$ *and there exists a permutation of the indices* $1, 2, \ldots, r$ *such that* $N_\alpha = \mathrm{Diag}(T_1 \times \cdots \times T_k) \times \mathrm{Diag}(T_{k+1} \times \cdots \times T_{2k}) \times \cdots \times \mathrm{Diag}(T_{r-k+1} \times \cdots \times T_r)$. *Moreover,* $n = |T_1|^{(k-1)l}$.

Proof. Suppose first that G has a minimal normal subgroup N with $C_G(N) \neq 1$. By Exercise 6.4, $C_G(N)$ is also a normal subgroup of G and its centralizer is nontrivial, since it contains N. Moreover, by Exercise 6.1, both N and $C_G(N)$ are transitive. However, by Lemma 6.1.4, N and $C_G(N)$ are semiregular as subgroups of the semiregular groups $C_{\mathrm{Sym}(\Omega)}(C_G(N))$ and $C_{\mathrm{Sym}(\Omega)}(N)$, respectively. Hence both N and $C_G(N)$ are regular.

If N is abelian then $N \leq C_G(N)$ and so $N = C_G(N)$. Moreover, since N is a minimal normal subgroup, it is the direct product of isomorphic simple groups, that is, an elementary abelian p-group for some prime p and $n = p^d$ for some $d \geq 1$. If N is nonabelian then N and $C_G(N)$ are two different subgroups. Both of them are minimal normal in G, since a proper subgroup of a regular group cannot be transitive. In both of the cases of abelian and nonabelian N, there are no other minimal normal subgroups of G because any two minimal normal subgroups must centralize each other. Hence we are in case I(i) or I(ii) of the theorem.

Suppose now that G has no minimal normal subgroup with nontrivial centralizer. Then, using again that any two minimal normal subgroups centralize each other, we obtain that G has only one minimal normal subgroup, which we may denote by N. Since N is minimal normal, there exists a nonabelian simple group T and an integer $r \geq 1$ such that $N = T_1 \times \cdots \times T_r$ with $T_i \cong T$ for all $i \in [1, r]$. The only minimal normal subgroups of N are the groups T_i for $i \in [1, r]$ (cf. Exercise 6.10), so G must permute these by conjugation. Hence $\mathrm{Aut}(N) \cong \mathrm{Aut}(T) \wr S_r$ and G can be identified with a subgroup of $\mathrm{Aut}(T) \wr S_r$. The minimality of N also implies that this conjugation action is transitive since any orbit corresponds to a normal subgroup of G.

Let $\alpha \in \Omega$. We first observe that G_α also acts by conjugation transitively on $\{T_1, \ldots, T_r\}$, since $G = NG_\alpha$ and the conjugation action of N is trivial. We consider the projections $\pi_i : N_\alpha \to T_i$.

If π_1 is surjective then all π_i are surjective since $\pi_i(N_\alpha) = \pi_1(N_\alpha)^g$ for some $g \in G_\alpha$ conjugating T_1 to T_i. In this case, N_α is a subdirect product of the T_i and, by Exercise 6.9, there is a partition $\Pi = (\Pi_1, \ldots, \Pi_l)$ of the index set $\{1, 2, \ldots, r\}$ such that N_α is the product of l diagonal subgroups, corresponding to this partition. Since $N_\alpha \lhd G_\alpha$, the parts Π_i are permuted by the conjugation action of G_α on $\{1, 2, \ldots, r\}$, and since the conjugation action is transitive, all parts must be of the same size. Hence we are in case II(iii).

If $1 \lneqq \pi_1(N_\alpha) \lneqq T_1$ then $1 \lneqq \pi_i(N_\alpha) \lneqq T_i$ for all $i \in [1, r]$ and we are in case II(ii). Finally, if $1 = \pi_1(N_\alpha)$ then $1 = \pi_i(N_\alpha)$ for all $i \in [1, r]$ and we are in case II(i). □

It is possible to describe the structure and action of primitive groups in much more detail than was done here, but we confine ourselves to the properties needed in the design and analysis of the composition series algorithm. A more thorough description can be found, for example, in [Dixon and Mortimer, 1996, Chap. 4]. Some further material is also listed in Exercise 6.7 concerning case I(ii) and in Exercise 6.8 concerning cases II(ii) and II(iii) with $l > 1$. We need more details only in the case II(iii) with $l = 1$. The proof of the following lemma can be found in [Dixon and Mortimer, 1996, Theorem 4.5A].

Lemma 6.2.6. *Let $G \le \mathrm{Sym}(\Omega)$ be primitive with a unique minimal normal subgroup $N = T_1 \times \cdots \times T_r$, such that all T_i are isomorphic to a nonabelian simple group T. Let $\alpha \in \Omega$ and suppose that $N_\alpha = \mathrm{Diag}(T_1 \times \cdots \times T_r)$. Then $G_\alpha = H \times P$, where $\mathrm{Inn}(T) \lesssim H \lesssim \mathrm{Aut}(T)$ and P is isomorphic to a primitive subgroup of S_r.*

The subgroup N_α defines an isomorphism among the T_i. So N can be identified with the set of sequences $(t_1, \ldots, t_r)$, $t_i \in T$. For $g \in P$, $(t_1, \ldots, t_r)^g$ is a sequence consisting of a permutation of the entries of $(t_1, \ldots, t_r)$, according to the permutation action of P in S_r.

Our second version of the O'Nan–Scott theorem follows.

Theorem 6.2.7. *Let $G \le \mathrm{Sym}(\Omega)$ be primitive and let $|\Omega| = n$. Then G satisfies at least one of the following properties:*

(A) *G has a minimal normal subgroup N with $C_G(N) \ne 1$.*
(B) *G has a proper normal subgroup of index less than n.*
(C) *G has a unique minimal normal subgroup $N = \mathrm{Soc}(G) = N_1 \times \cdots \times N_m$, where the N_i are isomorphic groups, $m!/2 \ge n$, and G acts by conjugation on $\{N_1, \ldots, N_m\}$ as the full alternating group A_m. Moreover, let $\alpha \in \Omega$. Then one of the following three cases occurs:*
    (i) *$N_\alpha = 1$;*
    (ii) *$N_\alpha = (N_1)_\alpha \times \cdots \times (N_m)_\alpha$ with $1 \ne (N_i)_\alpha \ne N_i$; or*

(iii) *each* N_i *is isomorphic to the same simple nonabelian group* T *and* $N_\alpha = \mathrm{Diag}(N_1 \times \cdots \times N_m)$.

(D) G *is simple.*

Proof. We have to show that the groups in case II of Theorem 6.2.5 are covered by cases B, C, and D of this theorem. Using the notation of Theorem 6.2.5, suppose that G is in case II, with minimal normal subgroup $N = T_1 \times \cdots \times T_r$ and all T_i isomorphic to the same simple nonabelian group T. Since $|T| \geq 60$ and $|T : H| \geq 5$ for any proper subgroup H of T, we have $r \leq \log_5 n$ in each subcase of case II.

If $r = 1$ then either G is simple (and it is in case D) or $N \cong \mathrm{Inn}(T) \lneqq G \leq \mathrm{Aut}(T)$. In the latter case, G/N is solvable by Schreier's Hypothesis (which is now a theorem, as a consequence of the classification of finite simple groups); hence G has a cyclic factor group of order p for some prime p. Clearly $p \leq n$, and so G is in case B.

Suppose now that $r \geq 2$. Let Σ be a minimal block system in the conjugation action of G on $\{T_1, , \ldots, T_r\}$ and let $m := |\Sigma|$. If G is in case II(iii) and $m < r$ then we also suppose that the groups T_i belonging to the same diagonal collection are in the same block of Σ. Then G acts primitively on Σ; let $P \leq S_m$ denote the image of this action. By [Praeger and Saxl, 1980], if $P \leq S_m$ is primitive and $P \neq A_m, S_m$ then $|P| < 4^m$; since $m \leq r \leq \log_5 n$, the kernel of action on Σ has index less than n and G is in case B. If $P = S_m$ then it has a normal subgroup of index 2 and G is in case B. This leaves us with the case that $P = A_m$. If $m!/2 < n$ then we are in case B, so we may suppose that $m!/2 \geq n$.

Now, if G is in case II(i) then G is in case C(i). If G is in case II(ii) then it is in case C(ii). If G is in case II(iii) with $l > 1$ then it is in case C(ii). Finally, if G is in case II(iii) with $l = 1$ then G is in case C(iii). □

The composition series algorithm that we shall describe in the next two sections uses different methods for solving the problem (6.7) from Section 6.2.1 for inputs belonging to the different cases of Theorem 6.2.7. If it is not known to which case of Theorem 6.2.7 the input primitive group G belongs, then the algorithm tries all methods. Note that if the output of (6.7) is a proper normal subgroup or a faithful permutation action on at most $n/2$ points then the correctness of the output can be checked in nearly linear time. Hence if all methods report failure or that G is simple then we can conclude that G is indeed simple.

The only regular primitive groups are cyclic of prime order, and these can be easily recognized. Hence, in the next two sections, we may assume that the input primitive group is not regular.

6.2.3. Normal Subgroups with Nontrivial Centralizer

In practice, solving the problem (6.7) is most time consuming in the case when the input primitive group G has a regular normal subgroup with a nontrivial centralizer (i.e., G has a regular abelian normal subgroup or G has two regular nonabelian normal subgroups). We shall present two approaches. The first method originated in [Luks, 1987] and subsequently was refined in [Babai et al., 1993] and [Beals and Seress, 1992] to a nearly linear-time algorithm. The second method was introduced in [Neumann, 1986], and it handles the case when G has a regular abelian normal subgroup. Here we give a nearly linear-time version. As we shall discuss in Remarks 6.2.11 and 6.2.14, both approaches can be extended to handle all primitive groups with regular normal subgroups. However, we cannot guarantee nearly linear running time of these extensions in the cases when G has a unique regular nonabelian normal subgroup or G has a nonabelian regular normal subgroup, respectively.

Let $G \le \mathrm{Sym}(\Omega)$ be primitive, and suppose that G has a regular normal subgroup N. For any $\alpha, \beta \in \Omega$, there is a unique element $x_{\alpha\beta} \in N$ such that $\alpha^{x_{\alpha\beta}} = \beta$. The first method finds generators for larger and larger subgroups of G that contain $x_{\alpha\beta}$, eventually constructing some maximal subgroup $K \le G$ such that the action of G on the cosets of K is not faithful and $|G:K| < n$. The action on the cosets of K is constructed, and its kernel is a solution for (6.7). The second method constructs smaller and smaller subsets containing $x_{\alpha\beta}$, eventually arriving to some subset $K \subseteq G$ that is small enough so that we can afford to take the normal closure of each element.

Computing the Socle of a Frobenius Group

As a preliminary step, both methods handle the case of Frobenius groups. We present Neumann's algorithm for this task; another algorithm is described in Section 6.2.5. Some background material on Frobenius groups can be found in Exercises 6.11 and 6.12.

Let $G \le \mathrm{Sym}(\Omega)$ be a primitive Frobenius group, and let N denote its unique minimal normal subgroup. Moreover, let $\alpha \in \Omega$ and $1 \ne z \in Z(G_\alpha)$. It is known (cf. Exercise 6.12(iii), (iv)) that such an element z exists. For any $g \in G \backslash G_\alpha$, we have $\alpha^{zz^g} = \alpha^{z^g} \ne \alpha^{z^g z}$, since the only fixed point of z is α and so, in particular, z moves the point α^{z^g}. Hence the commutator $[z, z^g] \ne 1$, but $[z, z^g] \in N$ since the coset Nz is in the center of G/N. Hence $\langle [z, z^g]^G \rangle = N$.

Given a base and SGS for G, the center of G_α, and thus a nontrivial element z in it, can be constructed by a nearly linear-time deterministic algorithm, as described in Section 6.1.3. The normal closure $\langle [z, z^g]^G \rangle$ can also be computed in nearly linear time by a deterministic algorithm.

From now on, we may suppose that the input primitive group has a normal subgroup with nontrivial centralizer, but it is not a Frobenius group. First we describe a solution for (6.7) based on Luks's approach.

Lemma 6.2.8. *Suppose that $G \le \mathrm{Sym}(\Omega)$ is primitive, $|\Omega| = n$, G has a regular normal subgroup with nontrivial centralizer, and there are two points $\alpha, \beta \in \Omega$ such that $G_{\alpha\beta} \ne 1$. If K is any maximal subgroup of G containing $N_G(G_{\alpha\beta})$ then K contains a regular normal subgroup of G and $|G : K| < n$.*

Proof. Let N be an arbitrary regular normal subgroup of G. It contains a unique element $x_{\alpha\beta}$ that maps α to β. For any $g \in G_{\alpha\beta}$, $x_{\alpha\beta}^g \in N$ and it maps α to β; hence $x_{\alpha\beta}^g = x_{\alpha\beta}$, which implies that $x_{\alpha\beta} \in C_G(G_{\alpha\beta})$ and $x_{\alpha\beta} \in N_G(G_{\alpha\beta})$.

Since K is a maximal subgroup of G, the action of G on the cosets of K is primitive. This action cannot be faithful, since G either has a regular abelian normal subgroup or two regular normal nonabelian subgroups, and in any primitive faithful action these minimal normal subgroups have to act regularly; however, in the G-action on the cosets of K, the element $x_{\alpha\beta}$ has a fixed point (the coset $K \cdot 1$).

Now, since the action of G on the cosets of K is not faithful, K contains a normal subgroup of G and so it contains a minimal normal subgroup M of G. Since K also contains $G_{\alpha\beta}$ and $M \cap G_{\alpha\beta} = 1$, we have $|G : K| \le |M||G : G_{\alpha\beta}| \le n - 1$. □

Luks's Algorithm

The algorithm for solving (6.7) is to compute $N_G(G_{\alpha\beta})$, embed it into a maximal subgroup K, construct the action of G on the cosets of K, and compute the kernel of the action. We have to show that these computations can be done in nearly linear time.

Lemma 6.2.9. *Given $G \le \mathrm{Sym}(\Omega)$ and $\alpha, \beta \in \Omega$ as in Lemma 6.2.8 and an SGS for G relative to some nonredundant base, $N_G(G_{\alpha\beta})$ can be computed in nearly linear time by a deterministic algorithm.*

Proof. Applying a base change (cf. Section 5.4), we may assume that α and β are the first two base points. Let R_1 and R_2 be the transversals for G mod G_α and G_α mod $G_{\alpha\beta}$, respectively, encoded in shallow Schreier trees (S_1, T_1) and (S_2, T_2). Let N denote a regular normal subgroup of G.

Let Δ be the fixed point set of $G_{\alpha\beta}$. For any $g \in G$, it is clear that $g \in N_G(G_{\alpha\beta})$ if and only if g fixes Δ setwise and g fixes Δ setwise if and only if $\alpha^g \in \Delta$ and

$\beta^g \in \Delta$. Moreover, for any $\gamma, \delta \in \Delta$, the unique element $x_{\gamma\delta} \in N$ moving γ to δ centralizes $G_{\alpha\beta}$ and so $x_{\gamma\delta} \in N_G(G_{\alpha\beta})$. Therefore $N_G(G_{\alpha\beta})$ acts transitively on Δ.

First, we compute Δ. This can be done in nearly linear time, using the generators for $G_{\alpha\beta}$ in the SGS for G. After that, generators for $N_G(G_{\alpha\beta})$ are computed in two steps. We initialize $H := G_{\alpha\beta}$ and the orbits $\alpha^H := \{\alpha\}$ and $\beta^H := \{\beta\}$.

In the first step, we compute $N_G(G_{\alpha\beta})_\alpha$. This is done by running through the elements of Δ. If for some $\gamma \in \Delta$ we have $\gamma \notin \beta^H$ but γ is in the fundamental orbit β^{G_α} then we multiply out the product r_γ of labels along the path from γ to β in the Schreier tree T_2, we replace H by $\langle H, r_\gamma \rangle$, and we recompute the orbit β^H. Since R_2 is encoded in a shallow Schreier tree, the computation of r_γ can be done in nearly linear time. Moreover, H is increased at most $\lceil \log(n-1) \rceil$ times, since $|N_G(G_{\alpha\beta})_\alpha : G_{\alpha\beta}| \leq n-1$.

In the second step, we embed $N_G(G_{\alpha\beta})_\alpha$ into a subgroup that acts transitively on Δ. By the discussion in the second paragraph of the proof, this subgroup will be $N_G(G_{\alpha\beta})$. Again, we run through the elements of Δ. If we encounter some $\delta \in \Delta$ that is not in α^H then we multiply out the product of labels r_δ in the Schreier tree T_1 along the path from δ to α and compute $\beta^{G_\alpha} \cap \Delta^{r_\delta}$. This intersection is nonempty, since there exist elements in G that map δ to α and fix Δ setwise (and so, in particular, map β to an element of Δ). Any such element x can be written in the form $\bar{x} r_\delta^{-1}$ for some $\bar{x} \in G_\alpha$ and then $\beta^{\bar{x}} \in \beta^{G_\alpha} \cap \Delta^{r_\delta}$. Therefore we can pick $\gamma \in \beta^{G_\alpha} \cap \Delta^{r_\delta}$, multiply out the product of labels r_γ in the Schreier tree T_2 along the path from γ to β, replace H by $\langle H, r_\gamma^{-1} r_\delta^{-1} \rangle$, and recompute the orbit α^H. Since H increases in the second step at most $\lceil \log |\Delta| \rceil$ times, the second step runs in nearly linear time as well. □

The next step in the solution of (6.7) is the embedding of $N_G(G_{\alpha\beta})$ into a maximal subgroup K of G. This part of the algorithm is of Monte Carlo type and requires $\Theta(n)$ group operations, but so many permutation multiplications are not allowed in the nearly linear-time context. Therefore, we consider the isomorphism $\psi : G \to H$ with the black-box group H consisting of the standard words, as described in Section 5.3. We embed the subgroup $\psi(N_G(G_{\alpha\beta}))$ into a maximal subgroup K^* of H, and we finally construct $K := \psi^{-1}(K^*)$. By Lemma 5.3.1, group operations in H (*including the construction of nearly uniformly distributed random elements in subgroups of* H) require only $O(\log^c |G|)$ time for an absolute constant c.

Lemma 6.2.10. *Let H be a black-box group, $L \leq H$, and suppose that we can test membership in L and that there exists a maximal subgroup of $K^* \leq H$ containing L that has index less than n. Then generators for a subgroup L^**

satisfying $L \lneqq L^* \lneqq H$ can be computed by a Monte Carlo algorithm, using $O(\log^2(\varepsilon^{-1}) n \log n)$ group operations and membership tests in L, where ε is the desired error probability.

Proof. Let $l := \lceil n \log((\varepsilon/2)^{-1}) \rceil$. We construct a sequence $(h_1, \ldots, h_l)$ of elements of H recursively. Let h_1 be a random element of $H \backslash L$ and, if h_{i-1} is already defined, let h_i be a random element of $\langle L, h_{i-1} \rangle \backslash L$. (We construct h_i as a random element of $\langle L, h_{i-1} \rangle$; if h_i happens to be in L, then we repeat the choice of h_i, up to $\lceil \log(2l/\varepsilon) \rceil$ times.)

We claim that with probability greater than $1 - \varepsilon$, the sequence $(h_1, \ldots, h_l)$ is constructed and $L \lneqq \langle L, h_l \rangle \lneqq H$. By the definition of the sequence, for each i we have $L \lneqq \langle L, h_i \rangle$. For a fixed i, if h_{i-1} is defined then the probability that a random element of $\langle L, h_{i-1} \rangle$ is not in L is at least $1/2$, so the probability that the algorithm fails to construct h_i with $\lceil \log(2l/\varepsilon) \rceil$ random selections in $\langle L, h_{i-1} \rangle$ is at most $\varepsilon/2l$. Therefore, the sequence $(h_1, \ldots, h_l)$ is constructed with probability at least $1 - \varepsilon/2$.

Suppose now that $(h_1, \ldots, h_l)$ is constructed. If $\langle L, h_{i-1} \rangle = H$ for some i then the probability that $h_i \in K^*$ is greater than $1/n$, and if $\langle L, h_i \rangle$ is a proper subgroup of H then $\langle L, h_j \rangle$ is a proper subgroup of H for all $j \geq i$. Hence the probability that $\langle L, h_l \rangle = H$ is less than $(1 - 1/n)^l < \varepsilon/2$. □

We compute a shallow Schreier tree data structure for $N_G(G_{\alpha\beta})$, compute $L := \psi(N_G(G_{\alpha\beta}))$, and apply Lemma 6.2.10 to obtain $L^* \leq H$ and the subgroup $\psi^{-1}(L^*) \leq G$ properly containing $N_G(G_{\alpha\beta})$. (Note that because we have a shallow Schreier tree data structure for $N_G(G_{\alpha\beta})$, we can test membership in L.) Then we repeat the same steps, starting with $\psi^{-1}(L^*) \leq G$. We stop when the subgroup returned by the algorithm of Lemma 6.2.10 is H (this fact is noticed when the next shallow Schreir tree data structure is computed); with high probability, the input subgroup of the last iteration is maximal in G. Since $|G : N_G(G_{\alpha\beta})| < n^2$, the number of iterations is $O(\log n)$.

The final stage of the solution of (6.7) is the construction of the action of G on the cosets of K and the computation of the kernel of this action. This can be done as described in Lemma 6.2.4.

Remark 6.2.11. A similar algorithm can be used to solve (6.7) in the case when G has a unique nonabelian regular normal subgroup. The construction of $N_G(G_{\alpha\beta})$ and its embedding into a maximal subgroup K can be done in nearly linear time, as described in Lemmas 6.2.9 and 6.2.10. However, we cannot guarantee that K contains N, and so the action of G on the cosets of K may be faithful and Lemma 6.2.4 may not be applied. We obtain a solution for

(6.7) even in the case when K does not contain N, since then $G = KN$ and $K \cap N \neq 1$, so $|G : K| \leq n/2$. However, it is not clear how we can construct the action on the cosets of K in nearly linear time. We shall see in Section 6.2.6 that this is possible if β is chosen from the smallest orbit of G_α in $\Omega\backslash\{\alpha\}$, but the proof of that uses consequences of the classification of simple groups. There is also an elementary nearly linear-time algorithm for the handling of groups with regular non-abelian socle, which we shall present in Section 6.2.4.

Neumann's Algorithm (the EARNS Subroutine)

Now we turn to the description of the **EARNS** (Elementary Abelian Regular Normal Subgroup) method. Suppose that $G \leq \mathrm{Sym}(\Omega)$ is primitive with a regular abelian normal subgroup N, $n := |\Omega| = p^d$ for some prime p, and there are two points $\alpha, \beta \in \Omega$ such that $G_{\alpha\beta} \neq 1$. We may suppose that β is from an orbit of G_α in $\Omega\backslash\{\alpha\}$ of the smallest size, since replacing any β with one from such an orbit does not decrease the size of $G_{\alpha\beta}$ and so the property $G_{\alpha\beta} \neq 1$ still holds.

As in the proof of Lemma 6.2.9, let Δ be the set of fixed points of $G_{\alpha\beta}$. For any two points $\gamma, \delta \in \Delta$, we have $G_{\alpha\beta} \leq G_{\gamma\delta}$ and so the maximality of $|G_{\alpha\beta}|$ implies $G_{\alpha\beta} = G_{\gamma\delta}$. Therefore the restriction $L := N_G(G_{\alpha\beta})|_\Delta$ is either regular or a Frobenius group. In either case, L has a regular normal subgroup R that consists of the restrictions of the elements $x_{\gamma\delta} \in N$ mapping γ to δ, for all (not necessarily different) $\gamma, \delta \in \Delta$. Moreover, all such $X_{\gamma\delta} \in N$ centralize $G_{\alpha\beta}$ (see the second paragraph of the proof of Lemma 6.2.9).

We start the algorithm by computing $M := N_G(G_{\alpha\beta})$, and then $C_G(G_{\alpha\beta}) = C_M(G_{\alpha\beta})$. As we have observed in the previous paragraph, $C_G(G_{\alpha\beta})|_\Delta$ contains R as a normal subgroup. We continue the algorithm by computing R, and the next step is the computation of the preimage D of R in $C_G(G_{\alpha\beta})$. The group D is abelian, because $D_{(\Delta)} = Z(G_{\alpha\beta})$ and so D is generated by $Z(G_{\alpha\beta})$ and by elements of N that commute with each other and with $Z(G_{\alpha\beta})$. Next, we construct the subgroup P of $Z(G_{\alpha\beta})$ consisting of the elements of $Z(G_{\alpha\beta})$ of order dividing p, and $x \in D$ of order p such that $x|_\Delta \neq 1$. Then the coset Px contains a nontrivial element of N. Therefore, we can construct N by taking the normal closure of each element of Px in G and pick the only abelian group among these normal closures.

By Lemma 6.2.9 and by Section 6.1.4, the computation of $C_G(G_{\alpha\beta})$ can be done in nearly linear time. We have seen earlier in this section that the regular normal subgroup in Frobenius groups can be computed in nearly linear time, so R can be obtained by a nearly linear-time algorithm. The group D is a preimage at a transitive constituent homomorphism, and $Z(G_{\alpha\beta})$ is the pointwise stabilizer

$D_{(\Delta)}$. Generators for P are obtained by taking the appropriate powers of the generators of $Z(G_{\alpha\beta})$, and x can be chosen as a power of an element in the transversal for D mod D_α. Hence the coset Px can be constructed in nearly linear time.

Next, we need an upper estimate for $|P|$. It is based on the following result [Neumann, 1986, Lemma 3.4], the proof of which we leave as Exercise 6.13.

Lemma 6.2.12. *Let M be any group, let A be an abelian p-subgroup of M of order q, let $C := C_M(A)$, and let $m := |M : C|$. If $O_p(M) = 1$ then $q \le m$.*

Corollary 6.2.13. $|P| < n$.

Proof. We claim that $O_p(G_\alpha) = 1$. Indeed, N can be identified with a d-dimensional vector space over GF(p), with α playing the role of the 0 vector; with this identification, G_α acts on N by conjugation as an irreducible subgroup of $\mathrm{GL}_d(p)$, and the fixed vectors of $O_p(G_\alpha)$ comprise a block of imprimitivity of G (cf. Exercise 6.14). Hence the only fixed vector of $O_p(G_\alpha)$ is the 0 vector and so $O_p(G_\alpha) = 1$.

We can apply Lemma 6.2.12 with $M := G_\alpha$ and $A := P$. Since $P \le Z(G_{\alpha\beta})$, we have $C_M(P) \ge G_{\alpha\beta}$ and so $|M : C_M(P)| \le |G_\alpha : G_{\alpha\beta})| \le n - 1$. Hence $|P| \le |M : C_M(P)| < n$. □

By the corollary, $|Px| < n$. In a permutation group, we can compute the normal closure of an element in nearly linear time; however, we cannot do $\Theta(n)$ such computations within the nearly linear time constraint. Therefore, as in Luks's approach, we consider the isomorphism $\psi : G \to H$ with the black-box group $H = \langle S \rangle$ consisting of the standard words. By Theorem 2.3.9 and Lemma 2.3.14, for any $h \in H$ the normal closure $\langle h^H \rangle$ can be computed, and then the commutativeness of $\langle h^H \rangle$ can be tested, using $O(\log |H|(|S| + \log |H|))$ group operations. By Lemma 5.3.1, group operations in H can be performed in $\log^c |G|$ time, for an absolute constant c. Hence we can compute the normal closure of all standard words representing elements of Px in nearly linear total time by a Monte Carlo algorithm. When an abelian normal subgroup of H is found, the image of its generators under ψ^{-1} can be computed in nearly linear time by a deterministic algorithm.

Remark 6.2.14. This method can also be extended to solve (6.7) for primitive groups with one or two nonabelian regular normal subgroups. Let N denote a regular normal subgroup of G. Using the notation of Neumann's algorithm, our observation that R consists of the restrictions of the elements of N to Δ is still valid, and R can be constructed in nearly linear time. For an element

$x \in D$ of prime order r satisfying $x|_\Delta \neq 1$ and for the subgroup P of $Z(G_{\alpha\beta})$ consisting of elements of order dividing r, it is still true that Px contains a nontrivial element of N. Corollary 6.2.13 is also true, since G_α has no nontrivial solvable normal subgroups (cf. Exercise 6.15 for the case of two regular normal subgroups and [Dixon and Mortimer, 1996, Theorem 4.7B(i)] for the case of a unique nonabelian regular normal subgroup). Hence we can compute the normal closures in H of all standard words representing elements of Px in nearly linear total time. However, we cannot decide which normal closure corresponds to a regular subgroup of G within the nearly linear time constraint.

6.2.4. Groups with a Unique Nonabelian Minimal Normal Subgroup

Given a primitive group $G \leq \mathrm{Sym}(\Omega)$, with $|\Omega| = n$, if the algorithm of Section 6.2.3 fails to solve the problem (6.7) then, with high probability, G has a unique nonabelian minimal normal subgroup. In this section we solve (6.7) for such groups. The solution again originated in [Luks, 1987] and was sped up to nearly linear time in [Babai et al., 1993] and [Beals and Seress, 1992], but in this case the bulk of the work was done in [Babai et al., 1993]. Groups in cases B, C(i), C(ii), and C(iii) of Theorem 6.2.7 are handled by different methods, but in each subcase of case C, a common feature is that we embed a two-point stabilizer $G_{\alpha\beta}$ into a maximal subgroup $K < G$ and compute the primitive permutation representation of G on the cosets of K.

The first step is to check whether G belongs to case B. If G has a proper normal subgroup of index less than n then, for a uniformly distributed random $g \in G\backslash\{1\}$, the normal closure $\langle g^G \rangle$ is a solution of (6.7) with probability greater than $1/n$. Given an SGS for G, a random element and its normal closure can be computed in nearly linear time; however, we have to repeat this process $\Theta(n)$ times, which is not possible in the nearly linear time frame. Hence, once again, we consider the isomorphism $\psi : G \to H$ with the black-box group H consisting of the standard words representing the elements of G.

Lemma 6.2.15. *Let $H = \langle S \rangle$ be a black-box group, and suppose that there exists a normal subgroup of H of index less than n. Then generators for a proper normal subgroup of H can be computed by a Monte Carlo algorithm, using $O(\log(\varepsilon^{-1})n(|S| + \log|H| + \log n + \log(\varepsilon^{-1}))^2)$ group operations, where ε is the desired error probability. (As in Lemma* 6.2.10, *we consider constructing a nearly uniformly distributed random element in a subgroup of H as one group operation.)*

Proof. Let $l := \lceil n \log((\varepsilon/3)^{-1}) \rceil$. We construct a sequence $(h_1, \ldots, h_l)$ of elements of H recursively. Let h_1 be a random element of $H \backslash \{1\}$ and, if h_{i-1} is already defined, let h_i be a random element of $\langle h_{i-1}^H \rangle \backslash \{1\}$. (We construct h_i as a random element of $\langle h_{i-1}^H \rangle$; if h_i happens to be 1, then we repeat the choice of h_i, up to $\lceil \log(3l/\varepsilon) \rceil$ times.)

We claim that with probability greater than $1 - \varepsilon$, the sequence $(h_1, \ldots, h_l)$ is constructed and $1 \lneqq \langle h_l^H \rangle \lneqq H$. By the definition of the sequence, for each i we have $1 \lneqq \langle h_i^H \rangle$. For a fixed i, if h_{i-1} is defined then the probability that a random element of $\langle h_{i-1}^H \rangle$ is not 1 is at least 1/2, so the probability that the algorithm fails to construct h_i with $\lceil \log(3l/\varepsilon) \rceil$ random selections in $\langle h_{i-1}^H \rangle$ is at most $\varepsilon/3l$. Therefore, the sequence $(h_1, \ldots, h_l)$ is constructed with probability at least $1 - \varepsilon/3$.

Suppose now that $(h_1, \ldots, h_l)$ is constructed. If $\langle h_{i-1}^H \rangle = H$ for some i then the probability that h_i is in a proper normal subgroup is greater than $1/n$, and if $\langle h_i^H \rangle$ is a proper subgroup of H then $\langle h_j^H \rangle$ is a proper subgroup of H for all $j \geq i$. Hence the probability that $\langle h_l^H \rangle = H$ is less than $(1 - 1/n)^l < \varepsilon/3$.

The analysis in the previous paragraph assumed that the normal closure computations give the correct result. If we ensure that each of the l normal closure computations are correct with probability at least $1 - \varepsilon/3l$ then the entire algorithm succeeds with probability greater than $1 - \varepsilon$.

The normal closures are computed by the algorithm of Theorem 2.3.9. Since for each i we want the computation of $\langle h_i^H \rangle$ to succeed with probability at least $1 - \varepsilon/3l$, we need to construct $O(\log |H| + \log n + \log 1/\varepsilon)$ generators for $\langle h_i^H \rangle$, at a cost of $O((\log |H| + \log n + \log 1/\varepsilon))(|S| + \log |H| + \log n + \log 1/\varepsilon)$ group operations (see also Remark 2.3.5 for the justification of these quantities and Exercise 6.16). □

The Algorithm for Case B of Theorem 6.2.7

Therefore, to check whether G belongs to case B, we compute generators for a normal subgroup L of $H := \psi(G)$ by the algorithm of Lemma 6.2.15. After that, we compute an SGS for $\psi^{-1}(L) \leq G$ in nearly linear time by a deterministic algorithm. If $\psi^{-1}(L) \neq G$ then we have a solution for (6.7); otherwise, with high probability, G has no proper normal subgroup of index less than n.

The treatment of primitive groups belonging to case C is deterministic. In each subcase of C, we need information about the orbits of G_α. The elements of Ω correspond to the cosets of G_α, and the action of G_α on Ω is permutation isomorphic to its action on these cosets by multiplication. Furthermore, an elementary but very important observation is that the action of G_α on the cosets by multiplication is permutation isomorphic to its action on the cosets by conjugation, since for $g \in G_\alpha$ and $h \in G$, we have $(G_\alpha h)g = (G_\alpha h)^g = G_\alpha h^g$. Moreover, since $N = \mathrm{Soc}(G)$ is transitive on Ω, the elements of Ω can be also identified with the cosets of N_α in N. Hence, from each G_α-coset we can choose a representative $h \in N$; the conjugation action of G_α carries $G_\alpha h$ to a coset with representative $h^g \in N$.

First, we consider the subcases C(i) and C(ii). In these subcases, $N_\alpha = (N_1)_\alpha \times \cdots \times (N_m)_\alpha$. Any $\beta \in \Omega$ corresponds to a sequence of cosets $((N_1)_\alpha h_1, \ldots, (N_m)_\alpha h_m)$ with $h_i \in N_i$ for $i \in [1, m]$. We call the sequence $(h_1, \ldots, h_m)$ the coordinates of β; a coordinate is called nontrivial if $h_i \notin (N_i)_\alpha$. Note that the number of nontrivial coordinates is independent of the choice of the coordinate sequence for β and, since G_α permutes the N_i and the $(N_i)_\alpha$ by conjugation, for any $g \in G_\alpha$ the number of nontrivial coordinates in β and in β^g is the same.

Lemma 6.2.16. *Let $G \le \mathrm{Sym}(\Omega)$ be primitive, with $|\Omega| = n$, and suppose that G belongs to case* C(i) *or* C(ii) *of Theorem* 6.2.7. *Then for any $\alpha \in \Omega$, the smallest orbits of G_α in $\Omega \backslash \{\alpha\}$ have size less than $\log^2 n$. Moreover, if G belongs to case* C(ii) *then points in these smallest orbits have only one nontrivial coordinate and even the sum of the sizes of the smallest orbits is less than* $\log^2 n$.

Proof. Since $m!/2 \ge n$ and $(\log n / \log\log n)^{\log n/\log\log n} < n$, we have $m > \log n / \log\log n$. Therefore, since $n = |N_1 : (N_1)_\alpha|^m$, we must have $|N_1 : (N_1)_\alpha| < \log n$. Hence the number of points with one nontrivial coordinate is at most $m|N_1 : (N_1)_\alpha| < \log^2 n$. So there are orbits of G_α in $\Omega \backslash \{\alpha\}$, namely the orbits containing all points with exactly one nontrivial coordinate, whose union has size less than $\log^2 n$.

Hence, to finish the proof, it is enough to show that, in case C(ii), the smallest orbits in $\Omega \backslash \{\alpha\}$ consist of points with one nontrivial coordinate. Let $K \lhd G_\alpha$ be the kernel of the conjugation action of G_α on the set $\{N_1, \ldots, N_m\}$. Note that $N_\alpha \le K$, $G_\alpha/K \cong A_m$, and K permutes the cosets of $(N_1)_\alpha$ in N_1. Let q denote the size of the smallest K-orbit L among the cosets of $(N_1)_\alpha$, excluding $(N_1)_\alpha$ itself. Then, taking $\beta \in \Omega$ with the first coordinate in L and the other $m-1$ coordinates trivial, we have $|\beta^{G_\alpha}| = mq$.

We claim that for any $\gamma \in \Omega$ with k nontrivial coordinates, $|\gamma^{G_\alpha}| \ge \binom{m}{k} q 2^{k-1}$, which is greater than mq if $k \ge 2$. Indeed, if $(h_1, \ldots, h_m)$ is a coordinate sequence for γ and the jth coordinate is nontrivial, then $(N_j)_\alpha h_j$ is not fixed by the conjugation action of $(N_j)_\alpha \le K$ (otherwise $(1, \ldots, 1, h_j, 1, \ldots, 1)$ is a fixed point of N_α, contradicting the primitivity of G by Lemma 6.1.10(i)). Thus $|\gamma^K| \ge q2^{k-1}$, since γ has at least q images differing pairwise in the first nontrivial coordinate of γ, and for each of these, the action of $(N_j)_\alpha$ in the other nontrivial coordinates j provides at least 2^{k-1} images. Also, since $G_\alpha/K \cong A_m$, the k nontrivial coordinates can be conjugated into any k positions, accounting for the $\binom{m}{k}$ term. □

Lemma 6.2.17. *Let $G = \langle S \rangle \le \mathrm{Sym}(\Omega)$ be transitive, with $\Omega = [1, n]$, and suppose that for some $\alpha, \beta \in \Omega$, $|G_\alpha : G_{\alpha\beta}| = k$. Then the action of any $g \in G$ on the cosets of $G_{\alpha\beta}$ can be computed in $O(|S|nk^2)$ time by a deterministic algorithm, using $O(nk)$ memory.*

Proof. The action of G on the cosets of $G_{\alpha\beta}$ is permutation isomorphic to the action of G on the orbit $L := (\alpha, \beta)^G$ of the sequence (α, β). As preprocessing, we compute this orbit, combining the approach of Remark 2.1.2 and Theorem 2.1.1(ii). The elements of L are stored in an $n \times k$ matrix M, where the rows are numbered by the elements of $\Omega = [1, n]$. In row γ, we store the elements of L with first coordinate equal to γ. (Formally, the hash function $h : L \to [1, n]$ is defined by the rule $h : (\gamma, \delta) \mapsto \gamma$.) When the orbit algorithm has to decide whether a newly computed image (γ, δ) already occurs in L, it is enough to compare (γ, δ) with the pairs in row γ of M. If (γ, δ) is new, we store it at the first empty slot in row γ. This computation requires $nk|S|$ image computations of pairs and $O(nk^2)$ comparison of pairs.

The elements of L can be numbered $1, 2, \ldots, nk$, according to their positions in M. For any $g \in G$ and $i \in [1, nk]$, the image i^g is computed by looking up the pair (γ, δ) in position i in M and searching for the position of (γ^g, δ^g) in row γ^g of M. This requires $O(k)$ time. □

The Algorithm for Case C(i)

Now we are ready to describe the solution of (6.7) in the case C(i). Given $G \le \mathrm{Sym}(\Omega)$ belonging to this case, together with an SGS for G, we pick the first base point as α and pick β from a smallest orbit of G_α in $\Omega \backslash \{\alpha\}$. We compute $N_G(G_{\alpha\beta})$ as described in Lemma 6.2.9. By Lemmas 6.2.16 and 6.2.17, the action of G on the cosets of $G_{\alpha\beta}$ (or, more precisely, the permutation isomorphic action on $(\alpha, \beta)^G$) can be computed in nearly linear time by a deterministic algorithm. The set $(\alpha, \beta)^{N_G(G_{\alpha\beta})}$ is a block of imprimitivity in this action and, by the algorithm of Section 5.5, it can be embedded in a maximal block Δ. The action on the block system Σ consisting of the images of Δ can be computed easily and, by Remark 6.2.11, either $|\Sigma| \le n/2$ or the kernel of this action is a nontrivial normal subgroup of G.

The case C(ii) requires one more preparatory step.

Lemma 6.2.18. *Let $G \le \mathrm{Sym}(\Omega)$, with $|\Omega| = n$, belong to case* C(ii)*, let $\alpha \in \Omega$, and let β be in a smallest orbit of G_α in $\Omega\backslash\{\alpha\}$. Then there exist $\gamma, \delta \in \Omega$ such that*

(i) *γ is in the same orbit of G_α as β.*
(ii) *δ is in a smallest orbit of G_γ in $\Omega\backslash\{\gamma\}$.*
(iii) *The subgroup $H := \langle G_{\alpha\beta}, G_{\gamma\delta}\rangle$ is not equal to G, and the action of G on the cosets of any maximal subgroup K containing H is not faithful or $|G : K| \le n/2$.*

Proof. By Lemma 6.2.16, β has exactly one nontrivial coordinate; we may suppose that it is the first, and so $\beta = \alpha^t$ for some $t \in N_1\backslash(N_1)_\alpha$ and β has the coordinate sequence $(t, 1, \ldots, 1)$. Let γ be chosen such that the nontrivial coordinate of γ is the second one. Then $\gamma = \alpha^u$ for some $u \in N_2\backslash(N_2)_\alpha$ and γ has the coordinate sequence $(1, u, 1, \ldots, 1)$. Finally, let $\delta := \gamma^t = \beta^u$. Then δ has the coordinate sequence $(t, u, 1, \ldots, 1)$ and, since δ is defined as an image of γ under only one of the N_j, it is in a smallest orbit of G_γ in $\Omega\backslash\{\gamma\}$.

We claim that α, β, γ, and δ satisfy (iii). We first show that $H = \langle G_{\alpha\beta}, G_{\gamma\delta}\rangle$ normalizes N_1, and so it must be a proper subgroup of G. Let $g \in G_{\alpha\beta}$, and suppose that $N_1^g = N_j$ for some $j \in [1, m]$. Then $t^g t^{-1} \in N_j N_1$. If $j \ne 1$ then $\alpha^{t^g t^{-1}} = \alpha$ would have a nontrivial first coordinate since $t \notin (N_1)_\alpha$, which is a contradiction. The same argument can be repeated for $g \in G_{\gamma\delta}$, in the coordinate system according to the cosets of N_γ, since $N_\gamma = N_\alpha^u$ for $u \in N_2$, which centralizes N_1, and so $(N_1)_\gamma = (N_1)_\alpha^u = (N_1)_\alpha$ and $t \notin (N_1)_\gamma$.

Let K be a maximal subgroup containing H, and suppose that the action of G on the cosets of K is faithful. We claim that $|G : K| \le n/2^m$. Note first that, similarly to the argument in the previous paragraph, $(N_2)_\beta = (N_2)_\alpha^t = (N_2)_\alpha$ and $(N_2)_\delta = (N_2)_\gamma^t = (N_2)_\gamma$ since t centralizes N_2. Therefore $(N_2)_\alpha \le G_{\alpha\beta}$ and $(N_2)_\gamma \le G_{\gamma\delta}$, and so $\langle (N_2)_\alpha, (N_2)_\gamma\rangle \le K$. This means that $K \cap N_2 \ne 1$ and the action of G on the cosets of K falls in case C(ii), implying $|G : K| = |N_2 : (N_2 \cap K)|^m$ (since $N_2 \cap K$ is the stabilizer of the "point" K in N_2, in the action on the cosets of K). We have $(N_2)_\gamma = (N_2)_\alpha^u \ne (N_2)_\alpha$, because otherwise $(N_j)_\gamma = (N_j)_\alpha$ for all $j \in [1, m]$, contradicting Lemma 6.1.10(i). Thus $|N_2 : (N_2 \cap K)| < |N_2 : (N_2)_\alpha|$ and $|G : K| = |N_2 : (N_2 \cap K)|^m \le (|N_2 : (N_2)_\alpha|/2)^m = n/2^m$, as claimed. □

The Algorithm for Case C(ii)

If the input group G belongs to case C(ii) then we can solve (6.7) by fixing $\alpha \in \Omega$, fixing β in a smallest orbit Δ of G_α in $\Omega\backslash\{\alpha\}$, and testing for all $\gamma \in \Delta$ and all δ in the smallest orbits of G_γ in $\Omega\backslash\{\gamma\}$ whether or not $H := \langle G_{\alpha\beta}, G_{\gamma\delta}\rangle$

is equal to G. When a pair (γ, δ) is found with $H \neq G$, we construct the action of G on the cosets of a maximal subgroup of G containing H.

By Lemma 6.2.16, the number of pairs (γ, δ) to be tested is less than $\log^4 n$. Note that we can test whether $H = G$ without computing an SGS for H by constructing the action of the generators of H on the set $L := (\alpha, \beta)^G$. By Lemma 6.2.17, this can be done by a nearly linear-time deterministic algorithm, and $H = G$ if and only if H acts transitively on L. Similarly, we do not need to embed the group $H \neq G$ we have found into a maximal subgroup; instead, it is enough to compute a maximal block Δ in L containing $(\alpha, \beta)^H$ and the action of G on the block system consisting of the images of Δ.

The last subcase we have to handle is C(iii). In this subcase, using the notation of Theorem 6.2.7, $N_\alpha = \mathrm{Diag}(N_1 \times \cdots \times N_m)$ defines an identification among the terms N_i of the direct product N, and the points of Ω correspond to equivalence classes of the sequences $(t_1, \ldots, t_m)$, with $t_i \in T$. Two sequences $(t_1, \ldots, t_m)$ and $(s_1, \ldots, s_m)$ are equivalent if and only if there exists some $t \in T$ such that $t_i = ts_i$ holds for all $i \in [1, m]$. Since $m!/2 \geq n$ and $n = |T|^{m-1}$, we have $|T| < \log n$ and so each equivalence class has less than $\log n$ elements.

Lemma 6.2.19. *Let $G \leq \mathrm{Sym}(\Omega)$ be primitive, with $|\Omega| = n$, and suppose that G belongs to case* C(iii) *of Theorem* 6.2.7. *Then, for any $\alpha \in \Omega$, the cardinality of the union of orbits of G_α of size less than $\log^2 n$ is $O(\log^6 n)$.*

Proof. Since the conclusion of the lemma is an asymptotic statement, we can suppose that n is large enough so that $\binom{m}{4} \geq \log^3 n$. We shall prove that, for such n, there are less than $\log^6 n$ points in Ω in G_α-orbits of size less than $\log^2 n$.

Again, we use the notation of Theorem 6.2.7. For $g \in N$, let $E(g)$ denote the equivalence class of g and let $\beta_g \in \Omega$ denote the element of the permutation domain corresponding to $E(g)$. If some $g \in N$ has at least $\log^3 n$ conjugates under G_α then $|\beta_g^{G_\alpha}| \geq \log^2 n$ since the conjugates of g belong to at least $\log^2 n$ equivalence classes. Hence it is enough to estimate the number of equivalence classes that contain elements of N with less than $\log^3 n$ conjugates under G_α.

We assign a nondecreasing sequence $S(g) = (m_1, \ldots, m_k)$ of positive integers to each $g = (t_1, \ldots, t_m) \in N$. The length k of $S(g)$ is the number of elements of T occurring in the sequence g, and the numbers m_i are the frequencies of how often these elements occur. It is clear from the definition of $S(g)$ that if $h \in E(g)$ then $S(h) = S(g)$. Moreover, by Lemma 6.2.6, $G_\alpha = H \times P$, where $\mathrm{Inn}(T) \lesssim H \lesssim \mathrm{Aut}(T)$ and $P \cong A_m$. Hence if $g, h \in N$ are conjugates under G_α then $S(g) = S(h)$.

Suppose that for some $g = (t_1, \ldots, t_m) \in N$ there are indices $i_1, \ldots, i_l$ such that $4 \leq m_{i_1} + \cdots + m_{i_l} \leq m - 4$ for the numbers m_i in $S(g)$. Then g has at least $\binom{m}{4} \geq \log^3 n$ conjugates under G_α, since the subgroup $P \leq G_\alpha$ can move the set of coordinates of g where the elements with frequency $m_{i_1}, \ldots, m_{i_l}$ occur to any set of $m_{i_1} + \cdots + m_{i_l}$ positions, and $\binom{m}{m_{i_1}+\cdots+m_{i_l}} \geq \binom{m}{4}$. The only way that no such indices $i_1, \ldots, i_l$ exist is that some element of T occurs in the sequence $g = (t_1, \ldots, t_m)$ at least $m - 3$ times. Therefore, the elements $g \in N$ with less than $\log^3 n$ conjugates must have $S(g) = (1, 1, 1, m - 3), (1, 2, m - 3), (3, m - 3), (1, 1, m - 2), (2, m - 2), (1, m - 1)$, or (m). The number of elements with these $S(g)$-values is $\binom{m}{3}|T|(|T|-1)(|T|-2)(|T|-3) + \binom{m}{3}\binom{3}{2}|T|(|T| - 1)(|T| - 2) + \binom{m}{3}|T|(|T| - 1) + \binom{m}{2}|T|(|T| - 1)(|T| - 2) + \binom{m}{2}|T|(|T| - 1) + \binom{m}{1}|T|(|T| - 1) + \binom{m}{0}|T| < |T| \log^6 n$ and the set of these elements is the union of equivalence classes, so there are less than $\log^6 n$ equivalence classes with corresponding orbits in Ω of size less than $\log^2 n$. □

Lemma 6.2.20. *Let $G \leq \mathrm{Sym}(\Omega)$, with $|\Omega| = n$, belong to case* C(iii), *and let $\alpha \in \Omega$. Then there exists $\beta \in \Omega$ such that*

(i) *β is in an orbit of G_α in $\Omega \backslash \{\alpha\}$ of size less than* $\log^2 n$.
(ii) *The action of G on the cosets of any maximal subgroup K containing the setwise stabilizer $G_{\{\alpha,\beta\}}$ is not faithful or $|G : K| \leq n/2$.*

Proof. As usual, we use the notation of Theorem 6.2.7. Let $t \in T$ be an involution (note that t exists since, by the Feit–Thompson theorem, every finite nonabelian simple group has even order). Let $g := (t, 1, \ldots, 1) \in N$, and let $\beta := \alpha^t$. Then β satisfies (i) since g has at most $m|T| < \log^2 n$ conjugates under G_α.

Let K be a maximal subgroup K containing the setwise stabilizer $G_{\{\alpha,\beta\}}$, and suppose that the action of G on the cosets of K is faithful. We have $\beta^g = \alpha$ since g is an involution, and so $g \in K$. This means that $K \cap N_1 \neq 1$ and therefore the primitive action on the cosets of K belongs to case C(ii) of Theorem 6.2.7. Hence $|G : K| = |N_1 : (N_1 \cap K)|^m \leq |T|^m/2^m \leq n|T|/2^m$. Since $m!/2 \geq n$, we have $|T| = n^{1/(m-1)} \leq (m!/2)^{1/(m-1)} \leq 2^{m-1}$ and $|G : K| \leq n/2$. □

The Algorithm for Case C(iii)

In case C(iii) the solution of (6.7) is to fix $\alpha \in \Omega$, and for all $\beta \in \Omega$ in orbits of G_α of size less than $\log^2 n$, construct the action of G on the cosets of a maximal subgroup containing $G_{\{\alpha,\beta\}}$, and compute the kernel of the action.

By Lemma 6.2.19, the number of points β to be processed is $O(\log^6 n)$. For a fixed β, we construct the action of G on the set $L := (\alpha, \beta)^G$. By Lemma 6.2.17, this can be done in nearly linear time by a deterministic algorithm. If $(\beta, \alpha) \notin L$ then we discard this β since it cannot be the one described in the proof of

Lemma 6.2.20. If $(\beta, \alpha) \in L$ then we construct a maximal block $\Delta \subseteq L$ containing (α, β) and (β, α), and the action of G on the block system consisting of the images of Δ. This action is permutation isomorphic to the action of G on the cosets of a maximal subgroup containing $G_{\{\alpha,\beta\}}$.

If none of the algorithms described in Sections 6.2.3 and 6.2.4 find a solution for (6.7) then, with high probability, the input primitive group is simple.

6.2.5. Implementation

Composition series computations are implemented in both GAP and MAGMA. Reduction to the primitive case proceeds almost as described in Section 6.2.1; however, primitive groups are handled somewhat differently, based on ideas from [Kantor, 1991]. We give some details about the GAP implementation; the version in MAGMA, as described in [Cannon and Holt, 1997], is similar. The current GAP implementation works for inputs of degree up to 10^6, whereas the MAGMA implementation works up to degree 10^7.

The GAP implementation starts with checking whether the input group is solvable. This can be done very efficiently by an SGS construction for solvable groups from [Sims, 1990]. We shall describe this algorithm in Section 7.1. If the input group G is solvable then the output is not only an SGS but a composition series as well; if G is not solvable then the SGS construction reports failure. For nonsolvable inputs, we apply one more reduction step, the computation of derived subgroups, besides the transitive constituent and block homomorphisms indicated in Section 6.2.1. This eventually reduces the composition series problem to one of finding a composition series in primitive perfect groups.

Kantor's idea is that there are categories of the O'Nan–Scott theorem such that no primitive perfect group $G \leq \mathrm{Sym}(\Omega)$, with $|\Omega| = n$, can belong to these categories if n is in the practical range $n \leq 10^7$. Moreover, just by looking at n and $|G|$, for most inputs G we can determine which category of the O'Nan–Scott theorem G belongs to. Therefore, we do not have to try all algorithms designed for the different categories. In particular, simple groups can be often recognized without any further work.

Lemma 6.2.21. *Let $G \leq \mathrm{Sym}(\Omega)$, with $|\Omega| = n$, be a primitive perfect group, and let $n \leq 10^7$. Then one of the following holds:*

(1) *$n = p^d$ for some prime p, G has a regular normal abelian subgroup, and $|G|$ divides $p^d|\mathrm{GL}_d(p)|$;*
(2) *n is the order of a finite simple group, $|G| = n^2$, and G is the direct product of two isomorphic single groups each acting regularly;*

(3) $n = n_1^r$ *for some values* n_1, r *occurring in the first and second table in* (6.8), *respectively; morever,* $|G| = ts^r u$ *for some value* s *occurring in line* n_1 *of the first table and for some value* u *occurring in line* r *of the second table,* $t \mid 12^r$, *and* G *falls in case* II(ii) *of Theorem* 6.2.5; *or*

(4) G *is simple.*

n_1	s
5	$\lvert A_5\rvert$
6	$\lvert A_5\rvert; \lvert A_6\rvert$
7	$\lvert \mathrm{PSL}_2(7)\rvert; \lvert A_7\rvert$
8	$\lvert \mathrm{PSL}_2(7)\rvert; \lvert A_8\rvert$
9	$\lvert \mathrm{PSL}_2(8)\rvert; \lvert A_9\rvert$
10	$\lvert A_5\rvert; \lvert A_6\rvert; \lvert A_{10}\rvert$
11	$\lvert \mathrm{PSL}_2(11)\rvert; \lvert M_{11}\rvert; \lvert A_{11}\rvert$
12	$\lvert \mathrm{PSL}_2(11)\rvert; \lvert M_{11}\rvert; \lvert M_{12}\rvert; \lvert A_{12}\rvert$
13	$\lvert \mathrm{PSL}_3(3)\rvert; \lvert A_{13}\rvert$
14	$\lvert \mathrm{PSL}_2(13)\rvert; \lvert A_{14}\rvert$
15	$\lvert A_6\rvert; \lvert A_7\rvert; \lvert A_8\rvert; \lvert A_{15}\rvert$
16	$\lvert A_{16}\rvert$
17	$\lvert \mathrm{PSL}_2(16)\rvert; \lvert A_{17}\rvert$
18	$\lvert \mathrm{PSL}_2(17)\rvert; \lvert A_{18}\rvert$
19	$\lvert A_{19}\rvert$
20	$\lvert \mathrm{PSL}_2(19)\rvert; \lvert A_{20}\rvert$
21	$\lvert A_7\rvert; \lvert \mathrm{PSL}_2(7)\rvert; \lvert \mathrm{PSL}_3(4)\rvert; \lvert A_{21}\rvert$
22	$\lvert M_{22}\rvert; \lvert A_{22}\rvert$
23	$\lvert M_{23}\rvert; \lvert A_{23}\rvert$
24	$\lvert \mathrm{PSL}_2(23)\rvert; \lvert M_{24}\rvert; \lvert A_{24}\rvert$
25	$\lvert A_{25}\rvert$

r	u
5	$\lvert A_5\rvert$
6	$\lvert A_5\rvert; \lvert A_6\rvert$
7	$\lvert \mathrm{PSL}_2(7)\rvert; \lvert A_7\rvert$
8	$\lvert \mathrm{PSL}_2(7)\rvert; \lvert \mathrm{AGL}_3(2)\rvert; \lvert A_8\rvert$
9	$\lvert \mathrm{PSL}_2(8)\rvert; \lvert A_9\rvert$
10	$\lvert A_5\rvert; 2^4 \cdot \lvert A_5\rvert; \lvert A_6\rvert; \lvert A_{10}\rvert$

(6.8)

Proof. We go through the cases of Theorem 6.2.5 and use the notation of that theorem. If G belongs to case I(i) then it belongs to case (1) of this lemma.

If G belongs to case I(ii) then the minimal normal subgroups N_1, N_2 of G are isomorphic to T^r for some simple group T (cf. Exercise 6.7) and $n = |T|^r$. The conjugation action of G permutes the r simple groups in N_1 and N_2, and so G has a factor group isomorphic to a subgroup $F \le S_r \times S_r$. We must have $r \le 3$ since $|T| \ge 60$ and $60^4 > 10^7$. Since G is perfect, $F = 1$, and so $G/\mathrm{Soc}(G)$ is isomorphic to a subgroup of $\mathrm{Out}(T)^{2r}$. By Schreier's hypothesis, $\mathrm{Out}(T)^{2r}$ is solvable, so we must have $G = \mathrm{Soc}(G) = N_1 \times N_2$. Our last observation is that $r = 1$ because otherwise, identifying Ω with $N_1 = T_1 \times \cdots \times T_r$, we have that the set 1^{T_1} is a nontrivial block of imprimitivity for G (cf. Exercise 6.7 again). Hence we are in case (2).

Case II(i) cannot occur since the perfectness of G implies $r \geq 5$ and $n \geq 60^5$ (in fact, the smallest degree of a primitive group belonging to case II(i) is 60^6, cf. [Dixon and Mortimer, 1996, Theorem 4.7B(iv)]).

If G belongs to case II(ii) and $r = 1$ then, by Schreier's hypothesis, G belongs to case (4) of this lemma. If $r > 1$ then the transitive permutation action of G on $\{T_1, \ldots, T_r\}$ is isomorphic to a perfect subgroup of S_r of order u, and $n = n_1^r$ for some integer n_1. Moreover, by Exercise 6.8, there is a *primitive* permutation representation of degree n_1 of a subgroup of $\mathrm{Aut}(T)$ containing $\mathrm{Inn}(T)$. Note the subtle point that the permutation representation of degree n_1 of $T \cong \mathrm{Inn}(T)$ may not be primitive; the smallest example, in fact the only one in our range $n \leq 10^7$, occurs when $n_1 = 21$ and $r = 5$, and $T = \mathrm{PSL}_2(7)$. We are in case (3), and the possibilities for n_1, r, u, and $s := |T|$ are listed in Table (6.8). The group orders s, u are given in a form that indicates the structure of the groups. The Mathieu groups are denoted by M_i ($i \in \{11, 12, 22, 23, 24\}$). The factor t in the order of G in this case is the contribution of the outer automorphisms of T.

Finally, we show that G cannot belong to case II(iii). Suppose the contrary. Then we must have $l = 1$ or $l \geq 5$ since G is perfect. Here $l \geq 5$ cannot occur because $n \leq 10^7$. If $l = 1$ then $k \geq 5$, using again that G is perfect; hence $n \geq 60^4 > 10^7$, a contradiction. □

Both algorithms described in Section 6.2.3 for the handling of case (1) are implemented in GAP. Neumann's method runs somewhat faster in practice, so it is the current default technique. The algorithm for Frobenius groups differs from the one given in Section 6.2.3. By Exercise 6.12(v), if $G \leq \mathrm{Sym}(\Omega)$, with $|\Omega| = p^d$, is a perfect Frobenius group then $G/\mathrm{Soc}(G) \cong \mathrm{SL}_2(5)$. Hence, taking random elements of G, we have a $1/120$ chance of finding an element of the socle. Note that it is very easy to test whether some $g \in G$ is in $\mathrm{Soc}(G)$: This happens if and only if $g^p = 1$. Alternatively, $g \in \mathrm{Soc}(G)$ if and only if $g = 1$ or g fixes no points of Ω.

Groups belonging to case (2) are handled by the extension of Neumann's method, described in Remark 6.2.14.

For groups belonging to case (3), the algorithm is different from the one given in Section 6.2.4. Let $\alpha \in \Omega$ and let K denote the kernel of the permutation action of G on $\{T_1, \ldots, T_r\}$. We compute the smallest orbit Δ of G_α in $\Omega\backslash\{\alpha\}$. It can be checked that in each group in case (3), using the terminology of Lemma 6.2.16, Δ consists of points with exactly one nontrivial coordinate. Next, we compute the (unique) minimal block system Σ for $G_\alpha|_\Delta$ and the kernel M of the action of G_α on Σ. With one exception, points in Δ belong to the same block of Σ if and only if their nontrivial coordinate is in the same position, and so $M = K_\alpha$. The exception is when $n_1 = 5, r = 10$, and $u = 2^4 \cdot |A_5|$ in Table (6.8), in which case the nontrivial coordinates can be in two positions for points in the same block, and M is the extension of K_α by an elementary abelian group of order 16. Therefore, we can obtain $\mathrm{Soc}(G)$ as $\langle M^G\rangle'$ or $\langle M^G\rangle''$ (we may need the second derived subgroup in the case $n_1 = 21, s = |\mathrm{PSL}_3(4)|$ as well).

6.2.6. An Elementary Version

The purpose of this section is twofold. First, we describe the nearly linear-time Monte Carlo composition series algorithm from [Beals, 1993b], which does not use consequences of the classification of finite simple groups. Then we describe how one can upgrade Monte Carlo composition series constructions to Las Vegas type, provided that a base and SGS, or the order, is given for the input group.

Beals's method solves the following basic problem:

$$\text{Given a group } G\text{, find generators for a simple subnormal subgroup } T \text{ of } G. \qquad (6.9)$$

If $1 \lneqq T \lneqq G$ and $T \lhd\!\lhd G$ then of course $\langle T^G\rangle$ is a proper normal subgroup of G. Hence, if we can solve (6.9) in nearly linear time then, by Lemma 6.2.2, we can also compute a composition series in nearly linear time.

The reduction to the primitive case proceeds as described in Section 6.2.1. If $G \le \mathrm{Sym}(\Omega)$ is not transitive or transitive but not primitive then either we obtain a faithful permutation representation on at most $|\Omega|/2$ points or we can replace G by the kernel N of a transitive constituent or block homomorphism, since a simple subnormal subgroup of N is subnormal in G as well. The handling of primitive groups with no regular normal subgroups is based on the following result, which is an extension of the well-known lemma of Iwasawa used in the textbook proofs of the simplicity of the groups $\mathrm{PSL}_d(q)$ (cf. [Huppert, 1967, Hilfssatz 6.12]).

Lemma 6.2.22. *Let $G \le \mathrm{Sym}(\Omega)$ be primitive and $\alpha \in \Omega$. Suppose further that $T \lhd\!\lhd G_\alpha$ and T is simple. Let $M := \langle T^{G_\alpha}\rangle$ and $N := \langle M^G\rangle$. Then*

(i) $N = M\,\mathrm{Soc}(G)$.
(ii) *If T is solvable then $N/\mathrm{Soc}(G)$ is solvable.*
(iii) *If T is nonabelian then either $N = \mathrm{Soc}(G)$ or $C_{N_\alpha}(M) = \mathrm{Soc}(G)_\alpha$.*

Proof. (i) Since $\mathrm{Soc}(G)$ is transitive, any $g \in G$ can be written in the form $g = g_1 h$ for some $g_1 \in G_\alpha$ and $h \in \mathrm{Soc}(G)$. Therefore, since $M \lhd G_\alpha$, we have $M^g = M^{g_1 h} = M^h \leq M\mathrm{Soc}(G)$ and so $N \leq M\mathrm{Soc}(G)$. Conversely, $M \leq N$ by definition and we claim that $\mathrm{Soc}(G) \leq N$. This is clear if $\mathrm{Soc}(G)$ is the unique minimal normal subgroup of G. If G has two minimal regular normal subgroups then $\mathrm{Soc}(G)_\alpha \leq M \leq N$ by Exercise 6.15, and N also contains a minimal normal subgroup N_1 of G. Hence $\mathrm{Soc}(G) = \langle \mathrm{Soc}(G)_\alpha, N_1 \rangle \leq N$.

(ii) Suppose that T is a cyclic group of order p for some prime p. Then M is a p-group by Exercise 6.17(ii) and so $N/\mathrm{Soc}(G) \cong M/(\mathrm{Soc}(G) \cap M)$ is also a p-group.

(iii) If T is nonabelian then M is a minimal normal subgroup of G_α by Exercise 6.17(i). Since $\mathrm{Soc}(G)_\alpha \lhd G_\alpha$, we must have $M \leq \mathrm{Soc}(G)_\alpha$ or $M \cap \mathrm{Soc}(G)_\alpha = 1$. If $M \leq \mathrm{Soc}(G)_\alpha$ then $N = \mathrm{Soc}(G)$. If $M \cap \mathrm{Soc}(G)_\alpha = 1$ then $\mathrm{Soc}(G)_\alpha$ centralizes M and so $\mathrm{Soc}(G)_\alpha \leq C_{N_\alpha}(M)$. Conversely, if $c \in C_{N_\alpha}(M)$ then $\mathrm{Soc}(G)c$ centralizes $M\mathrm{Soc}(G)/\mathrm{Soc}(G)$. However, $M\mathrm{Soc}(G)/\mathrm{Soc}(G) \cong M$, so it has trivial center, implying $c \in \mathrm{Soc}(G)$, and so $c \in \mathrm{Soc}(G) \cap N_\alpha = \mathrm{Soc}(G)_\alpha$. □

Beals's Algorithm for Groups with Regular Normal Subgroups

If a primitive group G has a regular normal subgroup then we can solve (6.9) in nearly linear time by constructing a minimal normal subgroup N of G by the methods of Sections 6.2.3 and 6.2.4. Note that these algorithms do not depend on the classification of finite simple groups. If N is elementary abelian then for any nonidentity $g \in N$, $\langle g \rangle$ is a subnormal simple subgroup of G. If $N = T_1 \times \cdots \times T_r$ with isomorphic, nonabelian simple T_i then we construct a minimal normal subgroup of N by applying recursively the composition series algorithm for N. Note that the only subroutines that may be called during this recursion are the transitive constituent and block homomorphism computations of Section 6.2.1 and the algorithm for handling primitive groups with two regular normal subgroups, since a direct product of simple groups cannot fall into other categories of the O'Nan–Scott theorem.

Beals's Algorithm for Groups with No Regular Normal Subgroups

If a primitive group $G \leq \mathrm{Sym}(\Omega)$ has no regular normal subgroups then we construct $\mathrm{Soc}(G)$ as follows: First, we recursively construct a simple subnormal subgroup $T \lhd\lhd G_\alpha$ for some $\alpha \in \Omega$ and compute $M := \langle T^{G_\alpha} \rangle$ and $N := \langle M^G \rangle$. If T is abelian then we compute the solvable residual N^∞ of N. By Lemma 6.2.22(ii), $\mathrm{Soc}(G) = N^\infty$. If T is nonabelian then we compute $C_{N_\alpha}(M)$ and $\langle (C_{N_\alpha}(M))^G \rangle$. Since $M = M_\alpha \lhd N_\alpha$, this can be done in nearly linear time by the algorithm of Section 6.1.4. By Lemma 6.2.22(iii), Soc(G) is the smaller of N and $\langle (C_{N_\alpha}(M))^G \rangle$. After Soc($G$) is constructed, we solve (6.9) for G by recursively computing a composition series for Soc(G). We note again that only the homomorphism computations of Section 6.2.1 and the algorithm handling primitive groups with two regular normal subgroups may occur during this recursion.

We claim that the running time of the algorithm for solving (6.9) is nearly linear for any input group G. Let $f(n, k)$ be the maximum time requirement of the algorithm for groups $G \le S_n$ with $|G| \le k$. For an input $G \le S_n$, the algorithm calls itself recursively only at most once, when the reduction subroutines of Section 6.2.1 constructed some $H \lhd\!\lhd G$ such that H acts primitively, with no regular normal subgroups on some permutation domain Δ with $|\Delta| \le n$. In this case, the recursive call is made for the subgroup H_δ for some $\delta \in \Delta$. The other steps of the algorithm are nearly linear, so $f(n, |G|) \le Cn \log^c |G| + f(|\Delta|, |H_\delta|)$ for some absolute constants c and C. Clearly, $|H_\delta| \le |G|/2$; thus

$$f(n, G) \le \sum_{i=0}^{\lfloor \log |G| \rfloor} Cn(\log |G| - i)^c \le Cn \log^{c+1} |G|.$$

Verification of a Composition Series

Our next goal is to show how Lemma 6.2.22 can be used to verify in nearly linear time the correctness of a composition series construction, provided that at least the order of the input group G is known. In this case, we can compute a nonredundant base and an SGS for G and subgroups of G by nearly linear-time Las Vegas algorithms, and all algorithms in Chapter 5 and Section 6.1 have Las Vegas versions. In fact, if the initial SGS computation for G succeeds (and we can check that since $|G|$ is given) then all further computations we require from Chapter 5 and Section 6.1 can be done in nearly linear time by deterministic algorithms. The verification of composition series will use the classification of finite simple groups.

Given $G = \langle S \rangle \le \mathrm{Sym}(\Omega)$, with $|\Omega| = n$, the output of the composition series algorithm is a sequence of subgroups $G = N_1 \rhd N_2 \rhd \cdots \rhd N_l = 1$ and homomorphisms $\varphi_i : N_i \to S_n$ with $\ker(\varphi_i) = N_{i+1}$ for $i \in [1, l-1]$. We can verify that $N_{i+1} \lhd N_i$ since we can test membership in the subgroups N_i; therefore, the only task left is to verify that $\varphi_i(N_i)$ is simple, for $i \in [1, l-1]$. Furthermore, we can assume that $\varphi_i(N_i)$ acts primitively on some set Δ_i since we can apply transitive constituent and block homomorphism algorithms to $\varphi_i(N_i)$ with guaranteed correct output and check with certainty that the kernel of these homomorphisms is trivial. Hence the problem of upgrading the composition series algorithm to Las Vegas type is reduced to the following task:

$$\text{Given a nonredundant base and SGS for a primitive group } G \le \mathrm{Sym}(\Delta)\text{, decide by a nearly linear-time Las Vegas algorithm whether } G \text{ is simple.} \tag{6.10}$$

Solving (6.10) is an important theoretical exercise. However, in the practical range $|\Delta| \le 10^7$, only a small part of the following discussion is needed. We leave the details to the reader (cf. Exercise 6.18).

First, we check that $|G|$ is the order of some simple group. We claim that the arithmetic operations required for this computation can be done in nearly linear time. Since all prime divisors of $|G|$ are at most $|\Delta|$, we can factor $|G|$ in nearly linear time. After that, we can identify $O(\log^c |G|)$ simple groups S such that if $|G|$ is the order of a simple group then $|G|$ must be the order of one of these candidates. A very rough estimate provides $c = 3$. There are $O(\log |G|)$ sporadic simple or alternating groups S with $|S| \le |G|$ and there is at most one cyclic simple group whose order is divisible by $|G|$. For each of the $O(\log |G|)$ prime divisors p of $|G|$ and each infinite series of Lie-type groups in characteristic p, there are $O(\log |G|)$ possibilities for the size of the defining field of S and $O(\log |G|)$ possibilities for the rank of S.

If $|G|$ is the order of a nonabelian simple group then we solve (6.10) by a recursive procedure. For some $\delta \in \Delta$, we compute a simple subnormal subgroup $T \lhd\lhd G_\delta$ by the nearly linear-time Monte Carlo algorithm for (6.9). Recursively, we solve (6.10) for T and verify that T is really simple. Then we compute $M := \langle T^{G_\delta} \rangle$ and $N := \langle M^G \rangle$.

If T is abelian then we compute the solvable residual N^∞ of N. By Lemma 6.2.22(ii), if G has a regular abelian normal subgroup then $N^\infty = 1$ and otherwise $N^\infty = \mathrm{Soc}(G)$. Hence, if $N^\infty \ne G$ then we declare with certainty that $G \ne \mathrm{Soc}(G)$ and so G is not simple. If $G = \mathrm{Soc}(G)$ then we declare that G is simple. This decision is correct, since $G = \mathrm{Soc}(G)$ is the direct product of isomorphic simple groups and so $|G| = s^r$, where s is the order of some simple group S and r is a positive integer. We also know that $|G|$ is the order of a simple group. By the Ph.D. thesis results of Teague (cf. [Kimmerle et al., 1990, Section 6]), if s, t are orders of finite simple groups and r is a positive integer satisfying $s^r = t$ then $s = t$ and $r = 1$.

If T is nonabelian then we first check that $G = N$. If $G \ne N$ then of course G is not simple, so we can suppose $G = N$. Next, we compute $C_{N_\delta}(M)$. By Lemma 6.2.22(iii), $C_{N_\delta}(M) \le \mathrm{Soc}(G)_\delta$, with equality if $N \ne \mathrm{Soc}(G)$. Hence, if $C_{N_\delta}(M) \ne 1$ then we compute $K := \langle (C_{N_\delta}(M))^G \rangle$. We have $1 \ne K \le \mathrm{Soc}(G)$. If $K \ne G$ then G is not simple since K is a proper normal subgroup; however, if $K = G$ then $G = \mathrm{Soc}(G)$ and, by Teague's result, G is simple.

We are left with the case that $G = N$ and $C_{N_\delta}(M) = 1$. We have to decide whether $N \neq \mathrm{Soc}(G)$ (which implies $\mathrm{Soc}(G)_\delta = 1$ by Lemma 6.2.22(iii)) or $N = \mathrm{Soc}(G) = G$ (which implies that G is simple by Teague's result). If $|\Delta|$ is not a power of the order of a simple group then the first of these cases cannot occur and so G is simple. Hence we can suppose that $|\Delta| = s^r$, where s is the order of some (not necessarily nonabelian) simple group.

Lemma 6.2.23. *Let $G \leq \mathrm{Sym}(\Delta)$ be primitive, where $|\Delta| = p^r$ for some prime p and integer r. Suppose that $|G|$ is the order of a simple group, T is a simple group such that $T \lhd\lhd G_\delta$ for some $\delta \in \Delta$, and for $M = \langle T^{G_\delta} \rangle$, we have $G = \langle M^G \rangle$ and $C_{G_\delta}(M) = 1$. Then G is not simple if and only if $M = T = G_\delta$ and $|G_\delta|$ divides $|\mathrm{GL}_r(p)|$.*

Proof. We have seen that if a primitive group G satisfies the conditions of the lemma then either $\mathrm{Soc}(G)$ is regular abelian or G is simple.

If $\mathrm{Soc}(G)$ is regular abelian then $G_\delta \cap \mathrm{Soc}(G) = 1$, so $G = N = M\mathrm{Soc}(G)$ implies that $M = G_\delta$. This also implies that $T = M$, since if $T \neq G_\delta$ then $\langle T^{G_\delta} \rangle$ is a proper normal subgroup of G_δ, contradicting $M = G_\delta$. The other property in the conclusion of the lemma, that $|G_\delta|$ divides $|\mathrm{GL}_r(p)|$, obviously holds in this case.

If G is simple then by [Guralnick, 1983] or [Kantor, 1985a], the pair (G, G_δ) is one of the following:

(i) $G \cong A_{p^r}$ and $G_\delta \cong A_{p^r-1}$;
(ii) $G \cong \mathrm{PSL}_d(q)$ with $(q^d - 1)/(q - 1) = p^r$, and G_δ is a maximal parabolic subgroup of G;
(iii) $G \cong \mathrm{PSL}_2(11)$, $p^r = 11$, and $G_\delta \cong A_5$;
(iv) $G \cong M_{23}$, $p^r = 23$, and $G_\delta \cong M_{22}$;
(v) $G \cong M_{11}$, $p^r = 11$, and $G_\delta \cong M_{10}$;
(vi) $G \cong \mathrm{PSU}_4(2)$, $p^r = 27$, and G_δ is a maximal parabolic subgroup of G of order 960.

Here M_{23}, M_{22}, and M_{11} denote Mathieu groups, and M_{10} is an index two extension of A_6. In cases (ii), (v), and (vi) G_δ is not simple, whereas in cases (i), (iii), and (iv) $|G_\delta|$ does not divide $|\mathrm{GL}_r(p)|$. □

Based on Lemma 6.2.23, we can easily finish the solution of (6.10) in the case when $|\Delta|$ is a prime power. The last remaining case is to decide whether $\mathrm{Soc}(G)$ is regular nonabelian. We shall do this by showing that, for primitive groups with regular nonabelian socle, (6.7) from Section 6.2.1 can be solved in nearly linear time by a *deterministic* algorithm.

Lemma 6.2.24. *Let $G \leq \mathrm{Sym}(\Omega)$ be primitive with regular nonabelian socle and let $\alpha \in \Omega$ and $|\Omega| = n$. Then the smallest orbit of G_α in $\Omega \backslash \{\alpha\}$ has size less than* $\log^6 n$.

Proof. Let $\mathrm{Soc}(G) = T_1 \times \cdots \times T_r$ with all T_i isomorphic to a simple nonabelian group T, and let $\nu(T)$ denote the degree of the smallest faithful permutation representation of T. By [Dixon and Mortimer, 1996, Theorem 4.7B(iv)], T is isomorphic to a section (i.e., a factor group of a subgroup) of S_{r-1}; therefore, by Corollary 6.2.3, $r - 1 \geq \nu(T)$.

Let s denote the size of the smallest conjugacy class of $\mathrm{Aut}(T)$ in $\mathrm{Inn}(T)\backslash\{1\}$. We claim that $s < \nu^5(T)$. For sporadic T, this can be checked from the Atlas [Conway et al., 1985]; for alternating groups, the number of 3-cycles satisfies this inequality; for classical groups of Lie type, $\nu(T)$ can be obtained from [Cooperstein, 1978], and the number of transvections (or number of root elements in the orthogonal case) satisfies the inequality; and for exceptional groups of Lie type, $|T| < \nu^5(T)$ (cf. [Kantor and Penttila, 1999]).

As we discussed before Lemma 6.2.16 in Section 6.2.4, the points in Ω correspond to the cosets of $\mathrm{Soc}(G)_\alpha = 1$ in $\mathrm{Soc}(G)$, and the permutation action of G_α on Ω is the same as the conjugation action of G_α on these cosets. Hence the smallest orbit of G_α in $\Omega\backslash\{\alpha\}$ has size at most $sr < r^6 < \log^6 n$. □

By Lemmas 6.2.24 and 6.2.17, if G has regular nonabelian socle then the action of G on the cosets of $G_{\alpha\beta}$ can be constructed by a nearly linear-time deterministic algorithm, provided that β is chosen from the smallest orbit of G_α in $\Omega\backslash\{\alpha\}$. After that, the action of G on the cosets of a maximal subgroup containing $N_G(G_{\alpha\beta})$ can be constructed as described in Section 6.2.4. This latter action either is not faithful or falls into a different case of Theorem 6.2.5 (in fact, into case II(ii)). If the first of these alternatives holds then G is not simple, and in the second case we have seen already how (6.10) can be solved.

The last observation we have to make is that the algorithm solving (6.10) can be organized such that it calls itself recursively only once, for a group of order at most half of the order of the input group. If this property holds then the same argument that we used earlier in this section in the case of (6.9) shows that the running time is nearly linear. As we described the algorithm, one recursive call is made for the simple group $T \lhd\!\lhd G_\delta$. The only way a second recursive call may occur is that the subroutine for the handling of groups with regular nonabelian socle returns a faithful primitive permutation representation on a smaller domain, and we check the simplicity of a subnormal subgroup of a point stabilizer in this new representation. To avoid two recursive calls, we postpone the verification of the simplicity of T as the last step of the algorithm, and if a new permutation representation is constructed then we do not perform the verification of T at all.

6.2.7. Chief Series

Given a nonredundant base and SGS, and a composition series $G = N_1 \rhd N_2 \rhd \cdots \rhd N_l = 1$ for some group $G \leq \mathrm{Sym}(\Omega)$, we can construct a chief series as follows: First, we compute the subgroup chain $G = H_1 \geq H_2 \geq \cdots \geq H_l = 1$, where $H_i := \langle N_i^G \rangle$. The subgroups H_i are normal in G and, by repeated applications of Exercise 6.17, we see that if $H_i \neq H_{i+1}$ then H_i/H_{i+1} is the direct product of simple groups isomorphic to N_i/N_{i+1} if N_i/N_{i+1} is nonabelian, and H_i/H_{i+1} is a p-group if $N_i/N_{i+1} \cong C_p$. Moreover, by the same exercise, the nonabelian groups H_i/H_{i+1} are chief factors, and so the only task left is the refinement of the solvable factors H_i/H_{i+1}.

If K is a p-group then K^pK' is a characteristic subgroup of K, where K^p denotes the subgroup of K generated by the pth power of elements of K, and the factor group K/K^pK' is elementary abelian. (The subgroup K^pK' is called the *Frattini subgroup* of K, and it is the intersection of all maximal subgroups of K.) Therefore, if $H_i = \langle S_i \rangle$ and H_i/H_{i+1} is a p-group then we compute $S_i^p := \{g^p \mid g \in S_i\}$ and $H_{i,1} := \langle S_i^p, H_i', H_{i+1} \rangle$. We have $H_i \geq H_{i,1} \geq H_{i+1}$ and $H_{i,1} \lhd G$, and the factor $H_i/H_{i,1}$ is an elementary abelian p-group. Repeating this procedure recursively in $H_{i,1}$, we can construct a chain of normal subgroups $H_i = H_{i,0} \geq H_{i,1} \geq \cdots \geq H_{i,m_i} = H_{i+1}$ of G, with elementary abelian factors $H_{i,j}/H_{i,j+1}$. So far, obviously everything could be done by nearly linear-time deterministic algorithms.

The only nontrivial step of the chief series construction is to decide whether in the resulting series of normal subgroups $G = M_1 \rhd M_2 \rhd \cdots \rhd M_m = 1$ the elementary abelian factors are chief factors. Suppose that M_i/M_{i+1} is an elementary abelian p-group, and $|M_i/M_{i+1}| = p^d$. The factor group M_i/M_{i+1} can be identified with the vector space $\mathrm{GF}(p)^d$, and the conjugation action of G on M_i/M_{i+1} is isomorphic to a subgroup $G_i \leq \mathrm{GL}_d(p)$. The factor M_i/M_{i+1} is a chief factor if and only if G_i acts irreducibly on $\mathrm{GF}(p)^d$. We can construct generators for G_i (as $d \times d$ matrices over $\mathrm{GF}(p)$) by the following procedure. Suppose that $M_i = \langle M_{i+1}, y_1, \ldots, y_d \rangle$, with $y_j^p \in M_{i+1}$ for $j \in [1, d]$. We construct an SGS S_0 with shallow Schreier trees for M_{i+1} and then, recursively for $j = 1, 2, \ldots, d$, we augment S_{j-1} to an SGS S_j for $\langle M_{i+1}, y_1, \ldots, y_j \rangle$ by the algorithm described in Lemma 7.1.2. By Exercise 7.2, there are $z_j \in M_{i+1}y_j$ for $j \in [1, d]$ such that any label in the Schreier tree data structure defined by S_d is from S_0 or is a power of some z_j. During the SGS constructions, we can easily keep track of the exponents of the labels as powers of the z_j. To construct the conjugation action of any $g \in G$ on M_i/M_{i+1} we write z_j^g as a word $w(j, g)$ in S_d by sifting z_j^g as a word in the Schreier tree data structure of M_i and then deleting the elements of S_0 from $w(j, g)$. The remaining word defines the image of the basis vector z_j at the linear transformation $\hat{g} \in G_i$, corresponding to g.

After the matrix generators for G_i are constructed, the irreducibility of the action of G_i on $\mathrm{GF}(p)^d$ can be decided by Norton and Parker's Meat-Axe procedure (cf. [Parker, 1984]). The version described in [Holt and Rees, 1994] and augmented in [Ivanyos and Lux, 2000] is very fast in practice, is of Las Vegas type, and runs in polynomial time in d, $\log p$, and the number of generators (and hence in nearly linear time in terms of the original permutation group input $G \le \mathrm{Sym}(\Omega)$). The output is either a certificate that the action is irreducible, and so M_i/M_{i+1} is a chief factor, or generators for a proper invariant subspace, which can be used to insert a normal subgroup between M_i and M_{i+1}.

6.3. Quotients with Small Permutation Degree

For iterative processes like computing the upper central series, it would be useful to construct permutation representations of quotient groups. However, as shown in [Neumann, 1986] (cf. Exercise 6.6), there are examples of $G \le S_n$, $N \lhd G$ such that G/N has no faithful permutation representation on less than $2^{n/4}$ points. On the positive side, it is proven in [Easdown and Praeger, 1988] and [Luks, 1990] that if N is abelian then there exists a homomorphism $\varphi : G \to S_n$ such that $N \le \ker(\varphi)$ and, in a sense, $\ker(\varphi)$ is not too big. Also, as observed by Luks, φ can be constructed efficiently. The method was also rediscovered in [Holt, 1997].

Theorem 6.3.1. *Let $G \le \mathrm{Sym}(\Omega)$, with $|\Omega| = n$, and let $N \lhd G$ be abelian. Then there exists a homomorphism $\varphi : G \to S_n$ such that $N \le \ker(\varphi)$ and the prime divisors of N and $|\ker(\varphi)|$ are the same. In particular, if N is a p-group then so is $\ker(\varphi)$; moreover, if N is elementary abelian then $\ker(\varphi)$ is elementary abelian. Given generators for G and N, φ can be constructed in linear time by a deterministic algorithm.*

Proof. Let $\Delta_1, \ldots, \Delta_k$ be the orbits of N and so $\Omega = \bigcup_{i=1}^k \Delta_i$. By Corollary 6.1.5, N acts regularly on each set Δ_i, so $N|_{\Delta_i}$ has $|\Delta_i|$ elements. Let Σ_i denote the set of permutations in $N|_{\Delta_i}$, and let $\Sigma := \bigcup_{i=1}^k \Sigma_i$. Clearly, $|\Sigma| = n$. We shall define an action of G on Σ, satisfying the conclusion of the theorem.

Given $g \in G$ and $\sigma_i \in \Sigma_i$, we define $\sigma_i^g \in \Sigma$ in the following way: By Lemma 6.1.7, G permutes the orbits of N; suppose that $\Delta_i^g = \Delta_j$. Take any

$x \in N$ with $x|_{\Delta_i} = \sigma_i$, and let $\sigma_i^g := x^g|_{\Delta_j}$. This map is well defined since if $x|_{\Delta_i} = y|_{\Delta_i}$ for some $x, y \in N$ then $x^g|_{\Delta_j} = y^g|_{\Delta_j}$. It is also clear that this construction defines a homomorphism $\varphi : G \to \mathrm{Sym}(\Sigma)$.

Now $N \leq \ker(\varphi)$, since N acts trivially on itself by conjugation. Moreover, if $g \in \ker(\varphi)$ then g fixes the orbits of N and $g|_{\Delta_i}$ centralizes $N|_{\Delta_i}$. By the second part of Corollary 6.1.5, $N|_{\Delta_i}$ is self-centralizing in $\mathrm{Sym}(\Delta_i)$, so $g|_{\Delta_i} \in N|_{\Delta_i}$. Hence the transitive constituents of N and $\ker(\varphi)$ are the same, implying the required relations between the prime factors and exponents.

To see the algorithmic feasibility of the construction, let us fix a representative $\delta_i \in \Delta_i$ in each orbit of N and build a (breadth-first-search) Schreier tree T_i with root δ_i for N mod N_{δ_i}. In this tree, contrary to the usual Schreier tree constructions, we direct the edges *away* from the root. Identifying the trivial permutation in Σ_i with δ_i, the elements of Σ_i naturally correspond to the elements of Δ_i. If $\sigma_i \in \Sigma_i$ corresponds to $\gamma \in \Delta_i$ then, to compute σ_i^g, it is enough to determine the image of δ_j under the permutation x_γ^g, where x_γ is the product of edge labels in T_i along the path P_γ from δ_i to γ. (Of course, we do not compute the permutation x_γ; instead, we work with the word consisting of the sequence of edge labels along P_γ.) This path P_γ may be $\Theta(|\Delta_i|)$ long, so computing one image $\delta_j^{x_\gamma^g}$ may require $\Theta(|\Delta_i|)$ time; however, noticing that if δ is the parent of γ in T_i and $\delta_j^{x_\delta^g}$ is already computed then we can obtain $\delta_j^{x_\gamma^g}$ in constant time, we can construct the entire image Σ_i^g in $O(|\Delta_i|)$ time. □

6.3.1. Solvable Radical and p-Core

For a group G, the *solvable radical* (in short, radical) $O_\infty(G)$ is defined as the largest solvable normal subgroup of G and, for prime numbers p, the *p-core* $O_p(G)$ is the largest normal p-group in G. Note that, since two solvable normal subgroups generate a solvable normal subgroup, there is a unique maximal solvable normal subgroup in G. Similarly, G has a unique maximal normal p-subgroup. By the same argument, we see that $O_\infty(G)$ and $O_p(G)$ are characteristic in G.

Based on Theorem 6.3.1, [Luks, 1990] gave polynomial-time algorithms for computing $O_\infty(G)$ and $O_p(G)$. For p-cores, another polynomial-time algorithm is described in [Neumann, 1986], and p-cores can be also obtained as the intersection of Sylow p-subgroups (cf. [Butler and Cannon, 1989]), although the computation of a Sylow subgroup is much more difficult than obtaining the p-core. Here we follow the nearly linear-time version from [Luks and Seress, 1997]. The basic idea is the same as in [Luks, 1990]. The radical and p-core are computed similarly, so we handle both of them simultaneously. Our primary description is the computation of the radical; the necessary changes for the computation of the p-core are indicated in double brackets.

First, let us consider the special case when G has a maximal normal subgroup N with $O_\infty(N) = 1[[O_p(N) = 1]]$. It turns out that this extra condition restricts severely the radical and p-core of G.

Lemma 6.3.2. *Suppose that $N \trianglelefteq G$ is a maximal normal subgroup and $O_\infty(N) = 1[[O_p(N) = 1]]$. Then, if G/N is not cyclic [[not a p-cycle]] then $O_\infty(G) = 1[[O_p(G) = 1]]$.*

Proof. Since the radical and p-core of G intersect N trivially, they must centralize N. In particular, if the radical [[p-core]] of G is nontrivial then $C_G(N)$ has nontrivial radical [[p-core]]. If $C_G(N) \leq N$ then its radical [[p-core]] is trivial. If $C_G(N) \not\leq N$ then $G = NC_G(N)$ and $C_G(N)/Z(N) \cong G/N$, so $C_G(N)$ has a chance to have nontrivial radical [[p-core]] only if G/N is cyclic [[is a p-cycle]]. □

Lemma 6.3.3.

(i) *Suppose that $N \trianglelefteq G$ is a maximal normal subgroup, $O_\infty(N) = 1$, and G/N is cyclic. Then $O_\infty(G) \neq 1$ if and only if $C_G(N) \neq 1$.*

(ii) *Suppose that $N \trianglelefteq G$ is a maximal normal subgroup, $O_p(N) = 1$, and G/N is a p-cycle. Then $O_p(G) \neq 1$ if and only if $C_G(N) \not\leq N$.*

Proof. (i) Since $Z(N) = 1$, we must have $C_G(N) = 1$ or $C_G(N) \cong G/N$. In the latter case, $O_\infty(G) = C_G(N) \neq 1$.

(ii) First, we observe that $O_p(Z(N)) = 1$, since $O_p(Z(N)) \lhd N$. If $C_G(N) \leq N$ then $C_G(N) = Z(N)$ and so $O_p(C_G(N)) = 1$. If $C_G(N) \not\leq N$ then $C_G(N)/Z(N) \cong C_p$ and $C_G(N) = \langle Z(N), g\rangle$ for any $g \in C_G(N)\backslash Z(N)$. Since g commutes with N, it commutes with $Z(N)$. Therefore $C_G(N)$ is abelian, which implies $O_p(C_G(N)) \cong C_p$ and $O_p(G) = O_p(C_G(N)) \neq 1$. □

The Algorithms

Now we are ready to describe the algorithms. First, we compute a composition series $1 = N_1 \lhd N_2 \lhd \cdots \lhd N_m = G$. By the results of Section 6.2, this can be done in nearly linear time. Then we find the smallest index i such that N_i has a nontrivial radical [[p-core]] by the following method: Suppose that $O_\infty(N_i) = 1[[O_p(N_i) = 1]]$. If N_{i+1}/N_i is not cyclic [[not a p-cycle]] then, by Lemma 6.3.2, we conclude that $O_\infty(N_{i+1}) = 1[[O_p(N_{i+1}) = 1]]$. Otherwise, we compute $C_{N_{i+1}}(N_i)$, using the algorithm of Section 6.1.4. If $C_{N_{i+1}}(N_i) \not\leq N_i$

then, by Lemma 6.3.3, we get a nontrivial radical [[p-core]]. We take its normal closure H in G; by Exercise 6.17(ii), it is a solvable normal subgroup [[normal p-group]] in G.

Next, we compute the derived series of H. The last nontrivial term in the derived series is an abelian normal subgroup [[abelian normal p-group]] N of G. We compute the homomorphism $\varphi : G \to S_n$ described in Theorem 6.3.1 and repeat the procedure in the image of φ. Note that the composition series computation need not be repeated, since $1 = \varphi(N_1) \lhd \varphi(N_2) \lhd \cdots \lhd \varphi(N_m) = \varphi(G)$, with $\varphi(N_{i+1})/\varphi(N_i) \cong N_{i+1}/N_i$ or $\varphi(N_{i+1})/\varphi(N_i) = 1$.

Theorem 6.3.1 and computing $O_\infty(G)$ are becoming more and more important. In any group G, we can define a series of characteristic subgroups $1 \le N_1 \le N_2 \le N_3 \le G$, where $N_1 = O_\infty(G)$, $N_2/N_1 = \mathrm{Soc}(G/N_1) \cong T_1 \times \cdots \times T_m$ is the direct product of nonabelian simple groups, $N_3/N_2 \lesssim \mathrm{Out}(T_1) \times \cdots \times \mathrm{Out}(T_m)$, and G/N_3 is a permutation group of degree m, corresponding to the conjugation action of G on $\{T_1, \ldots, T_m\}$. Constructing these subgroups is one of the main approaches for matrix group computations (cf. [Babai and Beals, 1999; Kantor and Seress, 2002]), but a number of recent permutation group algorithms utilize them as well, for example, for computing conjugacy classes (cf. [Cannon and Souvignier, 1997; Hulpke, 2000]), maximal subgroups (cf. [Eick and Hulpke, 2001; Cannon and Holt, 2002]), or automorphism groups (cf. [Holt, 2001]).

In the permutation group setting, during the construction of $O_\infty(G)$ we obtain faithful permutation representations for the factor groups G/K_i for a sequence of normal subgroups $1 = K_r \le K_{r-1} \le \cdots \le K_1 = N_1 = O_\infty(G)$, with elementary abelian factors K_i/K_{i+1}. First we solve the problem at hand (for example, the computation of conjugacy classes) for $G/O_\infty(G)$. The bulk of the work in this step is the handling of the simple nonabelian groups T_j in $\mathrm{Soc}(G/O_\infty(G))$, which can be done by identifying these groups with a standard copy of the T_j (cf. Section 8.3 for details and a precise formulation of the identification). In the standard copy, the problems indicated here can be solved more easily than in arbitrary permutation representations. After that, we lift the result recursively from G/K_i to G/K_{i+1}, for $i \in [1, r-1]$. For that task, techniques for solvable permutation groups are used. We shall give examples of such lifting procedures in Section 7.3.

Exercises

6.1. Let $G \le \mathrm{Sym}(\Omega)$ be primitive and $1 \ne N \lhd G$. Prove that N is transitive.

6.2. Finish the proof of Lemma 6.1.9.

6.3. Construct a permutation group $G \le \mathrm{Sym}(\Omega)$, with $|\Omega| = n$, and with $\Theta(n/2^{\sqrt{\log n}})$ orbits that are of the same size but pairwise not equivalent in the equivalence relation defined in Section 6.1.2. *Hint:* We may choose G to be an elementary abelian 2-group.

6.4. Let $H, G \le \mathrm{Sym}(\Omega)$ such that G normalizes H. Prove that G normalizes $C_{\mathrm{Sym}(\Omega)}(H)$.

6.5. Prove that if $H \le G$ then $H \lhd\lhd G$ if and only if the subgroup chain defined recursively as $H_0 := G$, $H_i := \langle H^{H_{i-1}} \rangle$ for $i > 0$ reaches H.

6.6. [Neumann, 1986] Let $\Omega = \{\alpha_1, \dots, \alpha_{4m}\}$. For $i \in [1, m]$, let G_i be a dihedral group of order 8, acting on $\{\alpha_{4i-3}, \alpha_{4i-2}, \alpha_{4i-1}, \alpha_{4i}\}$ and fixing the other $4m - 4$ points of Ω. Let z_i be the generator of $Z(G_i)$. Prove that the subgroup $N := \langle z_1 z_2, z_1 z_3, \dots, z_1 z_m \rangle$ is normal in $G := G_1 \times \cdots \times G_m$ and G/N has no faithful permutation representation of degree less than 2^{m+1}.

6.7. Let $G \le \mathrm{Sym}(\Omega)$ be primitive, with two regular normal subgroups N_1 and N_2. Prove that $N_1 \cong N_2 \cong N$ for some characteristically simple non-abelian group N and that Ω can be identified with N such that N_1 and N_2 are the right-regular and left-regular representations of N on itself.

6.8. Let $1 \ne K \le \mathrm{Sym}(\Delta)$ and $1 \ne H \le S_m$. We define an action of $K \wr H$ on the set Δ^m by the rule that if $g := (k_1, \dots, k_m; h) \in K \wr H$ and $\delta := (\delta_1, \dots, \delta_m) \in \Delta^m$ then $\delta^g := (\delta_{1^{\bar{h}}}^{k_{1^{\bar{h}}}}, \dots, \delta_{m^{\bar{h}}}^{k_{m^{\bar{h}}}})$ for $\bar{h} := h^{-1}$. (This is called the *product action* of the wreath product.)

Prove that this product action is primitive if and only if K acts primitively but not regularly on Δ and H is a transitive subgroup of S_k.

6.9. Let $T_1, \dots, T_r$ be nonabelian simple groups and let H be a subdirect product of the T_i, $i = 1, 2, \dots, r$. Prove that there is a partition $(\Sigma_1, \dots, \Sigma_l)$ of $\{1, 2, \dots, r\}$ such that, for each fixed $j \in [1, l]$, all groups T_k with $k \in \Sigma_j$ are isomorphic, and there is a diagonal subgroup $D_j \le \prod_{k \in \Sigma_j} T_k$ such that $H = \prod_{j=1}^{l} D_j$.

6.10. Let $T_1, \dots, T_r$ be nonabelian simple groups. Prove that if M is a minimal normal subgroup of $T_1 \times \cdots \times T_r$ then $M = T_i$ for some $i \in [1, r]$.

6.11. For any group G, prove that the following are equivalent:
   (i) G has a faithful transitive permutation representation that is not regular, but the stabilizer of any two points is trivial (i.e., G is a Frobenius group).
   (ii) G has a nontrivial subgroup H such that $N_G(H) = H$ and any two conjugates of H are identical or their intersection is 1 (such a subgroup H is called a *Frobenius complement*).

6.12. This is not an exercise but rather a compilation of results about Frobenius groups. Proofs can be found, for example, in Sections 17 and 18 of [Passman, 1968]. Let $G \le \mathrm{Sym}(\Omega)$ be a Frobenius group and let $\alpha \in \Omega$. Then $H := G_\alpha$ is a Frobenius complement for G.

(i) (Frobenius) The set $\{1\} \cup (G \backslash \bigcup_{g \in G} H^g)$ is a regular normal subgroup of G, which is called the *Frobenius kernel*.
(ii) (Thompson) The Frobenius kernel is nilpotent.
(iii) If $|H|$ is even then H has a unique element of order 2, which is therefore central.
(iv) (Zassenhaus) If $|H|$ is odd then for every prime divisor p of $|H : H'|$, elements of order p in H are central.
(v) (Zassenhaus) If H is nonsolvable then H has a normal subgroup H_0 of index 1 or 2 such that $H_0 \cong \mathrm{SL}_2(5) \times S$ for some solvable group S of order relative prime to 30.

6.13. Prove Lemma 6.2.12. *Hint:* Use induction on m.

6.14. Let p be a prime number.
(i) Let P be the set of $d \times d$ lower triangular matrices over GF(p), with all diagonal entries equal to 1. Prove that P is a Sylow p-subgroup of $\mathrm{GL}_d(p)$.
(ii) For $H \leq \mathrm{GL}_d(p)$, let fix(H) denote that the set of vectors in $\mathrm{GF}(p)^d$ fixed by each element of H. Prove that if H is a p-group then fix(H) is a nonzero subspace of $\mathrm{GF}(p)^d$.
(iii) Prove that if $H \leq \mathrm{GL}_d(p)$ then $\mathrm{fix}(O_p(H))$ is an H-invariant subspace.

6.15. Let $G \leq \mathrm{Sym}(\Omega)$ be primitive with two regular normal subgroups and let $\alpha \in \Omega$. Prove that the only minimal normal subgroup of G_α is $\mathrm{Soc}(G)_\alpha$.

6.16. Design a faster version of the algorithm described in Lemma 6.2.15, where the intermediate groups $\langle h_i^H \rangle$ for $1 \leq i < l$ are not computed; instead, guarantee that a subgroup of $\langle h_i^H \rangle$ of order greater than n is constructed with sufficiently high probability and that h_{i+1} is chosen from this subgroup.

6.17. Suppose that $T \lhd\lhd G$ and T is simple. Prove that
(i) if T is nonabelian then $\langle T^G \rangle$ is a direct product of simple groups isomorphic to T and it is a minimal normal subgroup of G;
(ii) if T is cyclic of order p then $\langle T^G \rangle$ is a p-group;
(iii) it is possible that $\langle T^G \rangle$ is not elementary abelian in part (ii).

6.18. Combining ideas from Sections 6.2.5 and 6.2.6, design a fast Las Vegas algorithm to decide whether a primitive group of degree at most 10^7 and of known order is simple.

7

Solvable Permutation Groups

7.1. Strong Generators in Solvable Groups

Exploiting special properties of solvable groups, [Sims, 1990] describes a method for constructing a strong generating set. Recall that a finite group G is solvable if and only if it is *polycyclic*, that is, there exists a sequence of elements $(y_1, \ldots, y_r)$ such that $G = \langle y_1, \ldots, y_r \rangle$ and for all $i \in [1, r-1]$, y_i normalizes $H_{i+1} := \langle y_{i+1}, \ldots, y_r \rangle$.

The main idea is that given an SGS for a group $H \leq \mathrm{Sym}(\Omega)$ and $y \in \mathrm{Sym}(\Omega)$ such that y normalizes H, an SGS for $\langle H, y \rangle$ can be constructed without sifting of Schreier generators. The method is based on the following observation.

Lemma 7.1.1. *Suppose that $G = \langle H, y \rangle \leq \mathrm{Sym}(\Omega)$ and y normalizes H. For a fixed $\alpha \in \Omega$, let $\Gamma = \alpha^H$, $\Delta = \alpha^G$, and $m = |\Delta|/|\Gamma|$. Then*

(i) *m is an integer and there exists $h \in H$ such that $z := y^m h$ fixes α;*
(ii) *z normalizes H_α; and*
(iii) *$G_\alpha = \langle H_\alpha, z \rangle$.*

Proof. (i) By Lemma 6.1.7, Δ is the disjoint union of G-images of Γ. Moreover, the G-images of Γ are cyclically permuted by y and m is the smallest integer such that $\Gamma^{y^m} = \Gamma$. In particular, $\alpha^{y^m} \in \Gamma$, so there exists $h \in H$ with the desired property.

(ii) Since y^m and h normalize H, so does their product z. Moreover, since $z \in G_\alpha$, it normalizes $H_\alpha = G_\alpha \cap H$.

(iii) Any $x \in G$ can be written in the form $x = h_x y^i$ for some $h_x \in H$ and some integer $i \geq 0$. If, in addition, $x \in G_\alpha$ then $\Gamma^x = \Gamma$ and so m divides i. Therefore $x = h_x(zh^{-1})^{i/m} = uz^{i/m}$ for some $u \in H$. Since both x and z fix α, so does u. This implies that $x \in \langle H_\alpha, z \rangle$. □

Lemma 7.1.2. *Suppose that $G = \langle H, y\rangle \le \mathrm{Sym}(\Omega)$, with $|\Omega| = n$, and y normalizes H. Moreover, suppose that a nonredundant base B and an SGS S relative to B is given for H, and let t denote the sum of depths of the Schreier trees defined by S. Then there is a deterministic algorithm that computes an SGS $S^* \supseteq S$ for G in $O(n(t + \log|G : H|))$ time, satisfying $|S^*| \le |S| + 2\log|G : H|$ and that the sum of depths of the Schreier trees defined by S^* is at most $t + 2\log|G : H|$.*

Proof. Let $B = (\beta_1, \dots, \beta_M)$ be the nonredundant base of H, and let $m_1 := |\beta_1^G|/|\beta_1^H|$. By Lemma 7.1.1(i), m_1 is an integer. First, we determine the value of m_1. This is done by computing the images $\beta_1^{y^i}$ for positive integers i. The smallest i such that $\beta_1^{y^i}$ is in the fundamental orbit $\Gamma := \beta_1^H$ gives the value of m_1. After that, we compute the permutations y^{2^i} for all i satisfying $2^i \le m_1$.

The Schreier tree (S_1^*, T_1^*) for G mod G_{β_1} is constructed by augmenting the Schreier tree (S_1, T_1) for H mod H_{β_1}, using the labels y^{2^i}. Let t_1 denote the depth of (S_1, T_1). Then points in Γ^{y^j} are of distance at most $t_1 + b(j)$ from the root in (S_1^*, T_1^*), where $b(j)$ denotes the number of 1s in the binary expansion of j. Hence the depth of (S_1^*, T_1^*) is at most $t_1 + \lceil\log m_1\rceil \le t_1 + 2\log m_1$. We finish processing the first level by computing y^{m_1} and the coset representative h defined by (S_1, T_1) such that $y^{m_1}h$ fixes β_1. By Lemma 7.1.1(iii) we have $G_{\beta_1} = \langle H_{\beta_1}, y^{m_1}h\rangle$.

The time requirement is $O(n)$ for computing m_1, $O(n\log m_1)$ for computing the y^{2^i} for $0 < i \le \log m_1$, $O(n\log m_1)$ for computing (S_1^*, T_1^*), and $O(n(t_1 + \log m_1))$ for computing $y^{m_1}h$.

After processing the first level, we apply recursively the same procedure for $G_{(\beta_1,\dots,\beta_j)}$, for $j = 1, 2, \dots, M$. Denoting the depth of the jth Schreier tree for H by t_j and $|\beta_j^{G_{(\beta_1,\dots,\beta_{j-1})}}|/|\beta_j^{H_{(\beta_1,\dots,\beta_{j-1})}}|$ by m_j, the total time requirement of the algorithm is $O(n\sum_{j=1}^{M}(t_j + \log m_j)) = O(n(t + \log|G : H|))$. □

Following Sims, we call the algorithm described in the proof of Lemma 7.1.2 **Normalizing_Generator**.

Corollary 7.1.3. *Given a polycyclic generating sequence $(y_1, \dots, y_r)$ for a solvable group $G \le S_n$, an SGS supporting membership testing in $O(n\log|G|)$ time per test can be computed in $O(rn\log|G|)$ time by a deterministic algorithm. The memory requirement is $O(n\log|G| + rn)$.*

In particular, if $G = \langle T\rangle$ is abelian then the given generators form a polycyclic generating sequence and an SGS can be computed in $O(|T|n\log|G|)$ time by a deterministic algorithm.

Unfortunately, the input generating set for a solvable group is frequently not a polycyclic one and often we do not even know whether the input group is solvable. Hence we use **Normalizing_Generator** as a subroutine of the following *generalized normal closure* (**GNC**) algorithm. The input of **GNC** is an SGS for some $N \lhd G$ and $y \in G$ such that $y \notin N$. The output is either an SGS for the normal closure $M = \langle \langle N, y \rangle^G \rangle$ or two conjugates u, v of y such that the commutator $w = [u, v] \notin N$.

Starting with $N = 1$, repeated applications of **GNC** construct an SGS for a solvable group. If we get an output of the first type, the next call is made with M and a generator of G not in M. When an output of the second type is returned, the next call is made with the same N as before and with w. In this case, the progress we made is that w is in the derived subgroup G'. All conjugates of w remain in G', so a new output of the second type will be in G''. Clearly, in a solvable group sooner or later an output of the first type occurs.

If the input group is not solvable then recursive calls of **GNC** continue indefinitely. To handle this possibility, a result from [Dixon, 1968] is used. Dixon showed that the derived length of a solvable permutation group of degree n is at most $(5 \log_3 n)/2$. Hence, if more than $(5 \log_3 n)/2$ consecutive outputs of **GNC** are of the second type then we stop and conclude that G is not solvable.

The Generalized Normal Closure Algorithm

To finish our description of the solvable group SGS construction, we indicate how the procedure **GNC** works. Given N and y, we start a standard normal closure algorithm (cf. Section 2.1.2), collecting the conjugates of generators in a set U. We add only those conjugates that increase $\langle N, U \rangle$. Each time a permutation u is added to U, we check whether the commutators of u with all previously defined elements of U are in N. If some commutator is not in N then we have found an output of the second type. If all commutators are in N then in particular u normalizes $\langle N, U \backslash \{u\} \rangle$ and **Normalizing_Generator** can be used for constructing an SGS for $\langle N, U \rangle$.

We analyze the version of the algorithm when the inverses of transversals are stored in Schreier trees using the method of Lemma 7.1.2. Let T denote the set of given generators for G. The procedure **GNC** is called $O(\log |G| \log n)$ times, because outputs of the first type define an increasing chain of subgroups that must have length $O(\log |G|)$, and between two outputs of the first type only $O(\log n)$ outputs of the second type occur.

Within one call of **GNC**, the set U increases $O(\log |G|)$ times since the groups $\langle N, U \rangle$ define an increasing subgroup chain. Hence $O(\log^2 |G|)$ commutators and $O(|T| \log |G|)$ conjugates of the elements of U must be computed and their membership must be tested in the groups N and $\langle N, U\rangle$, respectively. By Corollary 7.1.3, one membership test costs $O(n \log |G|)$. By Lemma 7.1.2, the total cost of $O(\log |G|)$ calls of **Normalizing_Generator** is $O(n \log^2 |G|)$. Combining these estimates, we obtain the following theorem.

Theorem 7.1.4. *An SGS for a solvable permutation group $G = \langle T\rangle \leq S_n$ can be computed in $O(n \log n \log^3 |G|(|T| + \log |G|))$ time by a deterministic algorithm.*

The solvable SGS construction belongs to the rare species of *deterministic* nearly linear-time algorithms. An additional bonus is that we also obtain a polycyclic generating sequence (and so a composition series). Moreover, in practice, the algorithm runs faster than other speedups of the Schreier–Sims procedure and the time lost on nonsolvable inputs is often insignificant. Hence, it may be beneficial to execute this algorithm for all input groups that are not yet known to be nonsolvable.

7.2. Power-Conjugate Presentations

Computing with solvable groups is one of the success stories of computational group theory. The representation most convenient to work with is a generator–relator presentation with special properties. Let $(y_1, \ldots, y_r)$ be a polycyclic generating sequence for a solvable group G, with y_i normalizing $G_{i+1} := \langle y_{i+1}, \ldots, y_r\rangle$ for $i \in [1, r-1]$. Any $g \in G$ can be written uniquely in the form $g = y_1^{e_1} y_2^{e_2} \cdots y_r^{e_r}$ for integers e_i satisfying $0 \leq e_i < |G_i : G_{i+1}|$. This product is called the *collected word* for g. We also have

$$\begin{aligned} y_i^{|G_i:G_{i+1}|} &= y_{i+1}^{\varepsilon_{i;i+1}} y_{i+2}^{\varepsilon_{i;i+2}} \cdots y_r^{\varepsilon_{i;r}}, \text{ for } 1 \leq i \leq r, \\ y_j y_i &= y_i y_{i+1}^{\varepsilon_{i,j;i+1}} y_{i+2}^{\varepsilon_{i,j;i+2}} \cdots y_r^{\varepsilon_{i,j;r}}, \text{ for } 1 \leq i < j \leq r. \end{aligned} \tag{7.1}$$

These are the relators for the so-called power-conjugate presentation (pcp) for G. By repeated applications of the relators in (7.1), any word in the y_i representing an element $g \in G$ can be rewritten to the collected word for g. This process is called a *collection*. In particular, given collected words for $g, h \in G$, we can compute the collected word of their product by concatenating the collected words for g and h and performing a collection. We note that no matter in which order the rules (7.1) are applied, the collection process terminates; however, the number of applications can vary significantly. A discussion of the different collection methods can be found in [Leedham-Green and Soicher, 1990] and in [Vaughan-Lee, 1990].

The pcps are very compact, but still relatively easily manageable representations of polycyclic groups. In some instances, groups G with $\log|G|$ in the thousands can be handled, which is currently not possible in a representation as a permutation or matrix group. There is a large library of algorithms to deal with polycyclic groups, but this is the topic for another book. Here we just give the references [Laue et al., 1984], [Celler et al., 1990], [Sims, 1994], and [Eick, 1997] as a starting point for the interested reader. Most algorithms work well in practice, but the complexity analysis is not as well developed as in the case of permutation groups.

It is often beneficial to work with a pcp of a special form. By inserting appropriate powers of the generators into the polycyclic generating sequence, we can assume that the indices $p_i := |G_i : G_{i+1}|$ are prime numbers. Moreover, it is useful if the tower of subgroups $G = G_1 \rhd G_2 \rhd \cdots \rhd G_r \rhd G_{r+1} = 1$ is a refinement of a chain of normal subgroups $G = N_1 \rhd N_2 \rhd \cdots \rhd N_m = 1$, with N_i/N_{i+1} elementary abelian. Then we can use induction to extend information from G/N_i to G/N_{i+1}. Since G/N_i acts as a group of linear transformations on N_i/N_{i+1}, often linear algebra methods can be utilized.

Conversion to a pcp

Because of the usefulness of pcps, we may want to compute one for a given solvable permutation group. Fortunately, the solvable SGS construction in Section 7.1 accomplishes this task, with very little extra work. By Exercises 7.1 and 7.2, we can construct a polycyclic generating sequence $(y_1, \ldots, y_r)$ such that the indices in the corresponding subgroup chain $G = G_1 \rhd G_2 \rhd \cdots \rhd G_r \rhd G_{r+1} = 1$ are prime numbers and the SGS S constructed by the algorithm of Section 7.1 consists of powers of the y_i. Moreover, by Lemma 7.1.2, $S \cap G_i$ is an SGS for G_i relative to the base of G (but this base may be redundant for indices $i > 1$). Therefore, sifting $y_i^{|G_i:G_{i+1}|}$ as a word in G_{i+1} for $1 \le i \le r-1$, and sifting $y_j^{y_i}$ as a word in G_{i+1} for $1 \le i < j \le r$, we can express the left-hand-sides of the relators in (7.1) as words in the y_k. Inductively for $i = r-1, r-2, \ldots, 1$, we can rewrite these words for $y_i^{|G_i:G_{i+1}|}$, $y_j^{y_i}$ in the form required in (7.1), utilizing the already constructed part of (7.1). Although we cannot give a complexity estimate of this procedure because the last step involves a collection process, the practical performance is very good.

7.3. Working with Elementary Abelian Layers

The main tool when working with groups given by a power-conjugate presentation is to construct a chain $G = G_1 \ge G_2 \ge \cdots \ge G_m = 1$ of normal subgroups with elementary abelian quotients G_i/G_{i+1}, identify G_i/G_{i+1} with a vector space, and recursively solve the problem at hand for the factors G/G_i, for $i = 1, 2, \ldots, m$. In almost all instances, the extension of the solution from G/G_i to G/G_{i+1} does not use the fact that G/G_i and G_{i+1} are solvable; the only property exploited by the algorithms is the conjugation action of G, considered as a group of linear transformations on the vector space G_i/G_{i+1}. Therefore, the same methods can be used in not necessarily solvable permutation groups to extend a solution from a factor group G/N to G/M, provided that N/M is elementary abelian and we can handle the conjugation action of G on N/M.

Given $M \leq N \leq G \leq \mathrm{Sym}(\Omega)$ with $M, N \lhd G$ and N/M elementary abelian, we have already described in Section 6.2.7 how to identify N/M with a vector space $V_{N,M}$ in nearly linear time. Recall that we construct an SGS S_M with shallow Schreier tree data structure for M and then repeatedly use the algorithm **Normalizing_Generator**, described in the proof of Lemma 7.1.2, to augment S_M to an SGS S_N for N. The elements in $S_N \backslash S_M$ can be considered as a basis for the vector space $V_{N,M}$, and any coset Mg can be written in this basis by sifting a representative of the coset as a word in N, and deleting the elements of S_M from the resulting word. We shall refer to this process as the *coordinatization* of g (or Mg). The linear transformation of $V_{N,M}$ corresponding to some $h \in G$ is constructed by coordinatizing the permutations z^h for $z \in S_N \backslash S_M$.

The coordinatization process is more efficient if we have a faithful permutation representation for G/M, since working in this permutation representation we have $S_M = 1$. In particular, extending a solution from $G/O_\infty(G)$ to G can be done using the normal subgroup chain between 1 and $O_\infty(G)$, which was constructed in Section 6.3.1, at the computation of $O_\infty(G)$.

In this section, we give two examples of lifting solutions from G/N to G/M. To simplify notation, we assume that $M = 1$ (or, equivalently, we have a faithful permutation representation for G/M). We emphasize again that the same algorithms can be used in G/M without constructing a faithful permutation representation for G/M; the coordinatization process just becomes less efficient.

7.3.1. Sylow Subgroups

Sylow subgroups play a central role in group theory, so constructing them is a natural and important task. However, although there are numerous proofs for the existence of Sylow subgroups, none of them translates easily to an efficient algorithm. Consequently, some of the most intricate algorithmic machinery for permutation groups was first developed in the context of Sylow subgroup constructions.

The theoretical approach originates in [Kantor, 1985b; 1990] where a polynomial-time algorithm is described for the construction of Sylow subgroups. Given $G \leq \mathrm{Sym}(\Omega)$, first we construct a chain $G = G_1 \geq G_2 \geq \cdots \geq G_m = 1$ of normal subgroups with characteristically simple quotients G_i/G_{i+1} (i.e., G_i/G_{i+1} is the product of isomorphic simple groups). Such normal subgroups can be obtained by the methods of Section 6.2 (see especially Section 6.2.7).

For a prime p dividing $|G|$, we construct a Sylow p-subgroup of G by constructing Sylow p-subgroups recursively in G/G_i, for $i = 1, 2, \ldots, m$. The extension of the solution from G/G_i to G/G_{i+1} requires algorithms for the following two problems:

(1) Find a Sylow p-subgroup in G_i/G_{i+1}.
(2) Given two Sylow p-subgroups of G_i/G_{i+1}, find an element of G_i/G_{i+1} conjugating one to the other.

These problems are trivial if G_i/G_{i+1} is elementary abelian, but if G_i/G_{i+1} is nonsolvable then their solution requires some form of constructive recognition of the simple groups in the direct product G_i/G_{i+1}. For the definition of constructive recognition, we refer to Section 8.3. Constructive recognition identifies the simple groups under consideration with a standard copy. For classical groups of Lie type, the standard copy is a group of matrices with "nice" generating sets, and problems (1) and (2) can be solved by linear algebra. However, for exceptional groups of Lie type, the current solutions of (1) and (2) are essentially by brute force. The running time is polynomial in the input length, but the algorithm is impractical. An important direction of future research is the efficient algorithmic handling of exceptional groups of Lie type, including the solutions of (1) and (2).

Later theoretical developments, namely the nearly linear-time Sylow subgroup computations of [Morje, 1995; 1997] in groups with no exceptional composition factors and the parallel machinery of [Mark, 1993] and [Kantor et al., 1999], all follow this pattern. In this section we shall only describe how to construct a Sylow subgroup of G/G_{i+1} in the case when G_i/G_{i+1} is elementary abelian. In particular, solving this special problem is enough for the construction of Sylow subgroups in solvable groups.

The first practical algorithm in [Cannon, 1971; Butler and Cannon, 1989] constructs a Sylow p-subgroup by finding larger and larger p-subgroups, based on Exercise 7.3. If a p-subgroup $H \leq G$ is already constructed but H is not a Sylow subgroup then the algorithm constructs the centralizers $C_G(h)$ for elements of $Z(H)$ until some h is found with the p-part of $C_G(h)$ exceeding $|H|$. Then, recursively, a Sylow subgroup of $C_G(h)$ is constructed. In the group

$C_G(h)$, we can factor out a normal p-group using Theorem 6.3.1 and work in a smaller quotient group. This algorithm uses centralizer computations and thus backtrack searches, which we shall discuss in Chapter 9.

Later practical versions gradually converged toward the theoretical approach just described. In [Atkinson and Neumann, 1990], [Butler and Cannon, 1991], and [Cannon et al., 1997], the method of [Butler and Cannon, 1989] is combined with a divide-and-conquer approach that utilizes block homomorphisms and transitive constituent homomorphisms. In the most recent implementations, a Sylow p-subgroup of $G/O_\infty(G)$ is lifted to a Sylow p-subgroup of G through the elementary abelian layers of $O_\infty(G)$, as in the theoretical solution.

After this lengthy introduction, we can now get down to work. We would like to solve the following problem:

$$\begin{gathered}\text{Given } Q \lhd G \leq \mathrm{Sym}(\Omega), \text{ with } |\Omega| = n, \text{ such that } Q \text{ is elementary abelian}\\ \text{and } G/Q \text{ is a } p\text{-group, construct a Sylow } p\text{-subgroup of } G\\ \text{in nearly linear time by a deterministic algorithm.}\end{gathered} \tag{7.2}$$

The approach we present for solving (7.2) is a straightforward sequential modification of the parallel procedure in [Kantor et al., 1999]. Although we describe a deterministic version, randomized methods are very well suited to speed up Sylow subgroup computations. If we make sure that the orders of the input group and of the output (the alleged Sylow subgroup) are computed correctly, then we can check whether the output is correct.

Let $Q \cong C_q^m$ for some prime q. If $q = p$ then G is a p-group and we are done. Hence we assume now that $q \neq p$. We compute an SGS $\{z_1, \ldots, z_m\}$ for Q that supports the coordinatization of the elements of Q, as described in the introduction of Section 7.3.

First, we solve (7.2) in the special case when $G/Q \cong C_p^r$ is an elementary abelian p-group. Note that in this case any Sylow p-subgroup P of G is elementary abelian. For $x \in G$, we denote by $\bar{x}$ the coset Qx in G/Q.

Lemma 7.3.1. *Let $Q \lhd G \leq \mathrm{Sym}(\Omega)$ be given, with $Q \cong C_q^m$ and $G/Q \cong C_p^r$ for two different primes p, q. Then a Sylow p-subgroup of G can be found in nearly linear time by a deterministic algorithm.*

Proof. Let $\{z_1, \ldots, z_m\}$ be the SGS of Q supporting coordinatization. First, we construct $x_1, \ldots, x_r \in G$ such that $\bar{x}_1, \ldots, \bar{x}_r$ generate G/Q. This can be done in the faithful permutation representation of G/Q provided by Theorem 6.3.1, or just by picking a subsequence of the given generators of G that define a subgroup chain with groups having orders with increasing p-parts. After that, we compute the matrices $T_1, \ldots, T_r$ of the linear transformations corresponding to the conjugation action of $x_1, \ldots, x_r$ on Q.

It is enough to find $q_1, \ldots, q_r \in Q$ such that $x_i q_i$ and $x_j q_j$ commute (in G) for all $i, j \in [1, r]$. Having accomplished that, the qth powers of the permutations $x_i q_i$ generate a Sylow p-subgroup of G.

Since $z_1, \ldots, z_m$ is a basis of the vector space Q, the q_k can be written in the form $q_k = \sum_{l=1}^{m} \alpha_{kl} z_l$, where the α_{kl} are (a priori unknown) elements of GF(q). For fixed $i, j \in [1, r]$, the condition $[x_i q_i, x_j q_j] = 1$ provides m linear equations for the α_{kl}, since

$$\begin{aligned} 1 = [x_i q_i, x_j q_j] &= q_i^{-1} x_i^{-1} q_j^{-1} x_j^{-1} x_i q_i x_j q_j \\ &= q_i^{-1} \cdot \left(q_j^{-1}\right)^{x_i^{-1}} \cdot [x_i, x_j] \cdot q_i^{x_j} \cdot q_j. \end{aligned} \tag{7.3}$$

All five terms on the second line of (7.3) are elements of Q. We can compute the coordinate vectors of these elements (as functions of the α_{kl}) since $[x_i, x_j]$ is a known element of Q, and conjugation by x_i and x_j are linear transformations of Q with known matrices T_i and T_j, respectively. Hence, for example, the coordinate vector of $q_i^{x_j}$ is the product of the coordinate vector of q_i and the matrix T_j. The sum of the five coordinate vectors corresponding to the terms of the second line of (7.3) is 0 and each coordinate provides a linear equation for the α_{kl}. Altogether, we get $\binom{r}{2} m \in O(\log^3 |G|)$ equations in $mr \in O(\log^2 |G|)$ unknowns. This system of equations has a solution (since Sylow subgroups exist), and we can find a solution by Gaussian elimination, using $O(\log^7 |G|)$ field operations in GF(q). □

If G/Q is not elementary abelian then we use an algorithmic form of the "Frattini argument" (cf. Exercise 7.4). First, we compute a homomorphism $\varphi : G \to S_n$ with kernel Q using Theorem 6.3.1, and then we compute an elementary abelian normal subgroup $\bar{A} \cong C_p^r$ of $\varphi(G)$ and its preimage $A := \varphi^{-1}(\bar{A})$. By Lemma 7.3.1, we can compute a Sylow p-subgroup $P = \langle x_1, \ldots, x_r \rangle$ of A. As a final preparatory step, we set up the data structure (cf. Section 5.1.2) for computing with the isomorphism $\psi : P = \langle x_1, \ldots, x_r \rangle \to \bar{A} = \langle \varphi(x_1), \ldots, \varphi(x_r) \rangle$, which is the restriction of φ to P. The point is that, given $\bar{x} \in \bar{A}$, we can compute $x \in P$ with $\varphi(x) = \bar{x}$. As before, let $\{z_1, \ldots, z_m\}$ be the SGS for $Q \cong C_q^m$ supporting coordinatization.

Now comes the Frattini argument. Since $N_G(P)A/A \cong G/A$, the normalizer $N_G(P)$ contains a Sylow p-subgroup of G. Our next goal is to construct generators for $N_G(P)$, by computing some $y_g \in A$ for each generator g of G such that $g y_g$ normalizes P. In fact, since $A = PQ$, we shall construct y_g in Q.

Let g be a fixed generator of G. Since $A \lhd G$, we have $x_j^g \in A$ for $1 \le j \le r$ and so there exist $x'_j \in P$ and $q_j \in Q$ such that $x_j^g = x'_j q_j$. In fact, x'_j and q_j can be constructed, since $x'_j = \psi^{-1}(\varphi(x_j)^{\varphi(g)})$ and $q_j = x_j^g / x'_j$. Hence we need $y_g \in Q$ such that $(x'_j q_j)^{y_g} = x'_j$ for $1 \le j \le r$. This last equation is equivalent to

$$\left(y_g^{-1}\right)^{x_j} \cdot y_g \cdot q_j = 1, \tag{7.4}$$

which, as in the proof of Lemma 7.3.1, leads to a system of linear equations for the coordinates of y_g in the basis $\{z_1, \ldots, z_m\}$. The system of equations we have to solve consists of $mr \in O(\log^2 |G|)$ equations.

Given $G = \langle T \rangle$, we compute y_g for each $g \in T$. Then the set $\{g y_g \mid g \in T\}$ generates $N_G(P)$. Next, using Theorem 6.3.1, we construct an action of $N_G(P)$ with P in the kernel; repeating the foregoing procedure in the image, eventually we can discard the p'-part of $N_G(P)$ and so obtain a Sylow p-subgroup of G. Note that the number of recursive calls in the procedure is bounded by the maximal length of subgroup chains in G.

We can also solve the conjugation problem for Sylow subgroups in solvable groups.

Lemma 7.3.2. *Suppose that $Q \lhd G \le \mathrm{Sym}(\Omega)$, $|\Omega| = n$, Q is an elementary abelian q-group, and G/Q is a p-group for distinct primes p and q. Given Sylow p-subgroups P_1, P_2 in G, we can construct an element $y \in G$ conjugating P_1 to P_2 in nearly linear time by a deterministic algorithm.*

Proof. We start by constructing an SGS $\{z_1, \ldots, z_m\}$ for Q supporting coordinatization and constructing the homomorphism $\varphi : G \to S_n$ with kernel Q, as described in Theorem 6.3.1. We also construct the isomorphisms $\psi_1 : P_1 \to \varphi(G)$ and $\psi_2 : P_2 \to \varphi(G)$, which are the restrictions of φ to P_1 and P_2, respectively. The maps ψ_i define an isomorphism between P_i and G/Q, for $i = 1, 2$.

Since $P_1 Q = G$, there exists $y \in Q$ with $P_1^y = P_2$, and our aim is to construct a conjugating element in Q. For any $g \in P_1$ and $y \in Q$, the conjugate g^y is in the coset Qg; therefore, if y conjugates P_1 to P_2 then $g^y = \psi_2^{-1}(\psi_1(g))$ for all $g \in P_1$.

Based on the observations in the previous paragraph, it is straightforward to write a system of linear equations for the coefficients of a conjugating element y in the basis $\{z_1, \ldots, z_m\}$. Let $P_1 = \langle T \rangle$. The condition $g^y = \psi_2^{-1}(\psi_1(g))$ is equivalent to

$$(y^{-1})^g \cdot y = g^{-1} \psi_2^{-1}(\psi_1(g)). \tag{7.5}$$

Writing (7.5) for all $g \in T$ and noticing that $g^{-1}\psi_2^{-1}(\psi_1(g)) \in Q$, we obtain $|T|m$ equations for the coefficients of y. □

Theorem 7.3.3. *Given a solvable group $H \le \mathrm{Sym}(\Omega)$, with $|\Omega| = n$, and two Sylow p-subgroups P_1, P_2 of H, an element of H conjugating P_1 to P_2 can be constructed by a deterministic nearly linear-time algorithm.*

Proof. We start with the computation of a series of normal subgroups $1 = K_r \lhd K_{r-1} \lhd \cdots \lhd K_1 = H$ with elementary abelian quotients K_i/K_{i+1} and homomorphisms $\varphi_i : H \to S_n$ with kernel K_i as in Section 6.3.1, at the construction of $O_\infty(H)$. Then, recursively for $i = 1, \dots, r$, we construct $g_i \in \varphi_i(H)$ such that $\varphi_i(P_1)^{g_i} = \varphi_i(P_2)$. We can start with $g_1 := 1$. Suppose that g_i is already constructed for some i, and let $h_i \in H$ be an arbitrary preimage of g_i under φ_i^{-1}. If K_i/K_{i+1} is a p-group then $g_{i+1} := \varphi_{i+1}(h_i)$ conjugates $\varphi_{i+1}(P_1)$ to $\varphi_{i+1}(P_2)$. If K_i/K_{i+1} is a q-group for some prime $q \ne p$ then in the group $G := \langle \varphi_{i+1}(K_i), \varphi_{i+1}(P_2)\rangle$, the subgroup $Q := \varphi_{i+1}(K_i)$ is an elementary abelian normal q-subgroup with G/Q a p-group, and $\varphi_{i+1}(P_1^{h_i})$, $\varphi_{i+1}(P_2)$ are two Sylow p-subgroups of G. Hence we are in the situation of Lemma 7.3.2 and we can construct $y \in G$ conjugating $\varphi_{i+1}(P_1^{h_i})$ to $\varphi_{i+1}(P_2)$. Then we define $g_{i+1} := \varphi_{i+1}(h_i)y$. At the end of the recursion, $g_r \in \varphi_r(H) \cong H$ is the element of H conjugating P_1 to P_2. □

7.3.2. Conjugacy Classes in Solvable Groups

We defer the general discussion of conjugacy class computations to Section 9.4. Here we only describe an algorithm from [Mecky and Neubüser, 1989] to solve the following problem:

Let $Q \lhd G \le \mathrm{Sym}(\Omega)$, with $|\Omega| = n$, and suppose that Q is elementary abelian. Given representatives of the conjugacy classes of G/Q and the centralizers of the representatives in G/Q, construct representatives for the conjugacy classes of G and the centralizers of the representatives. (7.6)

Let $\{Qh_1, \dots, Qh_k\}$ be the set of representatives for the conjugacy classes of G/Q, and let $\bar{C}_i := C_{G/Q}(Qh_i)$ for $i \in [1, k]$. We denote the complete preimage of $\bar{C}_i$ in G by C_i.

Computation of the Class Representatives

For any $g \in G$, the conjugacy class of g intersects nontrivially exactly one of the cosets Qh_i (namely, the representative of the conjugacy class of Qg in G/Q). Hence we can find representatives for the conjugacy classes of G in the cosets $Qh_1, \ldots, Qh_k$. For fixed $i \in [1, k]$, two elements $g_1, g_2 \in Qh_i$ are conjugate in G if and only if $g_2 \in g_1^{C_i}$, since any element of $G \backslash C_i$ conjugates g_1 outside of the coset Qh_i. Therefore, we can obtain representatives for the conjugacy classes of G that intersect Qh_i nontrivially by computing the orbits of the conjugation action of C_i on Qh_i and taking a representative from each orbit.

The argument in the previous paragraph is valid for any normal subgroup Q of G. However, we can exploit the fact that Q is elementary abelian and so can be considered as a vector space, allowing us to make two possible shortcuts in the foregoing computation. To simplify notation, we drop the index i and consider an arbitrary coset $Qh \in \{Qh_1, \ldots, Qh_k\}$, with $\bar{C} := C_{G/Q}(Qh)$. We denote the complete preimage of $\bar{C}$ in G by C.

For $qh \in Qh$ and $c \in C$, we have $(qh)^c = q^c[c, h^{-1}]h$. For fixed c (and h), the map $A_c : Q \to Q$ defined by the rule $A_c : q \mapsto q^c[c, h^{-1}]$ is an affine map of the vector space Q, since $q \mapsto q^c$ is a linear transformation and $[c, h^{-1}]$ is a fixed element of Q, independent of q. Therefore the set of permutations of Qh, defined by the conjugation actions of the elements of C, correspond to a set of permutations defined by the affine transformations $\{A_c \mid c \in C\}$ of Q. We claim that $A_C := \{A_c \mid c \in C\}$ is in fact a group, and with the natural identification $\varphi : Qh \to Q$ defined as $\varphi : qh \mapsto q$, the conjugation action of C on Qh is permutation isomorphic to the action of A_C on Q. Indeed,

$$q^{c_1 c_2}[c_1 c_2, h^{-1}] = (q^{c_1}[c_1, h^{-1}])^{c_2}[c_2, h^{-1}],$$

since $[c_1 c_2, h^{-1}] = [c_1, h^{-1}]^{c_2}[c_2, h^{-1}]$. This last identity is a special case of Lemma 2.3.10, and it can also be checked directly.

The first shortcut is to compute an SGS $\{z_1, \ldots, z_m\}$ of Q supporting coordinatization and matrices in this basis for the affine transformations $A_T := \{A_c \mid c \in T\}$ for a generating set T of C. Then, instead of the orbits of the conjugation of C on Qh, we compute the orbits of $A_C = \langle A_T \rangle$ on Q and take the preimages of representatives of these orbits under φ^{-1}. The affine images of elements of the vector space Q can be computed much more quickly than the conjugates of the corresponding elements of Qh.

The second shortcut goes even further and computes the orbits of $A_C = \langle A_T \rangle$ on a potentially smaller factor space of Q instead of Q. Since $Q \lhd C$, Lemma 6.1.7 implies that the orbits of the subgroup $A_Q := \{A_c \mid c \in Q\}$ of A_C

are permuted by A_C. For $c \in Q$, we have $q^c = q$ for all $q \in Q$ and so A_c is just a translation of the vector space Q by the vector $[c, h^{-1}]$. Hence, for the 0 vector of Q, we have $0^{A_Q} = [Q, h^{-1}]$. This subspace $[Q, h^{-1}]$ can be computed easily, and the orbits of A_C on Q are the complete preimages of the orbits of A_C on the factor space $Q/[Q, h^{-1}]$. Let $\bar{A}_C$ denote the permutation action of A_C on $Q/[Q, h^{-1}]$. Note that $\bar{A}_C$ acts on $Q/[Q, h^{-1}]$ as a group of affine transformations as well, since for any $q, r \in Q$ and $c \in C$ we have

$$\begin{aligned}[q, h^{-1}]^c &= (q^{-1})^c h^c q^c (h^{-1})^c = (q^{-1})^c h[h, c] q^c (h^{-1})^c \\ &= (q^{-1})^c h q^c [h, c](h^{-1})^c = (q^{-1})^c h q^c h^{-1} = [q^c, h^{-1}]\end{aligned}$$

and so $([Q, h^{-1}]r)^c [c, h^{-1}] = [Q, h^{-1}] r^c [c, h^{-1}]$.

Computation of the Centralizers

To apply (7.6) recursively, we also have to compute the centralizers of the representatives of the conjugacy classes in G. Let $[Q, h^{-1}]q_1, \ldots, [Q, h^{-1}]q_l$ be the representatives of the orbits of $\bar{A}_C$ on $Q/[Q, h^{-1}]$ (and so $q_1 h, \ldots, q_l h$ are the representatives of those conjugacy classes of G that intersect Qh nontrivially). During the orbit computations in $Q/[Q, h^{-1}]$, we can build Schreier trees S_j that code elements of $\bar{A}_C$ carrying the representatives $[Q, h^{-1}]q_j$, $j \in [1, l]$, to other elements of their orbits (cf. the end of Section 2.1.1). The elements coded by S_j comprise a transversal for $\bar{A}_C$ mod $(\bar{A}_C)_{[Q,h^{-1}]q_j}$, and so some form of point stabilizer subgroup constructions can be used to obtain generators for $\bar{D}_j := (\bar{A}_C)_{[Q,h^{-1}]q_j}$. Note that we know $|\bar{D}_j|$ in advance, so the randomized methods from Chapter 4 give guaranteed correct output as well.

Let D_j denote the complete preimage of $\bar{D}_j$ in A_C. We have $A_Q \leq D_j$, since A_Q acts trivially on $Q/[Q, h^{-1}]$. In terms of the action of A_C on Q, the group D_j is the setwise stabilizer of $[Q, h^{-1}]q_j$. We still have to construct $C_G(q_j h)$, which is permutation isomorphic to the point stabilizer subgroup of D_j fixing q_j. For that, we do not need another orbit computation on $[Q, h^{-1}]q_j$; instead, we can use linear algebra.

As a first step, we construct $(A_Q)_{q_j}$. For $q \in Q$, we have $A_q(q_j) = q_j^q [q, h^{-1}]$, which is equal to q_j if and only if $[q, h^{-1}] = 1$, that is, $q \in C_Q(h) = C_Q(q_j h)$. Hence $(A_Q)_{q_j}$ can be obtained as the fixed point space of A_h, which is a linear transformation of Q.

Next, we claim that, for any $c \in D_j$,

(i) there is an element $r(c) \in Q$ such that $A_{cr(c)}$ fixes q_j; and
(ii) if A_{cr_1}, A_{cr_2} fix q_j for some $r_1, r_2 \in Q$ then $r_1 r_2^{-1} \in C_Q(h)$.

To show (i), observe that

$$A_c(q_j) = q_j^c[c, h^{-1}] = [q, h^{-1}]q_j \tag{7.7}$$

for some $q \in Q$, since A_c fixes $[Q, h^{-1}]q_j$. For that q, using (7.7) and using repeatedly that Q is abelian and that $[c, h^{-1}] \in Q$, we have

$$\begin{aligned} A_{cq^{-1}}(q_j) &= q_j^{cq^{-1}}[cq^{-1}, h^{-1}] = q_j^c[c, h^{-1}]^{q^{-1}}[q^{-1}, h^{-1}] \\ &= q_j^c[c, h^{-1}][q^{-1}, h^{-1}] = [q, h^{-1}]q_j[q^{-1}, h^{-1}] = q_j. \end{aligned}$$

Hence $r(c) := q^{-1}$ satisfies (i). Moreover, if $A_{cr_1}(q_j) = A_{cr_2}(q_j) = q_j$ then $[cr_1, h^{-1}] = [cr_2, h^{-1}]$, implying $[r_1, h^{-1}] = [r_2, h^{-1}]$ and $h^{r_1 r_2^{-1}} = h$. Therefore, (ii) holds.

We can construct $C_G(q_j h)$ as follows: For each generator c of $D_j = \langle T \rangle$, we construct $r(c)$ satisfying (i), by solving the system of linear equations

$$q_j^c[c, h^{-1}]q_j^{-1} = q^{-1} \cdot q^{h^{-1}}$$

for the coordinates of q in the basis $\{z_1, \ldots, z_m\}$ of Q supporting coordinatization. This equation was obtained by rewriting (7.7). The left-hand side is a known vector in Q, and the right-hand side can be expressed as a linear function of the coordinates of q. As we have seen in the previous paragraph, $r(c) := q^{-1}$ satisfies (i). Let $C_Q(h) = \langle S \rangle$. By (ii), the set $S \cup \{cr(c) \mid c \in T\}$ generates $C_G(q_j h)$.

The running time and memory requirement of the algorithm is proportional to $|Q/[Q, h^{-1}]|$, which may be exponentially large in terms of the input length. In general, an exponential blowup is unavoidable, since the length of the output may be this large. This happens, for example, if $Q \leq Z(G)$. Hence, the method should be used judiciously. For example, when computing the conjugacy classes of a solvable group, we may try to arrange the series of normal subgroups with elementary abelian factors guiding the recursion so that central extensions come at the end of the recursion. Moreover, we may have to restrict ourselves only to the counting of the conjugacy classes instead of listing representatives.

7.4. Two Algorithms for Nilpotent Groups

In this section, we describe two algorithms that use quite different methods than the top-down extension of information to larger and larger factor groups in Section 7.3.

7.4.1. A Fast Nilpotency Test

There are numerous characterizations of finite nilpotent groups (cf. Exercise 7.9), and quite a few of them can be utilized for testing whether a given permutation group G is nilpotent. Here we present a nilpotency test from [Rákóczi, 1995; 1997], which is extremely fast even in the case when $\log|G|$ is proportional to the permutation degree.

First, we describe an algorithm that decides whether a group $G = \langle S\rangle \le \mathrm{Sym}(\Omega)$, with $|\Omega| = n$, is a p-group for a given prime p. A possibility is to compute $|G|$ by the algorithm of Section 7.1, but we can do better. A permutation group is a p-group if and only if all of its transitive constituents are p-groups; hence we start with constructing the transitive constituents of G. This amounts to computing the orbits of G and can be done in $O(|S|n)$ time. If G has an orbit Δ such that $|\Delta|$ is not a power of p then we can conclude immediately that G is not a p-group.

For the timing of the next theorem, recall that the inverse Ackerman function $A^{-1}(m, n)$ was defined in Section 5.5.2.

Theorem 7.4.1. *Let $G = \langle S\rangle \le \mathrm{Sym}(\Omega)$ be transitive, with $|\Omega| = n = p^k$ for some prime p. It can be decided whether G is a p-group by a deterministic algorithm in $O(|S|n \log n A^{-1}(2|S|n, n))$ time.*

Proof. The algorithm is based on the following simple characterization of transitive p-groups. A transitive group $G \le \mathrm{Sym}(\Omega)$, with $|\Omega| = p^k$ for some $k \ge 1$, is a p-group if and only if

(i) G has a block system $\Sigma = \{\Sigma_1, \dots, \Sigma_p\}$;
(ii) G acts on Σ as the cyclic group C_p; and
(iii) if G_1 is the setwise stabilizer of Σ_1 in G then the restriction $G_1|_{\Sigma_1}$ is a p-group.

If G satisfies (i) and (ii) then G_1 is the kernel of action on Σ, since G acts regularly on Σ. Hence $|G| = p|G_1|$ and (iii) implies that G is a p-group. Conversely, Theorem 6.2.5 and Exercise 6.14 imply that the only primitive p-group is C_p in its natural action on p points, so if G is a transitive p-group then it has a block system satisfying (i) and (ii), and (iii) holds obviously. Therefore, we can test whether $G = \langle S\rangle$ is a p-group by checking whether it has a block system satisfying (i)–(iii).

The first step is a subroutine that constructs a block system of size p if G is a p-group and returns a block system of size p or $\{\Omega\}$ if G is not a p-group. We use the version of block computations described in Section 5.5.2. To this end, we need to construct some subset $\Delta \subseteq \Omega$, with $|\Delta| \geq 2$, such that if G is a p-group then there is a nontrivial block system with a block containing all points of Δ.

One way of finding such a set Δ is based on the observation that if G is a p-group and $g, h \in G$ are not commuting then $[g, h]$ fixes all blocks in any block system of size p for G, and so any nontrivial orbit of $[g, h]$ can be used as Δ. Another way is to take a nontrivial element g of $Z(G)$. By Lemma 6.1.7, the orbits of $\langle g \rangle$ are permuted by G. Therefore, if g has more than one orbit on Ω then any nontrivial orbit of g is suitable as Δ, whereas if g has only one orbit then any orbit of g^p suffices.

Let g be the first generator of G given in the generating set S. We compute the commutator of g with all other elements $h \in S$. If one of these commutators is nontrivial then the first method described in the previous paragraph for finding Δ can be applied, whereas if all commutators are equal to 1 then $g \in Z(G)$ and the second method can be used. Applying the algorithm of Section 5.5.2, we construct a block system Σ^* such that some block of Σ^* contains all points of Δ. If $\Sigma^* = \{\Omega\}$ then we conclude that G is not a p-group, and if $|\Sigma^*| > p$ then we construct the action of G on Σ^* and repeat the procedure recursively for this action. Eventually, we obtain a block system Σ for G satisfying (i), or we terminate with the conclusion that G is not a p-group.

If Σ is constructed then the second step of the algorithm checks whether the action of G on Σ satisfies (ii). Any element of S acting nontrivially on Σ must define a cyclic ordering of the blocks in Σ and it is straightforward to check whether the other generators respect this cyclic ordering.

The third step of the algorithm is the construction of generators for $G_1|_{\Sigma_1}$. We find some $g \in S$ acting nontrivially on Σ and construct the bijections $\varphi_j : \Sigma_1 \to \Sigma_j$ and their inverses for $j \in [2, p]$, defined by the powers $g, g^2, \ldots, g^{p-1}$ of g. These bijections can be used to construct the restrictions of the permutations in the set

$$T := \{g^i h g^{-j} \mid i \in [0, p-1],\ j \in [0, p-1],\ h \in S,\ g^i h g^{-j} \in G_1\}$$

to Σ_1. By Lemma 4.2.1, these restrictions generate $G_1|_{\Sigma_1}$.

After that, the same procedure as we just described can be applied recursively to determine whether $G_1|_{\Sigma_1}$ is a p-group. Note that $G_1|_{\Sigma_1}$ acts transitively on Σ_1, since for any $\alpha, \beta \in \Sigma_1$, there exists $x \in G$ with $\alpha^x = \beta$ and obviously $x|_{\Sigma_1} \in G_1|_{\Sigma_1}$.

Now we analyze the time requirement of the algorithm. In the first step of the algorithm, the set Δ is computed in $O(|S|n)$ time. By Corollary 5.5.9, the block system Σ^* is obtained in $O(|S|nA^{-1}(2|S|n, n))$ time. If $|\Sigma^*| > p$

then further block computations are required; however, these recursive calls of block computations are on sets of size decreasing by at least a factor p, so they contribute only a constant multiplier to the time complexity. The second step, checking the action on Σ, is in $O(|S|p)$ time. In the third step, the bijections φ_j and their inverses are constructed in $O(n)$ total time, because for each of the n/p elements $\alpha \in \Sigma_1$ and for each $j \in [2, p-1]$, $\varphi_{j+1}(\alpha) = \varphi_j(\alpha)^g$ can be obtained in constant time from $\varphi_j(\alpha)$. The $|S|p$ Schreier generators for $G_1|_{\Sigma_1}$ are computed in $O(|S|n)$ total time, since we construct permutations only on a domain of degree n/p. Hence the total time requirement of the three steps is $O(|S|nA^{-1}(2|S|n, n))$.

The recursive call is for the permutation group $G_1|_{\Sigma_1}$ of degree n/p with $|S|p$ generators, so the three steps of the algorithm for $G_1|_{\Sigma_1}$ require $O((|S|p)(n/p)A^{-1}(2|S|n, n/p)) = O(|S|nA^{-1}(2|S|n, n))$ time. Similarly, further recursive calls are on domains of degree decreasing by a factor p and with the number of generators increasing by a factor p, so the time requirement remains the same. Since the number of recursive calls is at most $\log n$, the total time requirement of the algorithm is $O(|S|n \log n A^{-1}(2|S|n, n))$, as claimed. □

The nilpotency test for permutation groups is based on the p-group test described in the proof of Theorem 7.4.1. A permutation group $G = \langle S \rangle \leq \mathrm{Sym}(\Omega)$, with $|\Omega| = n$, is nilpotent if and only if all of its transitive constituents are nilpotent, so again we start with the construction of the transitive constituents, in $O(|S|n)$ time. From now on, we suppose that G is transitive.

Let p be a prime divisor of n. If G is nilpotent then $G = P \times Q$, where P is a nontrivial p-group and Q is a nilpotent group of order not divisible by p. By Lemma 6.1.7, the orbits of P constitute a block system Σ^* for G. Since P is a p-group, the size of blocks in Σ^* is p^k for some $k \geq 1$. Moreover, as P is in the kernel of the action of G on Σ^*, the action of G on Σ^* has order relative prime to p. So, in particular, the divisor $|\Sigma^*|$ of this group order is relative prime to p and p^k is the largest power of p that divides n. The orbits of Q constitute another block system Σ^{**} for G. By the same argument we used for Σ^*, we obtain that $|\Sigma^{**}|$ is relative prime to each prime divisor of $|Q|$. Hence $|\Sigma^{**}| = p^k$.

Finally, we describe how can we get generators for P and Q. For a generator $s \in S$, let $|s| = p^l r$ with $(p, r) = 1$. Then $s_p := s^r \in P$ and $s_q := s^{p^l} \in Q$, and $s \in \langle s_p, s_q \rangle$. Hence $P = \langle s_p \mid s \in S \rangle$ and $Q = \langle s_q \mid s \in S \rangle$. Note that $|s|$ divides n since obviously $p^l \leq p^k$, and by induction on the number of different prime divisors of $|Q|$ we can prove that r divides n/p^k.

Lemma 7.4.2. *Let $G = \langle S \rangle \leq \mathrm{Sym}(\Omega)$, with $|\Omega| = n$, be transitive, let p be a prime divisor of n, and let $P := \langle s_p \mid s \in S \rangle$, $Q := \langle s_q \mid s \in S \rangle$ with s_p, s_q as* *defined in the previous paragraph. Let p^k be the largest power of p dividing n. Then G is nilpotent if and only if it satisfies all of the following properties:*

(i) *For each $s \in S$, $|s|$ divides n.*
(ii) *The orbits of P constitute a block system Σ^* for G.*
(iii) *The orbits of Q constitute a block system Σ^{**} for G.*
(iv) *$|\Sigma^*| = n/p^k$ and $|\Sigma^{**}| = p^k$.*
(v) *P acts as a p-group on its orbits.*
(vi) *Q acts as a nilpotent group on its orbits.*

Proof. We have seen that a transitive nilpotent group satisfies (i)–(vi). Conversely, suppose that a transitive group G satisfies (i)–(vi), and let N_1 and N_2 be the kernels of actions of G on Σ^* and Σ^{**}, respectively. We have $s \in \langle s_p, s_q \rangle$ for each $s \in S$, so $G = \langle P, Q \rangle$. Since $P \leq N_1$ and $Q \leq N_2$, we also have $G = \langle N_1, N_2 \rangle$. Moreover, $N_1 \cap N_2 = 1$, since the orbits of $N_1 \cap N_2$ constitute a block system that is a common refinement of Σ^* and Σ^{**}, and so by (iv) the orbits must have size 1. Hence $G = N_1 \times N_2$. Using again that $G = \langle P, Q \rangle$ and $P \leq N_1$, $Q \leq N_2$, we obtain $G = P \times Q$. By (v) and (vi), G is nilpotent. □

Corollary 7.4.3. *Let $G = \langle S \rangle \leq \mathrm{Sym}(\Omega)$, with $|\Omega| = n$, be transitive. Then we can test whether G is nilpotent in $O(|S|n \log n A^{-1}(2|S|n, n))$ time by a deterministic algorithm.*

Proof. We have to show that properties (i)–(vi) in Lemma 7.4.2 can be checked in the given time bound. The prime factorization of n can be obtained in $O(\sqrt{n}\log^2 n)$ time by brute force. For each $s \in S$, we can check whether $s^n = 1$ and, if the answer is yes, we can compute s_p and s_q in $O(n \log n)$ time. The orbits of P and Q can be determined, and properties (ii), (iii), and (iv) can be checked, in $O(|S|n)$ time. By Theorem 7.4.1, (v) can be checked in $O(|S|n \log n A^{-1}(2|S|n, n))$ time. Finally, (vi) is checked by a recursive call. Since the orbits of Q have size $n/p^k \leq n/2$, and further recursive calls are for groups of degree decreasing at least by a factor 2, the recursive calls contribute only a constant multiplier to the overall running time. □

7.4.2. The Upper Central Series in Nilpotent Groups

As we have seen in Section 6.1.3, the center of a group is easily computable. To obtain the upper central series, this computation should be iterated in factor groups. However, as we pointed out repeatedly, it is possible that even factor groups of p-groups $G \leq S_n$ can have no faithful permutation representations on less than $2^{n/4}$ points (cf. Exercise 6.6). Hence a completely new approach is necessary. In this section we present the nearly linear-time upper central series algorithm for nilpotent groups from [Seress, 1998]. The method is based on ideas from [Kantor and Luks, 1990].

As in Section 6.1.1, given $G, H \leq \mathrm{Sym}(\Omega)$, let Ω_1, Ω_2 be disjoint copies of the set Ω and introduce the subgroups $D = \mathrm{Diag}(G \times G) := \{(g, g) \mid g \in G\}$, $1 \times H := \{(1, h) \mid h \in H\}$, and $K := \langle D, 1 \times H \rangle$. The following lemma is a special case of the discussion in [Kantor and Luks, 1990, Section 6].

Lemma 7.4.4. *If $H \lhd G$ then $\mathrm{Core}_{G \times G}(K)$ acts as $H \cdot Z(G/H)$ on Ω_1.*

Proof. Denote the restriction of $\mathrm{Core}_{G\times G}(K)$ to Ω_1 by C. Then C contains H because $H \times H \lhd G \times G$ and $H \times H \leq K$. Also, $K = \{(g, gh) \mid g \in G, h \in H\}$ since G normalizes H. Now, for an arbitrary $(g, gh) \in K$,

$$
\begin{aligned}
&(g, gh) \in \mathrm{Core}_{G\times G}(K) \\
&\quad \Leftrightarrow (g, gh) \in K^{(y_1, y_2)} \text{ for all } y_1, y_2 \in G \\
&\quad \Leftrightarrow (g, gh)^{(y_1, y_2)} \in K \text{ for all } y_1, y_2 \in G \\
&\quad \Leftrightarrow (g, gh)^{\left(1, y_2 y_1^{-1}\right)} \in K \text{ for all } y_1, y_2 \in G \text{ (since } K^{(y_1, y_1)} = K) \\
&\quad \Leftrightarrow (g, g^y h^y) = (g, g(g^{-1} g^y) h^y) \in K \text{ for all } y \in G \\
&\quad \Leftrightarrow g^{-1} g^y \in H \text{ for all } y \in G.
\end{aligned}
$$

The last condition means that g and g^y are in the same coset of H for all $y \in G$; in other words, $Hg \in Z(G/H)$. □

Lemma 7.4.4 reduces the construction of $Z(G/H)$ to a core computation. By the algorithm of Section 6.1.5, we can compute the core of a subnormal subgroup in nearly linear time; hence, we need to determine in which cases is K subnormal in $G \times G$. The necessary and sufficient condition for that is provided by the following lemma.

Lemma 7.4.5. *$K \lhd\!\lhd G \times G$ if and only if G/H is nilpotent.*

Proof. Define $K_0 := G \times G$ and $K_i := \langle K^{K_{i-1}} \rangle$ for $i > 0$. Also, let $L_0 := G$, $L_i := [G, L_{i-1}]$ for $i > 0$ denote the lower central series of G. We claim that, for all $i \geq 0$,

$$K_i = \langle D, 1 \times H L_i \rangle. \tag{7.8}$$

We prove (7.8) by induction on i. The case $i = 0$ is trivial. The inductive step is based on the identity that for $g \in G$, $h_1, h_2 \in H$, and $l \in L_i$,

$$\begin{aligned}(g, gh_1)^{(1,h_2l)} &= \left(g, g^{h_2l} h_1^{h_2l}\right) \\ &= \left(g, g\left(g^{-1}l^{-1}h_2^{-1}gh_2l\right)h_1^{h_2l}\right) = \left(g, g\left([g,l]l^{-1}g^{-1}h_2^{-1}gh_2l\right)h_1^{h_2l}\right) \quad (7.9) \\ &= \left(g, g[g,l][g,h_2]^l h_1^{h_2l}\right) = (g,g)\left(1, [g,l][g,h_2]^l h_1^{h_2l}\right).\end{aligned}$$

Suppose that (7.8) holds for some $i \geq 0$. Since $K^d = K$ for all $d \in D$, the group $K_{i+1} = \langle K^{K_i} \rangle$ is generated by elements of the form $(g, gh_1)^{(1,h_2l)}$ for $g \in G$, $h_1, h_2 \in H$, and $l \in L_i$. By (7.9), all generators are in $\langle D, 1 \times HL_{i+1} \rangle$ since $[g,l] \in L_{i+1}$ and $[g,h_2]^l, h_1^{h_2l} \in H$. Conversely, $\langle D, 1 \times HL_{i+1} \rangle$ is generated by D and elements of the form $(1, [g,l]h)$ for $g \in G, h \in H$, and $l \in L_i$. Applying (7.9) backward with the notation $h_1 := h^{l^{-1}}$ and $h_2 := 1$, we obtain that

$$(1, [g,l]h) = \left(1, [g,l][g,1]^l \left(h^{l^{-1}}\right)^l\right) = (g^{-1}, g^{-1})\left(g, gh^{l^{-1}}\right)^{(1,l)} \in K_{i+1}.$$

Hence (7.8) holds for $i + 1$.

To finish the proof of the lemma, observe that HL_i is the ith term in the lower central series of G/H. Hence, if G/H is nilpotent then $HL_i = H$ for some i and (7.8) implies that $K = K_i$ for the same i. By Exercise 6.5, this means that $K \lhd\lhd G \times G$. Conversely, if $K_i = K$ for some i then (7.8) implies that for all $x \in HL_i$ we have $(1, x) = (g, gh)$ for some $g \in G$ and $h \in H$. This means $g = 1$ and $h = x$, so $HL_i = H$ and G/H is nilpotent. □

In particular, if G is nilpotent then we can compute its upper central series in nearly linear time by a deterministic algorithm. A practical concern is that some of the computations have to be carried out on a permutation domain that is four times larger than the original one, since the core construction doubles the permutation domain of $G \times G$. Fortunately, numerous shortcuts are possible. For example, (7.8) implies that generators for the subnormal series between K and $G \times G$ can be computed on the original permutation domain via the lower central series of G.

Exercises

7.1. Modify the solvable SGS construction of Section 7.1 such that the output polycyclic generating sequence defines a subgroup chain with prime indices.

7.2. Prove that if $G = \langle H, y\rangle \le \mathrm{Sym}(\Omega)$, y normalizes H, and S is an SGS for H then the algorithm of Lemma 7.1.2 constructs an SGS S^* for G relative to some base $(\beta_1, \ldots, \beta_M)$ such that for all $i \in [1, M]$ the ratio $|\beta_i^{G_{(\beta_1,\ldots,\beta_{i-1})}}|/|\beta_i^{H_{(\beta_1,\ldots,\beta_{i-1})}}|$ of the fundamental orbit sizes divides $|G : H|$. In particular, if $|G : H|$ is a prime number then
   (i) the fundamental orbits of H and G differ for exactly one base point;
   (ii) all elements in $S^*\backslash S$ are powers of some $z \in G\backslash H$ belonging to the coset Hy.

7.3. Let $H \le G$ and suppose that $|H| = p^m$ for a prime p. Prove that if H is not a Sylow p-subgroup of G then there exists $h \in Z(H)\backslash\{1\}$ such that $|C_G(h)|$ is divisible by p^{m+1}.

7.4. Let $N \lhd G$ and let P be a Sylow p-subgroup of N. Prove that $G = N_G(P)N$.

7.5. [Kantor and Taylor, 1988] Modify the methods of Section 7.3.1 to construct and conjugate Hall subgroups of solvable groups.

7.6. Use the techniques of Section 7.3 to design a nearly linear-time deterministic algorithm to construct a Sylow p-subgroup of a solvable group $G \le \mathrm{Sym}(\Omega)$ that contains a given p-subgroup of G.

7.7. Use the techniques of Section 7.3 to design a nearly linear-time constructive version of one half of the Schur–Zassenhaus theorem. Namely, given $N \lhd G \le \mathrm{Sym}(\Omega)$ with $(|N|, |G/N|) = 1$ and N solvable, construct a complement H for N in G. Furthermore, given two complements H_1, H_2 of N, construct some $g \in G$ that conjugates H_1 to H_2.

7.8. How can random subproducts speed up the algorithms for computing strong generating sets in solvable groups (Section 7.1), cores of subnormal subgroups (Section 6.1.5), and Sylow subgroups of solvable groups (Section 7.3.1)?

7.9. Show that the following are equivalent for a finite group G.
   (i) The lower central series of G terminates at 1.
   (ii) Every subgroup of G is subnormal.
   (iii) Every proper subgroup of G is properly contained in its normalizer.
   (iv) Every maximal subgroup of G is normal.
   (v) The Sylow subgroups of G are normal in G.
   (vi) G is the direct product of p-groups.
   (vii) The upper central series of G terminates at G.

8

Strong Generating Tests

The fast constructions of strong generating sets are randomized and in practice we often use heuristics; therefore, it is necessary to test the correctness of the output. Given an input group $G = \langle S \rangle \leq \mathrm{Sym}(\Omega)$, suppose that we have constructed an alleged base $B = (\beta_1, \ldots, \beta_m)$, generating sets S_i for each $G^{[i]} := G_{(\beta_1,\ldots,\beta_{i-1})}$, and transversals R_i for $\langle S_i \rangle$ mod $\langle S_i \rangle_{\beta_i}$. The correctness of the computation can be checked by forming all Schreier generators from S_i, R_i for $i = 1, 2, \ldots, m$ and sifting them through our data structure. However, this method is too slow, and trying to avoid the formation of all Schreier generators was the primary reason why we resorted to randomized constructions.

So far, we have seen two methods to check the result of an SGS computation. The most simple-minded way, which is very useful in practice, is to compare $\prod_i |R_i|$ and $|G|$ if the order of G is known in advance. Errors in SGS computations are one-sided in the sense that we never put permutations into S_i that are not in $G^{[i]}$, and the possible error is that S_i generates only a proper subgroup of $G^{[i]}$. Therefore, if $\prod_i |R_i| = |G|$ then the construction is correct.

The second method we have encountered is the Monte Carlo strong generating test described in Lemma 4.5.6. Although this procedure is reasonably fast in practice if the SGS construction used one of the shallow Schreier tree construction methods of Section 4.4, the drawback is that we cannot guarantee correctness with absolute certainty.

In this chapter, we describe three further methods to check the correctness of an SGS computation. By Lemma 4.2.3, it is enough to check that $\langle S_i \rangle_{\beta_i} = \langle S_{i+1} \rangle$ holds for each i. To this end, the first two methods we present solve the problem (4.6) in Section 4.3. Recall that (4.6) asks for deciding whether $H = G_\beta$, given the input $G = \langle S \rangle$, a transversal for G mod G_β, and $H \leq G_\beta$. The third method departs radically from the traditional point stabilizer approach and is based on structural properties of G. By the time it confirms the correctness of the SGS, it also computes a composition series and a short presentation for G.

8.1. The Schreier–Todd–Coxeter–Sims Procedure

One possibility for solving the problem (4.6) in Section 4.3 is a combination of coset enumeration and the Schreier generator approach. The idea was originally proposed in some unpublished notes of Sims in 1974 and in [Sims, 1978], and the algorithm was developed in [Leon, 1980b]. A brief summary of the method is in [Leon, 1980a].

We start with a crash course on coset enumeration from [Seress, 1997], copyright © 1997 American Mathematical Society. Reprinted by Permission. A very accessible and more thorough introduction to the subject can be found in [Neubüser, 1982], and the monograph [Sims, 1994] contains a comprehensive treatment of finitely presented groups.

8.1.1. Coset Enumeration

Let $G = \langle E \mid \mathcal{R} \rangle$ be a presentation for a group G, where $E = \{g_1, \ldots, g_n\}$ is a finite set of generators, and $\mathcal{R} = \{r_1 = 1, \ldots, r_m = 1\}$ is a set of defining relations. Each r_i is a word, using the generators in E and their inverses. The basic questions are to decide whether G is finite and to determine whether a given word represents the identity of G.

By the celebrated result of Novikov and Boone (cf. [Rotman, 1995, Chap. 12]), these questions are *undecidable*: They cannot be answered by a recursive algorithm. Nevertheless, because of the practical importance of the problem, considerable effort has been devoted to the development of methods for investigating finitely presented groups.

One basic method is the *Todd–Coxeter coset enumeration procedure*. Given $G = \langle E \mid \mathcal{R} \rangle$ and $H = \langle h_1, \ldots, h_k \rangle$, where $H \leq G$ and each h_j is a word in the generators of G and their inverses, our goal is to compute the permutation representation of G on the right cosets of H.

We set up a *coset table*. This is a matrix M, where the rows are labeled by positive integers, representing cosets of H, and the columns are labeled by the elements of $\bar{E} := \{g_1, \ldots, g_n, g_1^{-1}, \ldots, g_n^{-1}\}$. The entries (if defined) are positive integers: $M(k, g) = l$ if we know that $kg = l$ for the cosets k, l and for $g \in \bar{E}$. Originally, we have a $1 \times |\bar{E}|$ table with no entries, where 1 denotes the coset $H \cdot 1$. As new cosets are defined, we add rows to the coset table.

Of course, we have to detect when two words, defining different rows of the table, actually belong to the same coset of H. To this end, for each relation $r_i = g_{i_1} g_{i_2} \cdots g_{i_t}$, we also maintain a *relation table*. This is a matrix M_i, with rows labeled by the cosets $1, 2, \ldots$ as defined in M and columns labeled by the elements of the sequence $(g_{i_1}, g_{i_2}, \ldots, g_{i_t})$. The entry $M_i(k, g_{i_j})$, if defined, is the number of the coset $kg_{i_1}\cdots g_{i_j}$. Initially, we have $M_i(k, g_{i_t}) = k$ for each row number k, since $r_i = 1$ in G. Whenever a new coset is defined, we fill all entries of the relation tables that we can.

Finally, for each generator $h_j = g_{j_1}\cdots g_{j_t}$ of H, we maintain a *subgroup table*. This is a matrix S_j with only one row, corresponding to the coset $H\cdot 1$, and columns labeled by the factors of h_j. The rule for filling entries is the same as for the M_i; originally, $S_j(1, g_{j_t}) = 1$, since $Hh_j = H$.

When the last entry is filled in a row of a relation table or a subgroup table, we also get an extra piece of information $kg = l$ for some cosets k, l and $g \in \bar{E}$. This is called a *deduction*. If the entry $M(k, g)$ is not yet defined then we fill the entries $M(k, g)$, $M(l, g^{-1})$ and all possible entries in the relation and subgroup tables; in this way, we may get further deductions. If $M(k, g)$ is already defined but $l' := M(k, g) \neq l$, then we realize that l, l' denote the same coset of H. This is called a *coincidence*. We replace all occurrences of l, l' by the smaller of these two numbers and fill the entries of the tables that we can. This may lead to further deductions and coincidences. The process stops when all entries of the coset table, the relation tables, and subgroup tables are filled.

We illustrate these ideas by enumerating $G = \langle g_1, g_2 \mid g_1^2 = 1, g_2^2 = 1, (g_1g_2)^3 = 1\rangle \cong S_3$ on the cosets of the subgroup $H = \langle g_1g_2g_1g_2\rangle$ of order 3. Since both generators are involutions, we have $\bar{E} = E$. Also, we maintain only one relation table, corresponding to $(g_1g_2)^3 = 1$; the other two relators tell us that, at the definition of new cosets, we should multiply previous cosets by g_1, g_2 alternatingly. The display

CT	g_1	g_2
1	2	3
2	1	4
3		1
4		2

RT	g_1	g_2	g_1	g_2	g_1	g_2
1	2	4			3	1
2	1	3			4	2
3			4	2	1	3
4			3	1	2	4

(8.1)

ST	g_1	g_2	g_1	g_2
1	2	4	3	1

shows the coset table CT, relation table RT, and subgroup table ST after the definition of the cosets $1 := H, 2 := 1g_1, 3 := 1g_2, 4 := 2g_2$. At that moment, the last entry (in the second column) of ST is filled and we get the deduction $4g_1 = 3$, which also implies $3g_1 = 4$. Then all entries in CT are known, and we can complete RT; this leads to the coincidences $1 = 4$ and $2 = 3$.

There are different strategies for coset enumeration. The version we presented here, where we fill all consequences of new definitions, deductions, and coincidences in all relation tables as soon as possible, is called the Felsch strategy. However, it is possible to postpone the filling of some relation tables: If we ensure that if a coset k is defined then sooner or later we define kg for all $g \in \bar{E}$, and sooner or later we deduce all consequences of these definitions, then it is guaranteed that the algorithm terminates if $|G : H| < \infty$. This remark plays an important role in the theoretical justification of the algorithm described in Section 8.1.2, where we do not have a full presentation at the start of coset enumeration and, as the coset enumeration progresses, we add relators to $\mathcal{R}$. From a theoretical point of view, we can suppose that all relators are already present at the beginning; we just postpone the filling of entries in some relation tables.

There is no recursive function of $|G : H|$ and the input length that would bound the number of cosets defined during the Todd–Coxeter procedure. It is easy to give presentations of the trivial group such that no commonly used variant of the Todd–Coxeter algorithm can handle them. This, and different coset enumeration strategies, are discussed, for example, in [Sims, 1994, Chap. 5]. Success mostly depends on the number of entries defined in the coset table rather than $|G : H|$. There are instances of successful enumeration with $|G : H| > 10^6$.

8.1.2. Leon's Algorithm

Now we are ready to present the strong generating test using coset enumeration. Leon calls his procedure the Schreier–Todd–Coxeter–Sims (STCS) method.

The factorization of Schreier generators as products of coset representatives gives a set of defining relations (cf. Exercise 5.2). Hence, shifting the Schreier generators can be interpreted as checking that each relation holds. However, this relator set is usually highly redundant. The idea behind STCS is to find a smaller set $\mathcal{R}$ of relations and check only the elements of $\mathcal{R}$.

In addition to the notation of (4.6) in Section 4.3, suppose that $|G : G_\beta| = n$ and we already have a presentation $\langle E_1 \mid \mathcal{R}_1 \rangle$ for H. The latter assumption is satisfied automatically as we process groups along a stabilizer chain.

The STCS Algorithm

We have to verify that $|G : H| = n$ or find an element of $G_\beta \backslash H$. To this end, we perform a coset enumeration of G on the cosets of H. If H has n cosets then $H = G_\beta$. We work with a presentation $\langle E \mid \mathcal{R} \rangle$, where E is a set of symbols containing E_1 and a symbol for each generator of G. However, at the beginning we do not have relators for G and initialize $\mathcal{R} := \mathcal{R}_1$. During the procedure, we add relators to $\mathcal{R}$. The algorithm verifies that for each relator added to $\mathcal{R}$, the product of the corresponding elements of G is 1. Let E^* denote the set of finite words over the alphabet E, and let $\varphi : E^* \to G$ be the map that evaluates words in E^* in G.

We fix a number $m > n$ and we interrupt the coset enumeration when m cosets are defined. There are three possibilities:

Case 1: The coset table closes with n cosets. In this case, $|G : H| = n$ or, equivalently, $H = G_\beta$.

Case 2: The coset table closes with more than n cosets. Then $H \neq G_\beta$ and an element of $G_\beta \backslash H$ can be obtained by choosing two words w_1, w_2 from the coset table that represent different cosets of H but $\beta^{\varphi(w_1)} = \beta^{\varphi(w_2)}$. Then $\varphi(w_1 w_2^{-1}) \in G_\beta \backslash H$.

Case 3: The coset enumeration was interrupted after defining m cosets. Then, as in Case 2, there are words w_1, w_2 in the coset table such that $\beta^{\varphi(w_1)} = \beta^{\varphi(w_2)}$. This means that $g := \varphi(w_1 w_2^{-1}) \in G_\beta$; we test whether $g \in H$. If the answer is "no" then we have found a witness for $G_\beta \neq H$. If the answer is "yes" then we sift g as a word in H and obtain a word t in the strong generators of H such that the product of permutations in t is g. Let w denote the word corresponding to t in E^*. We add the relator $w_1 w_2^{-1} w^{-1} = 1$ to $\mathcal{R}$ and perform the collapse $w_1 = w_2$ in the coset table. Then we resume the coset enumeration.

Owing to the elusive nature of coset enumeration, it is hard to analyze what happens. The finiteness of the procedure can be ensured by a careful choice of the words w_1, w_2 in Step 3. Namely, at each interruption of the coset enumeration, we build a breadth-first-search tree T that contains the already defined cosets as vertices. The tree is built the same way as at orbit calculations (cf. Section 2.1.1). The root is coset 1, corresponding to H. After the ith level L_i is defined, let L_{i+1} consist of those cosets k that do not occur on some already constructed level and were defined as $k := lg$ for some $l \in L_i$, $g \in E \cup E^{-1}$. This construction can be easily done using the (partial) coset table we have at the interruption. Since there are m cosets, the depth of T is at most m and each coset k is represented by a word w_k of length at most m in $E \cup E^{-1}$. Also, for each vertex k of T, we compute $\beta^{\varphi(w_k)}$. Recall that we choose cosets k_1, k_2 such that $\beta^{\varphi(w_{k_1})} = \beta^{\varphi(w_{k_2})}$ and add a relator of the form $w_{k_1} w_{k_2}^{-1} w$ to $\mathcal{R}$, where w is a word in $E_1 \cup E_1^{-1}$ and w is determined uniquely by w_{k_1} and w_{k_2}. Hence there are only finitely many possibilities to add a relator to $\mathcal{R}$ and so the procedure must terminate. In the implementation, whenever a coset k is added to T, $\beta^{\varphi(w_k)}$ is computed immediately and the construction of T can be abandoned as soon as we encounter two words with $\beta^{\varphi(w_{k_1})} = \beta^{\varphi(w_{k_2})}$.

Of course, we need much more than finiteness of the procedure. Leon experimented extensively with the value of the parameter m and the coset enumeration technique to be used. He chose the value $m = 1.2n$ and developed a generalization of one of the basic strategies for coset enumeration, the Haselgrove–Leech–Trotter method. Note that if $m < 2n$ then Case 2 cannot occur, since $|G : H| > n$ means $|G_\beta : H| \geq 2$ and so $|G : H| = |G : G_\beta| \cdot |G_\beta : H| \geq 2n$.

When the entire procedure terminates in Case 1, we have a presentation $\langle E \mid \mathcal{R} \rangle$ for a group $\bar{G}$ that has a subgroup $\bar{H}$ of index n, and G satisfies this presentation. Hence there is a homomorphism $\psi : \bar{G} \to G$ such that $\psi(\bar{H}) = H$. However, since relators for H are a subset of $\mathcal{R}$, we must have $\bar{H} \cong H$, $\ker(\psi) = 1$, and $\bar{G} \cong G$. Hence we obtain a presentation for G, which is usually much shorter than the one described in Exercise 5.2.

The STCS algorithm is used in MAGMA.

8.2. Sims's Verify Routine

We present another solution for (4.6) in Section 4.3. This method was developed by Sims in the early 1970s when he constructed a permutation representation for Lyon's simple group, but it has not been published. Lecture notes from Sims have circulated at least since 1972, and the procedure is implemented both in GAP and MAGMA. Again, we use the notation of (4.6).

Lemma 8.2.1. *Suppose that for all $\delta \in \beta^G$ there exist sets $U(\delta) \subseteq G$ satisfying the following properties:*

(i) $(\forall u \in U(\delta))\,(\beta^u = \delta)$ *and* $U(\beta) \subseteq H$*;*
(ii) $(\forall u, v \in U(\delta))\,(uv^{-1} \in H)$*; and*
(iii) $(\forall x \in S)\,(\exists u \in U(\delta))\,(\exists v \in U(\delta^x))\,(uxv^{-1} \in H)$.

Then $H = G_\beta$.

Proof. We use the idea of the proof of Lemma 4.2.1. Let $g \in G_\beta$ be arbitrary. Since $G_\beta \leq G$, g can be written in the form $g = x_1 x_2 \cdots x_k$ for some nonnegative integer k and $x_i \in S$ for $i \leq k$. Let $\delta_i := \beta^{x_1 \cdots x_i}$, for $0 \leq i \leq k$.

By (iii), for each $i \in [1, k]$ there exist $u_{i-1} \in U(\delta_{i-1})$ and $v_i \in U(\delta_i)$ such that $u_{i-1} x_i v_i^{-1} \in H$. Also, $v_i u_i^{-1} \in H$ by (ii) and $u_0^{-1}, v_k \in H$ by (i). Hence

$$\begin{aligned} g &= x_1 x_2 \cdots x_k \\ &= u_0^{-1}\left(u_0 x_1 v_1^{-1}\right)\left(v_1 u_1^{-1}\right)\left(u_1 x_2 v_2^{-1}\right) \cdots \left(v_{k-1} u_{k-1}^{-1}\right)\left(u_{k-1} x_k v_k^{-1}\right) v_k \in H. \end{aligned}$$

Since $g \in G_\beta$ was arbitrary, we have $H = G_\beta$. □

Let $H = \langle S_1 \rangle$. We consider the special case when $S = S_1 \cup \{z\}$; that is, G is generated by H and one additional element. We would like to define sets $U(\delta)$ that can be used in (i)–(iii) of Lemma 8.2.1.

Let $\Delta_1 = \{\beta\}$ and let $\beta^G = \Delta_1 \cup \cdots \cup \Delta_l$ be the decomposition of the G-orbit of β into orbits of H. For each $i \in [1, l]$, we fix a representative $\delta_i \in \Delta_i$ and $u(\delta_i) \in G$ such that $\beta^{u(\delta_i)} = \delta_i$. We may choose $u(\delta_1) := 1$. After that, for $\delta \in \Delta_i$, we define

$$U(\delta) := \left\{u(\delta_i)h \mid h \in H, \delta_i^h = \delta\right\}. \tag{8.2}$$

It is clear that the sets $U(\delta)$ satisfy (i).

Recall that we consider the special case when G is generated by H and one additional element z. Let $\gamma := \beta^{(z^{-1})}$ and, for $1 \le i \le l$, let $\Delta_i = \Gamma_{i1} \cup \cdots \cup \Gamma_{il_i}$ be the decomposition of Δ_i into orbits of H_γ. We fix a representative γ_{ij} of each H_γ-orbit Γ_{ij}. Since δ_i and γ_{ij} are in the same H-orbit, we can also fix $y(\gamma_{ij}) \in H$ such that $\delta_i^{y(\gamma_{ij})} = \gamma_{ij}$. We also define $u(\gamma_{ij}) := u(\delta_i)\, y(\gamma_{ij})$. Finally, if the representative of the H_γ-orbit of γ_{ij}^z is γ_{kj_k} then we fix $v(\gamma_{ij}^z) := u(\gamma_{kj_k})y$ for some $y \in H_\gamma$ such that $\beta^{v(\gamma_{ij}^z)} = \gamma_{ij}^z$.

Lemma 8.2.2. *If the following conditions* (*a*), (*b*), *and* (*c*) *hold then the sets* $U(\delta)$ *defined in* (8.2) *satisfy* (ii) *and* (iii) *of Lemma* 8.2.1.

(*a*) $H_{\delta_i}^{u(\delta_i)^{-1}} \le H$ *for all* $i \in [1, l]$;
(*b*) $u(\gamma_{ij})\, z\, v(\gamma_{ij}^z)^{-1} \in H$ *for all* $i \in [1, l]$ *and for all* $j \in [1, l_i]$; *and*
(*c*) $H_\gamma^z \le H$.

Proof. Let us fix $\delta \in \Delta_i$. First we check (ii). If $u(\delta_i)h_1,\ u(\delta_i)h_2 \in U(\delta)$ then ${\delta_i}^{h_1} = {\delta_i}^{h_2} = \delta$ and so $h_1 h_2^{-1}$ fixes δ_i. Hence the ratio

$$u(\delta_i)h_1\,(u(\delta_i)h_2)^{-1} = u(\delta_i)\big(h_1h_2^{-1}\big)\,u(\delta_i)^{-1} \in u(\delta_i)H_{\delta_i}u(\delta_i)^{-1},$$

and by (a) it is in H.

For (iii), we have to consider two cases: $x \in S_1$ or $x = z$. If $x \in S_1$ then let $u = u(\delta_i)h \in U(\delta)$ be arbitrary, and $v := ux$. Note that δ^x is in the same H-orbit Δ_i as δ, so $v = u(\delta_i)\,(hx) \in U(\delta^x)$. Moreover, $uxv^{-1} = 1 \in H$.

Finally, let $x = z$. Let γ_{ij} be the representative of the H_γ-orbit of δ, and fix $w \in H_\gamma$ such that $\gamma_{ij}^w = \delta$. Define $u := u(\gamma_{ij})\,w$. Then $u = u(\delta_i)\,(y(\gamma_{ij})w) \in U(\delta)$. We also define $v := v(\gamma_{ij}^z)\,w^z$. By (c), $w^z \in H$, so, if the representative of the H_γ-orbit of γ_{ij}^z is γ_{kj_k}, then $v = u(\delta_k)\,h$ for some $h \in H$. We have $\beta^v = \beta^{v(\gamma_{ij}^z)\,w^z} = (\gamma_{ij}^z)^{z^{-1}wz} = \delta^z$, and so $v \in U(\delta^z)$. By (b),

$$uzv^{-1} = u(\gamma_{ij})\,wz(z^{-1}w^{-1}z)\,v\big(\gamma_{ij}^z\big)^{-1} \in H,$$

so (iii) holds. □

The Verify Routine

To solve (4.6), we have to check whether (a), (b), and (c) of Lemma 8.2.2 are satisfied. On one hand, it is clear that $H_{\delta_i}^{u(\delta_i)^{-1}} \le G_\beta$, $u(\gamma_{ij})\, z\, v(\gamma_{ij}^z)^{-1} \in G_\beta$, and $H_\gamma^z \le G_\beta$, so if (a), (b), and (c) do not hold then we have constructed a witness for $H \ne G_\beta$. On the other hand, if (a), (b), and (c) are satisfied then, by Lemmas 8.2.2 and 8.2.1, we have $H = G_\beta$.

The required orbit computations of this algorithm are deterministic and can be done in nearly linear time. If the transversals for the point stabilizer chain of H are stored in shallow Schreier trees then each sifting required in (b), and each base change required in (a) and (c), can also be executed by a nearly linear-time deterministic or by a nearly linear-time Las Vegas algorithm. Hence the time requirement of the algorithm is $O(nM \log^c |G|)$, where n is the size of the permutation domain and M is the number of H_γ-orbits in β^G.

Now we handle the general case. Let $S = \{s_1, \ldots, s_k\}$ and $G(i) := \langle H, s_1, \ldots, s_i \rangle$. Recursively for $i = 1, 2, \ldots, k$, we shall check whether $G(i)_\beta = H$.

The case $i = 1$ is discussed in the previous paragraph. Suppose that we have verified that $G(i-1)_\beta = H$ for some $i \ge 2$ and have constructed a Schreier tree data structure for $G(i-1)$. Our next aim is to decide whether $G(i)_\beta = H$.

We sift s_i in $G(i-1)$. If $s_i \in G(i-1)$ then $G(i) = G(i-1)$ and we are done. If s_i has a nontrivial siftee g then there are two cases. The first one is that $\beta^g = \beta$, which means that $g \in G(i)_\beta \backslash G(i-1)$ and so it is a witness for $G(i)_\beta \ne H$. The second case is that $\beta^g \ne \beta$. This can happen only if the sifting procedure failed on the first level because $\beta^g \in \beta^{G(i)} \backslash \beta^{G(i-1)}$. In this case, let $\Delta := \beta^{G(i)}$ and $\Gamma := \beta^{G(i-1)}$. If indeed $G(i)_\beta = H$ then $G(i) \gneqq G(i-1) \ge G(i)_\beta$ and Γ is a block of imprimitivity for the G-action on Δ, so our first task is to check whether Γ is a block. This is exactly the situation we have encountered in Section 5.5.1, in Step 4 of the construction of a block system. By Lemma 5.5.5, in nearly linear time it is possible to check whether Γ is a block and, if it is not, to construct some $h \in G(i)_\beta \backslash H$. (We leave to the reader to check the details that the algorithm described in Lemma 5.5.5(b) works in the situation considered here, with $G(i-1)$ playing the role of $\langle H, r_\lambda \rangle$ in Lemma 5.5.5.)

If Γ is a block of imprimitivity then let Σ denote the block system consisting of the $G(i)$-images of Γ. In the $G(i)$-action on Σ, the "point" Γ is stabilized by $G(i-1)$; therefore, since $G(i)$ is generated by $G(i-1)$ and one additional element, we can use the special case of Sims's **Verify** routine discussed earlier in this section to check whether $G(i)_\Gamma = G(i-1)$. If indeed $G(i)_\Gamma = G(i-1)$ then $|G(i)| = (|\Delta|/|\Gamma|)|G(i-1)| = (|\Delta|/|\Gamma|)|\Gamma||H| = |\Delta||H|$ and $|G(i) : G(i)_\beta| = |\Delta|$, so $G(i)_\beta = H$. However, if $G(i)_\Gamma \ne G(i-1)$ then the algorithm constructs some $h_1 \in G(i)_\Gamma \backslash G(i-1)$ to witness this fact. Taking $h_2 \in G(i-1)$ such that $\beta^{h_1} = \beta^{h_2}$, we obtain $h_1 h_2^{-1} \in G_\beta \backslash H$.

Versions of Sims's **Verify** routine are implemented both in MAGMA and GAP. The bottleneck to nearly linear running time and to good practical performance is that in the special case $G = \langle H, z\rangle$, the running time is proportional to the number of orbits of H_γ, for $\gamma = \beta^{(z^{-1})}$. To alleviate this problem, the **Verify** routine is combined with the Schreier–Todd–Coxeter–Sims method (cf. Section 8.1) in MAGMA. This combination is referred to as the Brownie–Cannon–Sims algorithm. Since this algorithm is unpublished, we do not know what criterion on the number of orbits of H_γ is used to choose between **Verify** and STCS.

The GAP implementation constructs a maximal block system Σ for the G-action on β^G, and $z \in G$ such that z fixes the block B containing β. Then it verifies that $\langle H, z\rangle_\beta = H$ and, recursively by the same method for computing block systems, that $G_B = \langle H, z\rangle$ for the point stabilizer G_B in the action on Σ. Block computations are so cheap that their cost seems to be more than offset by the decrease of the number of orbits of H_γ, which we achieve with that choice of z. In essence, this modification inserts a chain of subgroups between G and G_β, which is in agreement with the general philosophy behind permutation group algorithms, namely, that we try to define a subgroup chain $G = G_1 \geq G_2 \geq \cdots \geq G_m = 1$ with small indices $|G_i : G_{i+1}|$.

As mentioned in Section 4.5.1 and discussed further at the end of Section 5.2, GAP employs a heuristic version of the nearly linear-time Monte Carlo SGS construction discussed in the above mentioned sections. The user can choose whether the output should be checked with a Monte Carlo algorithm by applying Lemma 4.5.6 or its correctness should be decided with certainty, applying Sims's **Verify** routine. Not only do both strong generating tests give a yes/no answer but, in the case of incorrect SGS, they construct an element of the input group that can be the starting point of the continuation of the SGS construction. However, the **Verify** routine is a quite expensive way to discover that an SGS is not correct, and GAP tries to call **Verify** only for correct inputs. Therefore, another heuristics is built in: If the user asks for **Verify** then, as a first step, ten random elements are tested, as described in Lemma 4.5.6; **Verify** is called only after these elements pass the test.

8.3. Toward Strong Generators by a Las Vegas Algorithm

In this section, we present a nearly linear-time Las Vegas algorithm that constructs a strong generating set in groups that satisfy some mild restrictions on their composition factors. The restriction on the composition factors is necessary because at present we do not have nearly linear-time constructive recognition algorithms, which we shall define shortly, for all finite simple groups. We shall follow the description of the algorithm from [Kantor and Seress, 1999], which makes clear how the class of groups covered by the algorithm extends when the nearly linear-time constructive recognition of further simple groups becomes available. In Theorem 8.3.2, we shall give the current list of simple groups that can be recognized constructively in nearly linear time.

The algorithm can be used as a strong generating test by comparing the order of the input group it computed with the order computed from the SGS to be tested. As we emphasized earlier, if the initial SGS computation is correct then the nearly linear-time algorithms described in Chapters 5 and 6 are deterministic or of Las Vegas type; therefore, for groups with restricted composition factors, *the entire nearly linear-time library of algorithms is upgraded to Las Vegas type*.

As promised, we give the definition of constructive recognition of simple groups. We formulate constructive recognition in the black-box group setting, since its applications extend beyond permutation group algorithms: Constructive recognition of simple groups is a central concept in matrix group algorithms as well. Let $\mathcal{F}$ be a family of simple groups and let $f : \mathcal{F} \to \mathbb{R}$ be a function taking positive values. We say that $\mathcal{F}$ is *black-box f-recognizable* if, whenever a group $H = \langle S \rangle$ isomorphic to a member of $\mathcal{F}$ is given as a black-box group encoded by strings of length at most N and, in the case of Lie-type H, the characteristic of H is given, there are Las Vegas algorithms for (i) and (ii) and a deterministic algorithm for (iii) of the following:

(i) Find the isomorphism type of H.

(ii) Find a new set S^* of size $O(\log |H|)$ generating H and a presentation of length $O(\log^2 |H|)$ such that S^* satisfies the presentation. (Note that $\log |H| \leq N$, and this presentation proves that H has the isomorphism type determined in (i).)

(iii) Given $h \in H$, find a straight-line program of length $O(N)$ from S^* to h. (Straight-line programs were defined in Section 1.2.3.)

Moreover,

(iv) The algorithms for (i)–(iii) run in time $O((\xi + \mu) f(H) N^c)$, where ξ is an upper bound on the time requirement per element for the construction of independent, (nearly) uniformly distributed random elements in subgroups of H, μ is an upper bound on the time required for each group operation in H, and c is an absolute constant.

As we have seen in Section 5.3, a permutation group $G \leq \mathrm{Sym}(\Omega)$ is isomorphic to a black-box group H over an alphabet consisting of the labels in a Schreier tree data structure of G. Let $\psi : G \to H$ denote this isomorphism.

By Lemma 5.3.1, the quantities ξ, μ, N for H in part (iv) of the definition of constructive recognition are bounded from above by $\log^{c'} |G|$ for some absolute constant c'. Moreover, for any $g \in G$, $\psi(g)$ can be computed in $O(\log^{c'} |G|)$ time, and for any $h \in H$, $\psi^{-1}(h)$ can be computed in $O(|\Omega| \log^{c'} |G|)$ time.

Therefore, a constructive recognition algorithm runs in nearly linear time for permutation group inputs if f is bounded from above by a nearly linear function of the degree of the input group. This fact motivates the consideration of the function $m : \mathcal{G} \to \mathbb{R}$, where $\mathcal{G}$ is the family of all finite simple groups and $m(G)$ is the degree of the smallest faithful permutation representation of G.

Theorem 8.3.1. *Given a permutation group $G = \langle T \rangle \leq \mathrm{Sym}(\Omega)$, with $|\Omega| = n$, such that all composition factors of G are m-recognizable, a base and strong generating set for G can be computed by a nearly linear-time Las Vegas algorithm.*

Proof. We compute an alleged base and strong generating set, and a composition series $G = N_1 \rhd N_2 \rhd \cdots \rhd N_l = 1$, by the nearly linear-time Monte Carlo algorithms in Sections 4.5 and 6.2. The composition series algorithm also provides homomorphisms $\varphi_i : N_i \to S_n$ with $\ker(\varphi_i) = N_{i+1}$, for $1 \leq i \leq l-1$. We also compute strong generating sets for all N_i relative to the base of G. We will verify the correctness of the base and strong generating sets for the subgroups N_i by induction on $i = l, l-1, \ldots, 1$.

Suppose that we already have verified an SGS for N_{i+1}. We compute a base, SGS, shallow Schreier-tree data structure, and an isomorphism $\psi_i : \varphi_i(N_i) \to H_i$ with a black-box group H_i for the image $\varphi_i(N_i) \leq S_n$, which is allegedly isomorphic to a simple group. Our first goal is to recognize H_i, and so $\varphi_i(N_i)$, constructively.

As a consequence of the classification of finite simple groups, we know that there are no three pairwise nonisomorphic simple groups of the same order. Therefore, since we know $|\varphi_i(N_i)|$, we have at most two candidate simple groups C for the isomorphism type of $\varphi_i(N_i)$, and in the ambiguous cases we try both possibilities. Also, if $|\varphi_i(N_i)| > 8!/2$ then $|\varphi_i(N_i)|$ determines whether $\varphi_i(N_i)$ is of Lie type, and if it is, its characteristic. Hence, using that $\varphi_i(N_i)$ is from an m-recognizable family, we can obtain by a nearly linear-time Las Vegas algorithm a generating set S_i^* of H_i of size $O(\log |H_i|)$ and a presentation $\langle E_i \mid \mathcal{R}_i \rangle$ of length $O(\log^2 |H_i|)$ satisfied by S_i^*. Moreover, for any given $h \in H_i$, we can write a straight-line program of length $O(\log |H_i|)$ from S_i^* to h in nearly linear time by a deterministic algorithm. By Lemma 5.3.1, the preimage $S_i^{**} = \psi_i^{-1}(S_i^*) \subseteq \varphi_i(N_i)$ can also be obtained in nearly linear time by a deterministic algorithm.

Now the correctness of the SGS for N_i can be proved in the following way: Let T_i be the set of generators of N_i computed by the composition series algorithm. We check that

(a) $N_{i+1} \lhd N_i$ and $N_i \neq N_{i+1}$;
(b) $\langle \varphi_i^{-1}(S_i^{**}) \rangle N_{i+1}/N_{i+1}$ satisfies the presentation $\langle E_i \mid \mathcal{R}_i \rangle$ computed for $\varphi_i(N_i) \cong H_i$; and
(c) $T_i \subset \langle \varphi_i^{-1}(S_i^{**}) \rangle N_{i+1}$, where $\varphi_i^{-1}(g)$ denotes a lift (i.e., an arbitrary preimage) of $g \in S_i^{**} \subseteq \varphi_i(N_i)$.

If (a), (b), and (c) hold then $|N_i| = |N_{i+1}||\varphi_i(N_i)|$. If $|N_i|$ is equal to the value for $|N_i|$ computed from the alleged SGS of N_i then the SGS construction is correct, since the alleged order of a group computed by a Monte Carlo SGS construction is not greater than the true order, with equality if and only if the SGS construction is correct.

We indicate how (a), (b), and (c) can be checked algorithmically. For (a), we conjugate the generators of N_{i+1} by the elements of T_i and check that the resulting permutations are in N_{i+1}. Because the correctness of the SGS for N_{i+1} is already known, membership testing giving guaranteed correct results is available for that group. Also, we check that not all elements of T_i are in N_{i+1}. For (b), we multiply out the relators that were written in terms of S_i^*, using the permutations in $\varphi_i^{-1}(S_i^{**}) = \varphi_i^{-1}\psi^{-1}(S_i^*)$; then we check that the resulting permutations are in N_{i+1}. Finally, for (c), for each $t_i \in T_i$ we write a straight-line program P from S_i^* to $\psi_i\varphi_i(t_i) \in H_i$ and evaluate P starting from $\varphi_i^{-1}\psi^{-1}(S_i^*)$ (recall that straight-line programs can be evaluated in any group, giving values to the symbols representing generators in the program). The result of the evaluation is some $t^* \in \langle \varphi_i^{-1}\psi_i^{-1}(S_i^*) \rangle$ and we check that $t^*t^{-1} \in N_{i+1}$. By (b) and (c), we have checked that the factor group $N_i/N_{i+1} \cong C$.

At the end of the induction, we have obtained a correct SGS for the group $N_1 = \langle T_1 \rangle$ that was output by the composition series algorithm. After that, we verify that $G = N_1$ by sifting the elements of the original generating set T in N_1.

To justify the nearly linear running time of the entire algorithm, note that $l \in O(\log |G|)$, so it is enough to show that the ith step of the induction for a fixed value $i \leq l$ runs in nearly linear time. We have already seen that the constructions of both S_i^* and of the presentation for $\varphi_i(N_i) \cong H_i$ are within this time bound. Since both $|T_i|$ and $|S_i^*|$ are $O(\log |G|)$, whereas the length of the presentation $\langle E_i \mid \mathcal{R}_i \rangle$ is $O(\log^2 |G|)$ and the Schreier-tree data structure of N_{i+1} is shallow, the number of permutation multiplications that occur while checking (a), (b), and (c) is bounded from above by a polylogarithmic function of $|G|$.

We note that, to achieve an algorithm failure with probability less than ε, we have to require that calls to the SGS construction algorithm and to constructive recognition fail with probability less than $\varepsilon/(c \log |G|)$ for some constant c, because, during the induction, $O(\log |G|)$ such calls may be made; however, this multiplies the running time only by a $O(\log \log |G|)$ factor. □

To apply the theorem to a wide class of groups, we need m-recognizable simple groups. The current state of the art is summarized in the following theorem.

Theorem 8.3.2. *The set of all finite simple groups, with the possible exceptions of the groups ${}^2F_4(q)$ and ${}^2G_2(q)$, comprises a black-box m-recognizable family.*

We do not prove Theorem 8.3.2; instead we just make some comments and give the appropriate references. Cyclic simple groups are m-recognizable by brute force, and the twenty-six sporadic simple groups can be handled with a constant number of black-box group operations.

The alternating groups comprise a 1-recognizable family (i.e., one can take $f(G) = 1$ in the definition of constructive recognition for all alternating groups G). This result was first proven in [Beals and Babai, 1993], and more efficient versions are described in [Bratus and Pak, 2000] and [Beals et al., 2002]. We shall return to the recognition of alternating and symmetric groups in Section 10.2.

Classical simple groups are treated in [Kantor and Seress, 2001]. We note that in that paper there is no algorithm for the constructive recognition of three-dimensional unitary groups, since at the time of writing it was not known whether the groups $G \cong \mathrm{PSU}_3(q)$ have presentations of length $O(\log^2 |G|)$, as required in part (ii) of the definition. Such presentations are described in [Hulpke and Seress, 2001]. A predecessor of [Kantor and Seress, 2001] is [Cooperman et al., 1997], where a constructive recognition algorithm for the groups $\mathrm{GL}_n(2)$ is given. In [Bratus, 1999], that method is extended to all special linear groups $\mathrm{PSL}_n(q)$.

Constructive recognition algorithms for the exceptional groups of Lie type are in [Kantor and Magaard, 2002]. Both [Kantor and Seress, 2001] and [Kantor and Magaard, 2002] far exceed the scope of this book.

Actually, in all these cited references, much more is done than parts (i)–(iii) of constructive recognition: Given a black-box group H isomorphic to some simple group S, an isomorphism $\lambda : H \to C$, defined by the images of the elements of S^*, is constructed between H and a standard copy C of S, together with procedures to compute the image of any element of H under

λ or of any element of C under λ^{-1}. These procedures are very useful for further computations with groups containing S as a composition factor, such as the construction of Sylow subgroups (cf. [Kantor, 1985b; Morje, 1995]) or the computation of maximal subgroups or conjugacy classes (cf. [Cannon and Souvignier, 1997; Hulpke, 2000; Eick and Hulpke, 2001]). Another possible application of these procedures is in computation with matrix groups.

The "standard copy" of a simple group S referred to in the previous paragraph is the natural permutation representation of degree n if S is the alternating group A_n, matrices modulo scalars in the correct dimension, over the field of definition if S is a classical group, and a presentation utilizing the Bruhat decomposition if S is an exceptional group of Lie type.

In [Kantor and Seress, 2001] and [Kantor and Magaard, 2002], there are also more precise timings of algorithms than required in Theorem 8.3.1. It is shown that the family of Lie-type simple groups, with the possible exception of the groups ${}^2F_4(q)$ and ${}^2G_2(q)$, is black-box f-recognizable for the function

$$f(G) = \begin{cases} q^{3/2} & \text{if } G \cong \mathrm{PSU}_d(q^{1/2}) \text{ for some } d \\ q & \text{for all other classical } G \text{ defined on a vector space over GF}(q) \\ q^2 & \text{if } G \cong {}^2B_2(q) \\ q^3 & \text{for other exceptional groups defined over GF}(q) \text{ except} \\ & G \cong {}^2F_4(q) \text{ and } {}^2G_2(q). \end{cases}$$

It seems very likely that the set of *all* groups of Lie type is a black-box f-recognizable family with some $f(G) \le m(G)$.

The algorithm described in Theorem 8.3.1 has not been implemented. Although the running time is nearly linear, computing the order of the input group by this algorithm involves a lot of superfluous work. However, it may be worthwhile to verify the correctness of an SGS computation in this way if the constructive recognition of the composition factors is needed in further computations. Besides the examples of computing Sylow subgroups, maximal subgroups, or conjugacy classes that we have mentioned earlier, this is the case at the construction of a short presentation, as described in the next section.

Constructive recognition of simple groups is even more important in the matrix group setting, where every attempt so far for finding the structure of a matrix group has led to some version of the constructive recognition problem (cf. [Babai and Beals, 1999; Leedham-Green, 2001; Kantor and Seress, 2002]).

Currently, constructive recognition of black-box groups isomorphic to alternating, special linear, symplectic, and unitary groups is implemented in GAP, but these algorithms are not in the standard distribution.

We finish this section with a lemma that will be needed in Section 8.4.

Lemma 8.3.3. *Let $G \le \mathrm{Sym}(\Omega)$, with $|\Omega| = n$, be a permutation group, such that all composition factors of G are m-recognizable. Suppose that the following have already been computed, as in the proof of Theorem* 8.3.1*: a composition series $G = N_1 \rhd N_2 \rhd \cdots \rhd N_l = 1$, homomorphisms $\varphi_i : N_i \to S_n$ with* $\ker(\varphi_i) = N_{i+1}$, *and presentations satisfied by generating sets $S_i^{**} \subset \varphi_i(N_i)$. Then any $g \in G$ can be reached from $\bigcup_i \varphi_i^{-1}(S_i^{**})$ by a straight-line program of length $O(\log |G|)$, and such a straight-line program can be computed in nearly linear time by a deterministic algorithm.*

Proof. By induction on $i = 1, 2, \ldots, l$, we construct a straight-line program P_i of length $O(\log(|G|/|N_i|))$ to some $g_i \in G$ such that $gg_i^{-1} \in N_i$. Let $g_1 := 1$. If g_i has already been obtained for some i, then we write a straight-line program L_i of length $O(\log |N_i/N_{i+1}|)$ from S_i^* to $\varphi_i(gg_i^{-1})$. In the case when N_i/N_{i+1} is cyclic or sporadic, this can be done by brute force. In the other cases, we use an isomorphism ψ_i between $\varphi_i(N_i)$ and a black-box group H_i, as in the proof of Theorem 8.3.1, and the fact that $\psi_i\varphi_i(N_i)$ is black-box m-recognizable. We evaluate this straight-line program starting from $\varphi_i^{-1}(S_i^{**})$, producing an element $h_i \in N_i$. Here, $gg_i^{-1}h_i^{-1} \in N_{i+1}$, and we can define $g_{i+1} := h_i g_i$. The straight-line program P_{i+1} reaching g_{i+1} is defined as the concatenation of P_i and L_i, with the added term (w_j, w_k) for the entries $w_j \in L_i$ and $w_k \in P_i$, which evaluate to h_i and g_i, respectively.

Finally, we notice that the entire procedure runs in nearly linear time, since $m(N_i/N_{i+1}) \le n$ for all i. □

8.4. A Short Presentation

In strong generating set constructions for a permutation group $G \le \mathrm{Sym}(\Omega)$, we defined a subgroup chain $G = G_1 > G_2 > \cdots > G_l = 1$, constructed generators S_i for the subgroups G_i, and tested that S_i indeed generates G_i. In all but one of the SGS constructions, the subgroups G_i comprised a point stabilizer chain relative to some nonredundant base, whereas in the nearly linear-time Las Vegas algorithm presented in Section 8.3, the subgroups G_i defined a composition series of G. In this section, we describe how the testing phase of these SGS constructions can be used to construct a presentation for G.

In the variants using a point stabilizer chain, the testing phase checks for $i \in [1, l-1]$ whether or not a suitable subset of Schreier generators obtained from S_i and a transversal R_i for G_i mod G_{i+1} are in G_{i+1}. This subset of Schreier generators defines a presentation for G in the following way. We introduce a set of symbols E_i corresponding to S_i, and we write the presentation using the generating set $E := \bigcup_i E_i$. Each element of the transversal R_i corresponds to a word in E_i, and hence each Schreier generator s_i constructed from S_i and R_i corresponds to a word $w(s_i)$ using the symbols in E_i. Since $s_i \in G_{i+1}$, it has a standard word decomposition (cf. Section 5.3), which corresponds to a word $\bar{w}(s_i)$ using the symbols in $\bigcup_{j>i} E_j$. We claim that $\langle E \mid \mathcal{R} \rangle$, where $\mathcal{R}$ is the set of relators $w(s)\bar{w}(s)^{-1} = 1$ for the Schreier generators s used in the checking phase in the SGS construction, is a presentation for G.

In the case of Sims's original SGS construction, when all Schreier generators composed from S_i and R_i are tested for all $i \in [1, l]$, our claim is just Exercise 5.2. In the case of the Schreier–Todd–Coxeter–Sims method described in Section 8.1.2, the SGS test itself constructs the presentation described in our claim, and we have seen in Section 8.1.2 that $\langle E \mid \mathcal{R} \rangle$ is indeed a presentation for G. Finally, in the case of Sims's **Verify** routine in Section 8.2, the Schreier generators used in the presentation can be obtained from Lemma 8.2.2, and the proof that all relations corresponding to other Schreier generators are consequences of the relators in $\mathcal{R}$ can be deduced from Lemma 8.2.1. We leave the details of this argument to the reader.

Another method for reducing the set of relations corresponding to Schreier generators, while of course maintaining that the remaining relators define a presentation for G, is in [Cannon, 1973]. A recent improvement of Cannon's method is described in [Gebhardt, 2000].

The disadvantage of presentations obtained from SGS constructions using a point stabilizer chain is that the number of relators, and so the length of the presentation, can be proportional to the degree n of the permutation domain. (The *length* of a presentation is simply the number of symbols needed to write down the presentation.) The main purpose of this section is to construct in nearly linear time a presentation of length polylogarithmic in $|G|$, in groups G where all composition factors are m-recognizable. The method is based on the strong generating test described in Theorem 8.3.1. The *existence* of a presentation of length $O(\log^C |G|)$, provided that all composition factors of G have presentations of length $O(\log^{C-1} |G|)$ for some absolute constant $C \geq 3$, was proven in [Babai et al., 1997a]. The following algorithmic version of this result for groups with m-recognizable composition factors is from [Kantor and Seress, 1999].

Theorem 8.4.1. *Let $G \leq \mathrm{Sym}(\Omega)$, with $|\Omega| = n$, be a permutation group, such that all composition factors of G are m-recognizable. Then it is possible to construct a presentation of length $O(\log^3 |G|)$ for G by a nearly linear-time Las Vegas algorithm.*

Proof. We use the Las Vegas algorithm of Theorem 8.3.1 to compute a composition series $G = N_1 \rhd N_2 \rhd \cdots \rhd N_l = 1$ and, for each $i \in [1, l-1]$, a homomorphism $\varphi_i : N_i \to S_n$ with $\ker(\varphi_i) = N_{i+1}$, together with a presentation $\langle E_i \mid \mathcal{R}_i \rangle$ of length $O(\log^2 |N_i/N_{i+1}|)$ for N_i/N_{i+1}, satisfied by generating sets $S_i^{**} \subset \varphi_i(N_i)$ of cardinality $O(\log |N_i/N_{i+1}|)$.

Next, for each $i < l$, we compute a set $T_i = \{\varphi_i^{-1}(g) \mid g \in S_i^{**}\} \subseteq N_i$, where $\varphi_i^{-1}(g)$ denotes an arbitrary preimage of g under the homomorphism φ_i. Then, corresponding to parts (a) and (b) of the SGS test in Theorem 8.3.1, we compute a set of permutations by the following two methods. The first method computes all permutations $p(t_i, t_j) = t_i^{-1} t_j t_i$, where $1 \le i < j \le l-1$ and $t_i \in T_i, t_j \in T_j$. Since $N_{i+1} \lhd N_i$, we have $p(t_i, t_j) \in N_{i+1}$. The second method computes all permutations $p(i, r)$, where $1 \le i \le l-1$ and $r = 1$ is a relator in $\mathcal{R}_i$. The permutation $p(i, r) \in G$ is computed by multiplying out r, using the appropriate elements of T_i. Since $r = 1$ is satisfied in N_i/N_{i+1}, we have $p(i, r) \in N_{i+1}$.

Since the number of generators is $O(\log |N_i/N_{i+1}|)$ and the total length of relators is $O(\log^2 |N_i/N_{i+1}|)$ in $\langle E_i \mid \mathcal{R}_i \rangle$, all permutations $p(t_i, t_j)$, $p(i, r)$ can be computed using $O(\log^2 |G|)$ permutation multiplications. Moreover, the total number of the permutations $p(t_i, t_j)$, $p(i, r)$ is $O(\log^2 |G|)$.

As a final step of preparation, for each $p = p(t_i, t_j)$ and $p = p(i, r)$ we compute a straight-line program $W(p)$ of length $O(\log |G|)$ reaching p from $\bigcup_{k \ge i+1} T_k$. By Lemma 8.3.3, this can also be done in nearly linear time. We may assume that the straight-line programs use the symbols from the sets E_i to denote the generators and that each element of each E_i occurs in at least one of the straight-line programs $W(p)$ (if this is not the case then we add the missing generators to one of the straight-line programs).

Now we are ready to write a presentation for G. More exactly, we define a group $\hat{G} = \langle \hat{E} \mid \hat{\mathcal{R}} \rangle$ and prove that $G \cong \hat{G}$. The generating set $\hat{E}$ is the union of symbols occurring in all straight-line programs $W(p)$, for $p = p(t_i, t_j)$ and $p = p(i, r)$ (i.e., if $W(p) = (w_1, w_2, \ldots, w_{l(p)})$ then $w_k \in \hat{E}$ for all $k \in [1, l(p)]$). In particular, $\bigcup_i E_i \subseteq \hat{E}$. The relator set $\hat{\mathcal{R}}$ consists of the following defining relations:

(i) If $w_j = (w_k, -1)$ for some w_j in a straight-line program $W(p)$ then we add the relator $w_j w_k = 1$ to $\hat{\mathcal{R}}$.
(ii) If $w_j = (w_k, w_m)$ then we add the relator $w_j (w_k w_m)^{-1} = 1$ to $\hat{\mathcal{R}}$.
(iii) For $1 \le i < j \le l-1$ and for $t_i \in T_i, t_j \in T_j$, let $e_i \in E_i$ and $e_j \in E_j$ be the symbols corresponding to t_i and t_j, respectively, and let $W(p(t_i, t_j)) = (w_1, w_2, \ldots, w_{l(p)})$. We add the relator $e_i^{-1} e_j e_i w_{l(p)}^{-1} = 1$ to $\hat{\mathcal{R}}$.
(iv) For $1 \le i \le l-1$ and $\{r = 1\} \in \mathcal{R}_i$, let $W(p(i, r)) = (w_1, w_2, \ldots, w_{l(p)})$. We add the relator $r w_{l(p)}^{-1} = 1$ to $\hat{\mathcal{R}}$.

Substituting the elements of $\bigcup_i T_i$ for the appropriate elements $\bigcup_i E_i \subseteq \hat{E}$ and substituting the evaluations starting from $\bigcup_i T_i$ for the elements of $\hat{E}$ representing nongenerator symbols in straight-line programs, it is clear that G satisfies the presentation $\hat{\mathcal{R}}$. Therefore there exists a homomorphism $\psi : \hat{G} \to G$, and it is enough to prove that $|\hat{G}| \leq |G|$.

For $i = 1, 2, \ldots, l-1$, let $\hat{G}_i$ be the subgroup of $\hat{G}$ generated by $\bigcup_{j \geq i} E_j$. Then $\hat{G}_1 = \hat{G}$, since (i) and (ii) ensure that all generators in $\hat{E}$ can be expressed as products of the elements of $\bigcup_{j \geq 1} E_j$.

We claim that $\hat{G}_{i+1} \lhd \hat{G}_i$ for $i = 1, 2, \ldots, l-1$. Indeed, by (iii), if $j \geq i+1$ then for any $e_i \in E_i$ and $e_j \in E_j$ we have $e_i^{-1} e_j e_i = w_{l(p)}$ for the last term $w_{l(p)}$ of the straight-line program $W(p(t_i, t_j))$. Since $t_i^{-1} t_j t_i \in N_{i+1}$ for the corresponding generators t_i, t_j of G, by (i) and (ii) again we have that all symbols in $W(p(t_i, t_j))$, in particular $w_{l(p)}$, are in $\hat{G}_{i+1}$.

Finally, we claim that $\hat{G}_i / \hat{G}_{i+1}$ satisfies the presentation $\langle E_i \mid \mathcal{R}_i \rangle$ for N_i / N_{i+1}. Indeed, for any relator $\{r = 1\} \in \mathcal{R}_i$, by (iv) we have $r \in \hat{G}_{i+1}$. Our last claim proves that $|\hat{G}_i / \hat{G}_{i+1}| \leq |N_i / N_{i+1}|$, and so $|\hat{G}| \leq |G|$ and $\hat{G} \cong G$.

Since the total number of permutations $p = p(t_i, t_j)$ and $p = p(i, r)$ is $O(\log^2 |G|)$ and each straight-line program $W(p)$ is of length $O(\log |G|)$, we have $O(\log^3 |G|)$ relators described in (i) and (ii) and each of these relators has length bounded by an absolute constant. Similarly, since there are $O(\log^2 |G|)$ pairs (t_i, t_j), the total length of relators described in (iii) is $O(\log^2 |G|)$. Finally, since the sum of lengths of the presentations $\langle E_i \mid \mathcal{R}_i \rangle$ is $O(\log^2 |G|)$, the total length of relators described in (iv) is also $O(\log^2 |G|)$. □

As we mentioned in Section 8.3, if H is a simple group not isomorphic to ${}^2G_2(q)$ then H has a presentation of length $O(\log^2 |H|)$ (cf. [Suzuki, 1962, p. 128; Babai et al., 1997a; Hulpke and Seress, 2001]). Hence if a group G has no composition factors isomorphic to ${}^2G_2(q)$ then the result of [Babai et al., 1997a] mentioned just before Theorem 8.4.1 implies that G has a presentation of length $O(\log^3 |G|)$. Moreover, if a permutation group G has no composition factors isomorphic to ${}^2G_2(q)$ or ${}^2F_4(q)$ then Theorems 8.3.2 and 8.4.1 imply that such a short presentation can be constructed in nearly linear time by a Las Vegas algorithm. Currently it is not known whether the groups ${}^2G_2(q)$ have presentations of length polylogarithmic in their order.

9

Backtrack Methods

In this chapter we consider computational problems that, at present, have no polynomial-time solutions. They can be formulated by the following general description: Given $G = \langle S \rangle \leq \mathrm{Sym}(\Omega)$ and a property $\mathcal{P}$, find the set $G(\mathcal{P})$ of elements of G satisfying $\mathcal{P}$. (We assume that for each $g \in G$, it is possible to decide whether g satisfies $\mathcal{P}$.) For certain properties, there seems to be no better general method than to run through a listing of G and examine each element until we find the ones satisfying $\mathcal{P}$. This happens, for example, when we search for the solutions of an equation in G. If a method checks each $g \in G$ separately then, of course, there is a severe limit on the orders of the groups it can handle in practice. However, in many cases of practical interest, the set $G(\mathcal{P})$, if not empty, is a subgroup of G or a coset of a subgroup. Examples when $G(\mathcal{P})$ is a subgroup are centralizers of elements or subgroups of $\mathrm{Sym}(\Omega)$, normalizers of subgroups, and setwise stabilizers of subsets of Ω. In these cases, we are usually interested in finding all $g \in G$ satisfying $\mathcal{P}$. The most important examples for properties satisfied by a coset of a subgroup (or no elements of G at all) are when $\mathcal{P}$ is the property that $g \in G$ conjugates some fixed $a \in \mathrm{Sym}(\Omega)$ to $b \in \mathrm{Sym}(\Omega)$ or $A \leq \mathrm{Sym}(\Omega)$ to $B \leq \mathrm{Sym}(\Omega)$, or $\Delta_1^g = \Delta_2$ for some $\Delta_1, \Delta_2 \subseteq \Omega$. In these cases, we are usually interested in finding only one element of $G(\mathcal{P})$, or in deciding that $G(\mathcal{P}) = \emptyset$. As these examples indicate, subgroup-type and coset-type problems come in pairs. To find all elements of $G(\mathcal{P})$ in a coset-type problem, we may search for one element of $G(\mathcal{P})$ and then solve the corresponding subgroup-type problem.

Throughout this chapter, *we assume that $G(\mathcal{P})$ is a subgroup, or a coset of a subgroup in G, or empty*. This assumption will enable us to search G without processing every element of G separately, thereby leading to practical algorithms.

9.1. Traditional Backtrack

The content of this section essentially was formulated in [Sims, 1971a, b]. Let $B = (\beta_1, \ldots, \beta_m)$ be a base of $G \leq \mathrm{Sym}(\Omega)$, and let $G = G^{[1]} \geq G^{[2]} \geq \cdots \geq G^{[m]} \geq G^{[m+1]} = 1$ be the corresponding point stabilizer subgroup chain. The images of base points uniquely determine the elements of G, and we may identify group elements with their sequence of base images. The image of an initial segment of B of length l defines a coset of $G^{[l+1]}$. The images of all initial segments define a partial order by inclusion, which is called the *search tree* $\mathcal{T}$ for G. The lth level $\mathcal{T}_l$ of $\mathcal{T}$ corresponds to the cosets of $G^{[l+1]}$; in particular, the root of $\mathcal{T}$ corresponds to G and the leaves of the search tree correspond to the elements of G. We shall denote the subtree of $\mathcal{T}$ rooted at $t \in \mathcal{T}$ by $\mathcal{T}(t)$. Formally, we can define a function φ from $\mathcal{T}$ to the power set of G by the rule

$$\varphi((\gamma_1, \ldots, \gamma_l)) := \left\{ g \in G \mid \beta_i^g = \gamma_i \text{ for } 1 \leq i \leq l \right\}.$$

Traditional backtrack methods systematically examine the elements of the search tree. It is especially valuable when a vertex t close to the root can be eliminated, by establishing either that $\varphi(t)$ does not contain any elements of $G(\mathcal{P})$ or that $\varphi(t) \cap G(\mathcal{P})$ is in the subset of $G(\mathcal{P})$ that is already found, because then all elements of G corresponding to leaves less than this vertex (i.e., elements of the coset $\varphi(t)$) can be excluded from the search at once. Of course, we do not construct and store $\mathcal{T}$ explicitly; rather, we traverse $\mathcal{T}$ and construct and discard vertices as required by the traversal.

Let $(\Omega, \prec)$ be an ordering of Ω with the property that $\beta_1 \prec \beta_2 \prec \cdots \prec \beta_m$ and $\beta_m \prec \beta$ for all $\beta \in \Omega \backslash B$. This ordering $(\Omega, \prec)$ induces a lexicographic ordering on G. Namely, for $g, h \in G$, we have $h \prec g$ if and only if there exists $l \in [1, m]$ such that $\beta_i^h = \beta_i^g$ for all $i < l$ and $\beta_l^h \prec \beta_l^g$. Similarly, $(\Omega, \prec)$ induces a lexicographic ordering of every level $\mathcal{T}_l$.

We require that our algorithm traverses $\mathcal{T}$ by a *depth-first-search*, or *backtrack*, traversal. This is defined by the following rule:

(i) The starting point of the traversal is the root of $\mathcal{T}$.
(ii) Suppose that the traversal arrived at a vertex $t \in \mathcal{T}$.
- If t is a leaf then examine whether the corresponding group element satisfies $\mathcal{P}$, and then go to the parent of t.
- If t is not a leaf then traverse the subtrees rooted at the children of t, considering these subtrees in the lexicographic ordering of their roots, and then go to the parent of t.

(iii) The algorithm ends when the previous step tries to go to the parent of the root.

In particular, the backtrack traversal considers the elements of G in increasing order and determines $G^{[i]} \cap G(\mathcal{P})$ before processing any element of $G \backslash G^{[i]}$.

There are many methods for eliminating a vertex t of the search tree, that is, for determining that no element of $G(\mathcal{P})$ is in the subtree rooted at t, or that we have already found all such elements. In this case, the traversal of this subtree can be skipped. These methods fall into two categories: *problem-independent* methods, which rely only on the fact that $G(\mathcal{P})$ is a subgroup or a coset of a subgroup, and *problem-dependent* ones, which use the specific features of the property $\mathcal{P}$.

9.1.1. Pruning the Search Tree: Problem-Independent Methods

Lemma 9.1.1. *Suppose that $G(\mathcal{P})$ is a subgroup of G, the subgroup $K := G(\mathcal{P}) \cap G^{[l+1]}$ is already computed for some $l \in [1, m]$, and currently the backtrack algorithm traverses the subtree $\mathcal{T}(t)$ with root $t = (\gamma_1, \gamma_2, \ldots, \gamma_l) \in \mathcal{T}_l$. Then, if any $g \in \varphi(t) \cap G(\mathcal{P})$ is found then $\langle g, K \rangle \supseteq \varphi(t) \cap G(\mathcal{P})$, so we can skip traversing the rest of $\mathcal{T}(t)$.*

Proof. Any $h \in \varphi(t) \cap G(\mathcal{P})$ is in the coset Kg. □

Note that in coset-type problems we terminate the entire search procedure as soon as one element of $G(\mathcal{P})$ is found.

In the following criteria, $G(\mathcal{P})$ may be a subgroup or a coset of a subgroup. Suppose that we already know subgroups $K, L \leq G$ such that any $g \in G$ belongs to $G(\mathcal{P})$ if and only if the double coset $KgL \subseteq G(\mathcal{P})$. If $G(\mathcal{P})$ is a subgroup then K and L may be chosen as the already constructed subgroup of $G(\mathcal{P})$. We may have nontrivial K and L at our disposal even when $G(\mathcal{P})$ is a coset of a subgroup, and we do not yet know any element of $G(\mathcal{P})$: For example, when $G(\mathcal{P})$ consists of the elements of G conjugating $a \in G$ to $b \in G$, we may choose $K = \langle a \rangle$ and $L = \langle b \rangle$.

Because it is enough to consider only one element from each double coset KgL and the backtrack traversing encounters the elements of G in increasing order, we can skip processing a subtree $\mathcal{T}(t)$ if we know that no $g \in \varphi(t)$ is the lexicographically first element of its double coset KgL. However, computing the first element of a double coset is, in general, an NP-hard problem (cf. [Luks, 1993]), so we have to settle for weaker, but easier computable criteria.

Lemma 9.1.2. *Let $t = (\gamma_1, \gamma_2, \ldots, \gamma_l) \in \mathcal{T}$. If $g \in \varphi(t)$ is the first element of KgL then γ_l is the $\prec$-minimal element of the orbit $\gamma_l^{L_{(\gamma_1, \gamma_2, \ldots, \gamma_{l-1})}}$.*

Proof. Suppose, on the contrary, that $\gamma \in \gamma_l^{L_{(\gamma_1,\gamma_2,\ldots,\gamma_{l-1})}}$ and $\gamma \prec \gamma_l$. Let $h \in L_{(\gamma_1,\gamma_2,\ldots,\gamma_{l-1})}$ with $\gamma_l^h = \gamma$. Then $gh \in KgL$ and $gh \prec g$, contradicting the minimality of g. □

Lemma 9.1.2 implies that when processing the successors of the vertex $t' := (\gamma_1, \gamma_2, \ldots, \gamma_{l-1})$, it is enough to consider extensions of t' by the $\prec$-minimal elements of the orbits of $L_{(\gamma_1,\gamma_2,\ldots,\gamma_{l-1})}$. The application of this criterion involves a base change in L, in which a base for L starting with $(\gamma_1, \gamma_2, \ldots, \gamma_{l-1})$ is computed. Because of the order in which the search tree is traversed, we already have $L_{(\gamma_1,\gamma_2,\ldots,\gamma_{l-2})}$ at hand when $L_{(\gamma_1,\gamma_2,\ldots,\gamma_{l-1})}$ is to be computed, so only γ_{l-1} has to be included in the new base. We have to find a balance between the cost of the base change and the gain in pruning the search tree. Hence, in some implementations, this criterion is applied only for small values of l.

[Leon, 1991] contains the following observation which, when an additional condition is satisfied, leads to a strengthening of Lemma 9.1.2.

Lemma 9.1.3. *Suppose that $\beta_l \in \beta_k^{K_{(\beta_1,\ldots,\beta_{k-1})}}$ for some $k \leq l$, and let $t = (\gamma_1, \gamma_2, \ldots, \gamma_l) \in \mathcal{T}$. If $g \in \varphi(t)$ is the first element of KgL then $\gamma_k \preceq \min(\gamma_l^{L_{(\gamma_1,\gamma_2,\ldots,\gamma_{k-1})}})$.*

Proof. Since $\beta_l \in \beta_k^{K_{(\beta_1,\ldots,\beta_{k-1})}}$, there exists $h_1 \in K_{(\beta_1,\ldots,\beta_{k-1})}$ such that $\beta_l = \beta_k^{h_1}$. Suppose, on the contrary, that there exists $\gamma \in \gamma_l^{L_{(\gamma_1,\gamma_2,\ldots,\gamma_{k-1})}}$ with $\gamma \prec \gamma_k$. Then $\gamma = \gamma_l^{h_2}$ for some $h_2 \in L_{(\gamma_1,\gamma_2,\ldots,\gamma_{k-1})}$, and $h_1 g h_2 \in KgL$. Moreover, $h_1 g h_2 \prec g$, since $\beta_i^{h_1 g h_2} = \beta_i^g$ for $i < k$, and $\beta_k^{h_1 g h_2} = \gamma \prec \gamma_k = \beta_k^g$. This contradicts the minimality of g in KgL. □

In particular, if $\beta_l \in \beta_k^{K_{(\beta_1,\ldots,\beta_{k-1})}}$ for some $k < l$ then we have to consider only vertices $t = (\gamma_1, \gamma_2, \ldots, \gamma_l) \in \mathcal{T}_l$ satisfying $\gamma_k \prec \gamma_l$. Sims's original criterion in Lemma 9.1.2 is the special case $k = l$ of Lemma 9.1.3.

Lemma 9.1.4. *Let s be the length of the lth fundamental orbit $\beta_l^{K_{(\beta_1,\ldots,\beta_{l-1})}}$ of K, and let $t = (\gamma_1, \gamma_2, \ldots, \gamma_l) \in \mathcal{T}$. If $g \in \varphi(t)$ is the first element of KgL then γ_l cannot be among the last $s - 1$ elements of its orbit in $G_{(\gamma_1,\ldots,\gamma_{l-1})}$.*

Proof. The set $\Gamma := \{\beta_l^{hg} \mid h \in K_{(\beta_1,\ldots,\beta_{l-1})}\}$ has s elements and $\gamma_l = \beta_l^g \in \Gamma$. All elements of Γ are in the same orbit of $G_{(\gamma_1,\ldots,\gamma_{l-1})}$, since for $\gamma = \beta_l^{hg} \in \Gamma$, we have $\gamma^{g^{-1}h^{-1}g} = \gamma_l$ with $g^{-1}h^{-1}g \in G_{(\gamma_1,\ldots,\gamma_{l-1})}$. Also, $hg \in KgL$, so the minimality of g implies $\gamma_l = \min \Gamma$. Hence $\gamma_l^{G_{(\gamma_1,\ldots,\gamma_{l-1})}}$ contains at least $s - 1$ elements after γ_l. □

Lemma 9.1.4 implies that when processing the successors of the vertex $t' := (\gamma_1, \gamma_2, \ldots, \gamma_{l-1})$, extensions by the last $s-1$ elements of each $G_{(\gamma_1,\ldots,\gamma_{l-1})}$-orbit can be skipped.

9.1.2. Pruning the Search Tree: Problem-Dependent Methods

The property $\mathcal{P}$ may imply restrictions on the base images of elements of $G(\mathcal{P})$. While processing the vertex $t = (\gamma_1, \ldots, \gamma_{l-1})$ of $\mathcal{T}$, we search for elements of $G(\mathcal{P})$ for which the images of the first $l-1$ base points are known. It is possible that all elements of $\varphi(t) \cap G(\mathcal{P})$ must satisfy additional properties, which enables us to eliminate some (or all) children of t from the backtrack traversal. A priori, we know that, for all children $(\gamma_1, \ldots, \gamma_l)$ of t, we must have $\gamma_l \in (\beta_l^{G_{(\beta_1,\ldots,\beta_{l-1})}})^g$, where g is an arbitrary element of $\varphi(t)$. We are looking for restrictions that define a subset $\Omega_{\mathcal{P}}(\gamma_1, \ldots, \gamma_{l-1})$ of $(\beta_l^{G_{(\beta_1,\ldots,\beta_{l-1})}})^g$, containing the possible extensions of t. These restrictions vary, depending on $\mathcal{P}$. As in the case of the problem-independent reductions, we try to satisfy two competing requirements: We are looking for restrictions that are easy to compute but preferably still eliminate a significant portion of the children of t.

Example 1: Centralizer Given $G \leq \mathrm{Sym}(\Omega)$ and $c \in G$, we search for a generating set of $C_G(c)$. The structure of $C_{\mathrm{Sym}(\Omega)}(c) = C_{\mathrm{Sym}(\Omega)}(\langle c \rangle)$ was described in Lemma 6.1.8. The equivalent orbits of $\langle c \rangle$ are the supports of cycles of c of equal length, so we know that the elements of $C_G(c)$ must respect the partition $\{\Omega_1, \Omega_2, \ldots\}$ of Ω, where Ω_i is the union of supports of cycles of c of length i. A further, quite severe, restriction is that for any cycle $C = (\omega, \omega^c, \ldots, \omega^{c^{k-1}})$ of c, the image ω^g uniquely determines C^g for any $g \in C_G(c)$. Namely, for all $l \in [1, k-1]$,

$$\left(\omega^{c^l}\right)^g = (\omega^g)^{c^l}.$$

Hence, if ω^g is decided for some $\omega \in B$ then we know the image of some additional elements of Ω. However, the elements of C are usually not among the base points. Therefore, to take full advantage of this criterion, we start the backtrack search with a base change. We compute a base according to an ordering of Ω, where elements from each cycle of c are consecutive. Then $\Omega_{\mathcal{P}}(\gamma_1, \ldots, \gamma_{l-1})$ consists of at most one point if $\beta_l = \beta_{l-1}^c$; otherwise, we can use

$$\Omega_{\mathcal{P}}(\gamma_1, \ldots, \gamma_{l-1}) := \left(\beta_l^{G_{(\beta_1,\ldots,\beta_{l-1})}}\right)^g \cap \Omega_i,$$

where g is an arbitrary element of $\varphi((\gamma_1, \ldots, \gamma_{l-1}))$ and i is the length of the c-cycle containing β_l. Because of this strong restriction, computing centralizers of elements is not considered to be difficult in practice, despite the exponential worst-case complexity of the procedure.

Example 2: Setwise Stabilizer Given $G \le \mathrm{Sym}(\Omega)$ and $\Delta \subseteq \Omega$, we are looking for generators of G_Δ. First, we compute a base for G according to an ordering of Ω where the elements of Δ precede all elements of $\Omega\backslash\Delta$. If the first base point in $\Omega\backslash\Delta$ is β_k then we know that $G^{[k]} = G_{(\Delta)} \le G(\mathcal{P})$. Therefore, by Lemma 9.1.1, $\varphi(t) \subseteq G(\mathcal{P})$ or $\varphi(t) \cap G(\mathcal{P}) = \emptyset$ for any $t = (\gamma_1, \dots, \gamma_{k-1})$, and it is enough to work with initial segments of base images of length at most $k-1$. We also have the restriction

$$\Omega_{\mathcal{P}}(\gamma_1, \dots, \gamma_{l-1}) := \left(\beta_l^{G_{(\beta_1,\dots,\beta_{l-1})}}\right)^g \cap \Delta$$

for all $l < k$ and all $g \in \varphi((\gamma_1, \dots, \gamma_{l-1}))$.

Example 3: Intersection of Groups Given $G, H \le \mathrm{Sym}(\Omega)$, we are looking for generators of $G \cap H$. We can suppose that $|G| \le |H|$. First, we compute a base $B = (\beta_1, \dots, \beta_m)$ for G and define the search tree $\mathcal{T}$ according to B. We also compute a base for H starting with B. The restriction on base images is

$$\Omega_{\mathcal{P}}(\gamma_1, \dots, \gamma_{l-1}) := \left(\beta_l^{G_{(\beta_1,\dots,\beta_{l-1})}}\right)^g \cap \left(\beta_l^{H_{(\beta_1,\dots,\beta_{l-1})}}\right)^h$$

for all $g \in \varphi((\gamma_1, \dots, \gamma_{l-1}))$ and all $h \in \varphi'((\gamma_1, \dots, \gamma_{l-1}))$, where φ' is the map from $\mathcal{T}$ to the power set of H, defined analogously to φ, namely, $\varphi'((\gamma_1, \dots, \gamma_{l-1})) := \{h \in H \mid \beta_i^h = \gamma_i, 1 \le i \le l-1\}$. When the traversing of $\mathcal{T}$ reaches $\mathcal{T}_m$, it means that we have constructed a sequence $t = (\gamma_1, \dots, \gamma_m)$ with $|\varphi(t)| = 1$ and $\varphi'(t) \ne \emptyset$. We have to check that the unique element of $\varphi(t)$ occurs in H as well.

Example 4: Conjugating Element Given $G \le \mathrm{Sym}(\Omega)$ and $a, b \in G$, we would like to find some $g \in G$ with $a^g = b$ or determine that no such element g exists. In this problem, the set $G(\mathcal{P})$ is empty or a right coset of $C_G(a)$. The restrictions on elements $g \in G(\mathcal{P})$ are similar to the ones we encountered at centralizer computations: g must map the cycles of a to cycles of b of the same length, and the image α^g of any $\alpha \in \Omega$ determines uniquely the image of the entire cycle of a containing α. Hence, we compute a base for G according to an ordering of Ω where elements from each cycle of a are consecutive. Then $\Omega_{\mathcal{P}}(\gamma_1, \dots, \gamma_{l-1})$ consists of at most one point if $\beta_l = \beta_{l-1}^a$; otherwise,

$$\Omega_{\mathcal{P}}(\gamma_1, \dots, \gamma_{l-1}) := \left(\beta_l^{G_{(\beta_1,\dots,\beta_{l-1})}}\right)^g \cap \Omega_i',$$

where g is an arbitrary element of $\varphi((\gamma_1, \ldots, \gamma_{l-1}))$, i is the length of the a-cycle containing β_l, and $\{\Omega'_1, \Omega'_2, \ldots\}$ is the partition of Ω where Ω'_i is the union of supports of cycles of b of length i.

Similarly to computing centralizers, finding conjugating elements is not considered a difficult problem in practice.

9.2. The Partition Method

An *ordered partition* of a set Ω is a sequence $\Pi = (\Pi[1], \ldots, \Pi[r])$ of pairwise disjoint, nonempty subsets of Ω, satisfying $\bigcup \Pi[i] = \Omega$. We call the sets $\Pi[i]$ the *cells* of the ordered partition; the cells are considered only as sets (i.e., we do not define orderings within the cells). We denote the set of all ordered partitions of Ω by $\mathrm{OP}(\Omega)$.

Ordered partitions in backtrack searches were first used in the highly successful program *nauty* (cf. [McKay, 1981]) for testing isomorphism of graphs and for computing automorphism groups of graphs. The basic idea is the following: If $\mathcal{X}(\Omega, E)$ is a graph with vertex set Ω and edge set E then any $x \in \mathrm{Aut}(\mathcal{X})$ must fix certain ordered partitions of Ω. Namely, vertices of valency k must be mapped by x to vertices of valency k, so x must respect the partition $\Pi = (\Pi[1], \ldots, \Pi[r])$, where the $\Pi[i]$ consist of the vertices of equal valency. This partition can also be *refined*: If $\alpha, \beta \in \Pi[i]$ have a different number of neighbors in some $\Pi[j]$ then there is no automorphism of $\mathcal{X}$ mapping α to β. Using this criterion, we can split the cells of Π such that $\mathrm{Aut}(\mathcal{X})$ must respect the finer partition as well. This refinement process can be iterated, by examining the number of neighbors in the cells of the finer partition. Let us call the final result of these refinements Σ.

Backtrack search comes into the picture when the refinement process in the previous paragraph bottoms out. We pick $\alpha_1 \in \Omega$ and look for automorphisms of $\mathcal{X}$ fixing α_1. The distance from α_1 in $\mathcal{X}$ defines a partition of Ω, which must be respected by $\mathrm{Aut}(\mathcal{X})_{\alpha_1}$. This partition can be intersected with Σ and refined by considering the number of neighbors in different cells, as described in the previous paragraph. We call the final result of the refinement process Σ_{α_1}. A backtrack search is started by constructing the analogous ordered partitions Σ_β for those $\beta \in \Omega$ that are in the same cell of Σ as α_1 and considering the cases that an automorphism x maps α_1 to β for all such β. If the lists of cell sizes in Σ_{α_1} and Σ_β are not identical then there is no $x \in \mathrm{Aut}(\mathcal{X})$ with $\alpha_1^x = \beta$. We build the next level of the search tree by picking some $\alpha_2 \in \Omega \backslash \{\alpha_1\}$ and computing a refinement $\Sigma_{(\alpha_1,\alpha_2)}$ of Σ_{α_1}, which must be fixed by $\mathrm{Aut}(\mathcal{X})_{(\alpha_1,\alpha_2)}$. Further vertices on this level are the partitions $\Sigma_{(\beta,\gamma)}$ for sequences (β, γ) such that the list of cell sizes does not exclude the existence of $x \in \mathrm{Aut}(\mathcal{X})$ with $\alpha_1^x = \beta$ and $\alpha_2^x = \gamma$.

Continuing, we arrive at a sequence $(\alpha_1, \alpha_2, \ldots, \alpha_m)$ of points in Ω, such that the base partition fixing the α_i is discrete (i.e., contains only one-element cells), and at other discrete partitions that at the appropriate positions contain the possible images of the α_i by elements of $\text{Aut}(\mathcal{X})$. Obviously, $\text{Sym}(\Omega)$ contains a unique permutation that maps a given discrete ordered partition to another given discrete ordered partition. Therefore, we constructed a base $(\alpha_1, \alpha_2, \ldots, \alpha_m)$ for $\text{Aut}(\mathcal{X})$ and the permutations in $\text{Aut}(\mathcal{X})$.

These ideas were adopted for searches in permutation groups G in [Leon, 1991]. A formal description and some examples can be found in [Leon, 1997] as well. For simplicity, we describe only the case when $G(\mathcal{P})$ is a subgroup and indicate the necessary modification for coset-type problems at the end of the section. Let $B = (\beta_1, \ldots, \beta_m)$ be a base for G. We consider the search tree $\mathcal{T}$ defined in Section 9.1 and use the sequences of base images as names for the vertices of $\mathcal{T}$ as in that section. However, the vertices themselves are ordered partitions of Ω. The vertex with name $t = (\gamma_1, \ldots, \gamma_l)$ is a partition Π_t, satisfying the following properties:

(p1) $\gamma_1, \ldots, \gamma_l$ occur in one-element cells of Π_t.
(p2) Any $g \in G(\mathcal{P}) \cap \varphi(t)$ must satisfy $(\Pi_{(\beta_1,\ldots,\beta_l)})^g = \Pi_t$.
(p3) If t' is the parent of t then Π_t is a refinement of $\Pi_{t'}$.

Here, the image of an ordered partition $\Pi = (\Pi[1], \ldots, \Pi[r])$ under $g \in \text{Sym}(\Omega)$ is defined as $\Pi^g := (\Pi[1]^g, \ldots, \Pi[r]^g)$. We say that $\Pi = (\Pi[1], \ldots, \Pi[r])$ is a *refinement* of $\Sigma = (\Sigma[1], \ldots, \Sigma[s])$ (in notation $\Pi \leq \Sigma$) if each $\Sigma[i]$ is the union of consecutive cells of Π. Formally,

$$\Sigma[i] = \bigcup_{j=k_{i-1}+1}^{k_i} \Pi[j],$$

for some sequence of indices $0 = k_0 < k_1 < \cdots < k_s = r$.

We shall also need the intersection of ordered partitions. Given $\Pi = (\Pi[1], \ldots, \Pi[r]) \in \text{OP}(\Omega)$ and $\Sigma = (\Sigma[1], \ldots, \Sigma[s]) \in \text{OP}(\Omega)$, the cells of their *intersection* $\Pi \wedge \Sigma$ are the nonempty subsets of Ω of the form $\Pi[i] \cap \Sigma[j]$ for $1 \leq i \leq r$ and $1 \leq j \leq s$, ordered by the following rule: $\Pi[i_1] \cap \Sigma[j_1]$ precedes $\Pi[i_2] \cap \Sigma[j_2]$ if and only if $i_1 < i_2$, or $i_1 = i_2$ and $j_1 < j_2$. Hence $\Pi \wedge \Sigma \leq \Pi$, but $\Pi \wedge \Sigma \leq \Sigma$ is not necessarily true. We listed some properties of intersections in Exercise 9.3.

Traversing a search tree of ordered partitions is a generalization of the original backtrack method. If the partitions are defined by the recursive rule $\Pi_{(\gamma_1,\ldots,\gamma_l)} := \Pi_{(\gamma_1,\ldots,\gamma_{l-1})} \wedge (\{\gamma_l\}, \Omega\backslash\{\gamma_l\})$, that is, the images of base points are in one-element cells and all further elements of Ω are in one cell, then we get back the traditional backtrack search tree. Ordered partitions have the advantage that they provide a convenient language to describe restrictions on the children of a vertex of the search tree. For example, defining a set of possible children $\Omega_{\mathcal{P}}(\gamma_1, \ldots, \gamma_l)$ of the vertex $(\gamma_1, \ldots, \gamma_l)$ as in Section 9.1.2 simply corresponds to intersecting $\Pi_{(\gamma_1,\ldots,\gamma_l)}$ with $(\Omega_{\mathcal{P}}(\gamma_1, \ldots, \gamma_l), \Omega\backslash\Omega_{\mathcal{P}}(\gamma_1, \ldots, \gamma_l))$. Even more importantly, as we shall see in some examples, it is possible to describe restrictions that cannot be expressed in terms of the base point images and so cannot be incorporated easily into a traditional backtrack search.

Given a partition $\Pi_{(\gamma_1,\ldots,\gamma_{l-1})}$ at vertex $(\gamma_1, \ldots, \gamma_{l-1})$ of $\mathcal{T}$, how do we define Π_t for $t := (\gamma_1, \ldots, \gamma_{l-1}, \gamma_l)$? To satisfy the properties (p1)–(p3), Π_t must be a refinement of $\Pi_t^* := \Pi_{(\gamma_1,\ldots,\gamma_{l-1})} \wedge (\{\gamma_l\}, \Omega\backslash\{\gamma_l\})$. Moreover, $\Omega_{\mathcal{P}}(\gamma_1, \ldots, \gamma_{l-1})$ provides the further restriction that Π_t can be chosen as a refinement of $\Pi_t^* \wedge (\Omega_{\mathcal{P}}(\gamma_1, \ldots, \gamma_{l-1}), \Omega\backslash\Omega_{\mathcal{P}}(\gamma_1, \ldots, \gamma_{l-1}))$. From property (p2), we may get further restrictions on Π_t. In general, we define a *$\mathcal{P}$-refinement process* $\mathcal{R} : \mathrm{OP}(\Omega) \to \mathrm{OP}(\Omega)$, incorporating the restrictions we can (or are willing to) compute, satisfying the following two rules:

(p4) $\mathcal{R}(\Pi) \leq \Pi$ for all $\Pi \in \mathrm{OP}(\Omega)$.
(p5) $\mathcal{R}(\Pi)^g = \mathcal{R}(\Pi^g)$ for all vertices Π of the search tree and for all $g \in G(\mathcal{P})$.

The partition Π_t is constructed as $\Pi_t := \mathcal{R}(\Pi_t^*)$, for the partition $\Pi_t^* = \Pi_{(\gamma_1,\ldots,\gamma_{l-1})} \wedge (\{\gamma_l\}, \Omega\backslash\{\gamma_l\})$ just defined. In practice, the refinement process usually consists of a sequence $(\mathcal{R}_1, \ldots, \mathcal{R}_l)$ of refinements. Given $\Pi \in \mathrm{OP}(\Omega)$, we initialize $\Pi' := \Pi$ and always replace Π' by $\mathcal{R}_i(\Pi')$ for the least index i such that $\mathcal{R}_i(\Pi') \neq \Pi'$, continuing this process until Π' becomes invariant for all $\mathcal{R}_i$. Then we define $\mathcal{R}(\Pi)$ as the current value of Π'.

The leftmost branch $(\emptyset, (\beta_1), (\beta_1, \beta_2), \ldots, (\beta_1, \beta_2, \ldots, \beta_m))$ of the traditional backtrack search tree corresponds to the point stabilizer subgroup chain $G = G^{[1]} \geq \cdots \geq G^{[m+1]} = 1$, where $G^{[i]} = G_{(\beta_1,\ldots,\beta_{i-1})}$, in the sense that for $t = (\beta_1, \ldots, \beta_{i-1})$ we have $\varphi(t) = G^{[i]}$. At partition backtrack, this leftmost branch consists of the ordered partitions $\Pi_1 \geq \cdots \geq \Pi_{m+1}$, where $\Pi_1 = \mathcal{R}((\Omega))$ and $\Pi_i = \mathcal{R}(\Pi_{i-1} \wedge (\{\beta_{i-1}\}, \Omega\backslash\{\beta_{i-1}\}))$ for $2 \leq i \leq m+1$. Some of these partitions may coincide: For example, in the case when $G(\mathcal{P}) = C_G(c)$ for some $c \in G$, we choose the base $(\beta_1, \ldots, \beta_m)$ such that an initial segment of base points, say $(\beta_1, \ldots, \beta_k)$, is contained in the same cycle of c (cf. Example 1 in Section 9.1.2). Then, as we saw in this example, the image β_1^g under an element $g \in C_G(c)$ determines uniquely $(\beta_1, \ldots, \beta_k)^g$. In particular, if $\beta_1^g = \beta_1$ then $(\beta_1, \ldots, \beta_k)^g = (\beta_1, \ldots, \beta_k)$. This restriction can be built into the refinement process $\mathcal{R}$, resulting in $\Pi_2 = \Pi_3 = \cdots = \Pi_{k+1}$. Deleting the repetitions from the chain of partitions $\Pi_1 \geq \cdots \geq \Pi_{m+1}$, we get a sequence of partitions that

Leon calls an $\mathcal{R}$-*base* for $G(\mathcal{P})$. The corresponding elements of $(\beta_1, \ldots, \beta_m)$ (i.e., those β_i for which Π_{i+1} was not deleted) form a base of $G(\mathcal{P})$. This base may be substantially shorter than the base for G, which speeds up the partition backtrack search. The first step of partition backtrack is the construction of an $\mathcal{R}$-base.

It would be enough to require that property (p5) holds when Π is an element of the $\mathcal{R}$-base, since the purpose of (p5) is to ensure that the refinement process does not violate (p2).

The problem-independent prunings of the search tree described in Section 9.1.1 can be applied for partition backtrack as well. There are further problem-independent restrictions, which show one of the advantages of partition backtrack over the traditional method: Instead of considering only the fundamental orbit $\beta_l^{G_{(\beta_1,\ldots,\beta_{l-1})}}$, we can work with all orbits of $G^{[l]} = G_{(\beta_1,\ldots,\beta_{l-1})}$. One such restriction is that if β_l is in the ith cell of $\Pi_{(\beta_1,\ldots,\beta_{l-1})}$ then (p2) implies that for any extension $(\gamma_1, \ldots, \gamma_l)$ of a vertex $(\gamma_1, \ldots, \gamma_{l-1})$, the extending element γ_l must come from the ith cell of $\Pi_{(\gamma_1,\ldots,\gamma_{l-1})}$. Also, a vertex $(\gamma_1, \ldots, \gamma_l)$ can be eliminated if the list of cell sizes of $\Pi_{(\beta_1,\ldots,\beta_l)}$ and $\Pi_{(\gamma_1,\ldots,\gamma_l)}$ differ.

Now we describe another restriction that uses all orbits of $G^{[l]}$. For $H \leq \mathrm{Sym}(\Omega)$, we define $\Theta(H) \in \mathrm{OP}(\Omega)$ to have as cells the orbits of H; orbit O_1 precedes orbit O_2 in $\Theta(H)$ if and only if $\min O_1 \prec \min O_2$.

Suppose that at the construction of an $\mathcal{R}$-base the partition $\Pi_l := \Pi_{(\beta_1,\ldots,\beta_{l-1})}$ is already defined and now we would like to construct $\Pi_{l+1} := \Pi_{(\beta_1,\ldots,\beta_l)}$. The elements of $G^{[l+1]}$ must leave $\Theta(G^{[l+1]})$ invariant, so it makes sense to define the first refinement of $\Pi^*_{l+1} := \Pi_l \wedge (\{\beta_l\}, \Omega\backslash\{\beta_l\})$ as $\mathcal{R}_1(\Pi^*_{l+1}) := \Pi^*_{l+1} \wedge \Theta(G^{[l+1]})$. Then we have to define $\mathcal{R}_1$ for partitions $\Pi^*_t := \Pi_{(\gamma_1,\ldots,\gamma_{l-1})} \wedge (\{\gamma_l\}, \Omega\backslash\{\gamma_l\})$, which will be placed on level T_l with the name $t = (\gamma_1, \ldots, \gamma_l)$, so that the compatibility conditions (p2) and (p5) are satisfied.

Let g be an arbitrary element of $\varphi(t)$ and define $\mathcal{R}_1(\Pi^*_t) := \Pi^*_t \wedge \Theta(G^{[l+1]})^g$. Note that $\mathcal{R}_1(\Pi^*_t)$ is well defined, because if $g_1, g_2 \in \varphi(t)$ then $g_1 = sg_2$ for some $s \in G^{[l+1]}$ and, since $\Theta(G^{[l+1]})^s = \Theta(G^{[l+1]})$, we have $\Theta(G^{[l+1]})^{g_1} = \Theta(G^{[l+1]})^{sg_2} = \Theta(G^{[l+1]})^{g_2}$. Since any $g \in \varphi(t)$ can be used at the definition of $\mathcal{R}_1(\Pi^*_t)$, we can compute $\mathcal{R}_1(\Pi^*_t)$ independently of whether we have at hand elements of $G(\mathcal{P})$ that map Π_l to $\Pi_{(\gamma_1,\ldots,\gamma_{l-1})}$. Note further that the cells of $\Theta(G^{[l+1]})^g$ are the orbits of $G_{(\gamma_1,\ldots,\gamma_l)}$, but not necessarily in the same order as in $\Theta(G_{(\gamma_1,\ldots,\gamma_l)})$.

We claim that the refinement $\mathcal{R}_1$ as just defined satisfies (p2) and (p5). To check (p2), let $g \in G(\mathcal{P}) \cap \varphi(t)$ be arbitrary. Then, in particular, $g \in G(\mathcal{P}) \cap \varphi((\gamma_1, \ldots, \gamma_{l-1}))$, so $\Pi_l^g = \Pi_{(\gamma_1,\ldots,\gamma_{l-1})}$. By Exercise 9.3, $\mathcal{R}_1(\Pi^*_{l+1})^g = (\Pi_l \wedge (\{\beta_l\}, \Omega\backslash\{\beta_l\}) \wedge \Theta(G^{[l+1]}))^g = \Pi_l^g \wedge (\{\gamma_l\}, \Omega\backslash\{\gamma_l\}) \wedge \Theta(G^{[l+1]})^g = \mathcal{R}_1(\Pi^*_t)$. To prove (p5), suppose that $\Pi = \Pi_{(\gamma_1,\ldots,\gamma_{l-1})} \wedge (\{\gamma_l\}, \Omega\backslash\{\gamma_l\})$ and $g \in G(\mathcal{P})$.

Let $g_1 \in \varphi((\gamma_1, \ldots, \gamma_l))$, and let $(\gamma_1', \ldots, \gamma_l') = (\gamma_1, \ldots, \gamma_l)^g$. Then $\Pi^g = \Pi_{(\gamma_1', \ldots, \gamma_{l-1}')} \wedge (\{\gamma_l'\}, \Omega \backslash \{\gamma_l'\})$ and $g_1 g \in \varphi((\gamma_1', \ldots, \gamma_l'))$, so

$$\begin{aligned} \mathcal{R}_1(\Pi)^g &= \mathcal{R}_1\big(\Pi_{(\gamma_1, \ldots, \gamma_{l-1})} \wedge (\{\gamma_l\}, \Omega \backslash \{\gamma_l\})\big)^g \\ &= \big(\Pi_{(\gamma_1, \ldots, \gamma_{l-1})} \wedge (\{\gamma_l\}, \Omega \backslash \{\gamma_l\}) \wedge \Theta\big(G^{[l+1]}\big)^{g_1}\big)^g \\ &= \Pi_{(\gamma_1', \ldots, \gamma_{l-1}')} \wedge (\{\gamma_l'\}, \Omega \backslash \{\gamma_l'\}) \wedge \Theta\big(G^{[l+1]}\big)^{g_1 g} = \mathcal{R}_1(\Pi^g). \end{aligned}$$

We finish this section by briefly indicating the necessary change in the partition backtrack method when $G(\mathcal{P})$ is a coset of a subgroup of G. We have to define *two* refinement processes $\mathcal{R}$ and $\mathcal{R}'$, connected by a modification of property (p5):

p(5)′ $\mathcal{R}(\Pi)^g = \mathcal{R}'(\Pi^g)$ for all vertices Π of the search tree and for all $g \in G(\mathcal{P})$.

For example, if $G(\mathcal{P})$ consists of the elements of G conjugating $a \in G$ to $b \in G$ then $\mathcal{R}$ can be chosen as the refinement process used at the computation of $C_G(a)$, which is based on the cycle structure of a, and $\mathcal{R}'$ can be chosen as the refinement process used at the computation of $C_G(b)$.

9.3. Normalizers

Given $G, H \le \mathrm{Sym}(\Omega)$, computing the normalizer $N_G(H)$ is considered to be a harder problem than the ones discussed in the examples in Section 9.1.2. All of those problems can be reduced to each other in polynomial time (cf. [Luks, 1993]), but no polynomial-time reduction of the normalizer computation to centralizer computations is known. In fact, there is no known "easy" way to compute even $N_{\mathrm{Sym}(\Omega)}(H)$; in contrast, we saw in Section 6.1.2 that $C_{\mathrm{Sym}(\Omega)}(H)$ is computable in polynomial time. Note that since $N_G(H) = N_{\mathrm{Sym}(\Omega)}(H) \cap G$, a polynomial-time algorithm for computing $N_{\mathrm{Sym}(\Omega)}(H)$ would imply the polynomial-time equivalence of normalizer computations with centralizer computations.

If some $g \in \mathrm{Sym}(\Omega)$ normalizes H then it must leave invariant certain structures associated with H. For example, Lemma 6.1.7 says that g must permute the orbits of H. The first algorithm for computing normalizers, which appeared in [Butler, 1983], was based on this observation. In this section we describe the method of [Theißen, 1997], which uses that g must permute the orbital graphs of H. Considering orbital graphs gives more efficient criteria for pruning the search tree, yet these criteria are still easy enough to compute.

Let $H \leq \text{Sym}(\Omega)$. Then H also acts on the ordered pairs $(\alpha, \beta) \in \Omega \times \Omega$ by the natural componentwise action: $(\alpha, \beta)^h := (\alpha^h, \beta^h)$, for all $h \in H$. The *orbital graphs* of H are directed graphs with vertex set Ω. The edge set of an orbital graph of H is one of the orbits in the action $\varphi : H \to \text{Sym}(\Omega \times \Omega)$. We denote the orbital graph with edge set $(\alpha, \beta)^H$ by $\mathcal{X}(\alpha, \beta)$. Note that H is a subgroup of the automorphism group of each orbital graph $\mathcal{X}(\alpha, \beta)$, since obviously H leaves invariant the set $(\alpha, \beta)^H$ of edges. Applying Lemma 6.1.7 to $\varphi(H)$, we obtain that any $g \in N_{\text{Sym}(\Omega)}(H)$ must permute the orbital graphs of H.

Let also $G \leq \text{Sym}(\Omega)$ be given; our goal is to compute $G(\mathcal{P}) := N_G(H)$. It is not necessary that $H \leq G$, although the condition $H \leq G$ makes the application of the criteria in Section 9.1.1 more efficient since we know a priori a nontrivial subgroup of $N_G(H)$.

The main tool of [Theißen, 1997] is a refinement process that can be applied to ordered partitions Π satisfying the condition that there are $\alpha, \beta \in \Omega$ belonging to the same H-orbit on Ω and that α and β occur in one-element cells in Π. The orbital graph $\mathcal{X}(\alpha, \beta)$ defines an ordered partition $\Delta := (\Delta[0], \Delta[1], \ldots, \Delta[k], \Delta[\infty])$, where $\Delta[i]$ consists of vertices of $\mathcal{X}(\alpha, \beta)$ of (directed) distance i from α for $0 \leq i \leq k$, and $\Delta[\infty]$ contains the elements of Ω that cannot be reached by directed paths from α. If $g \in G(\mathcal{P})$ maps Π to Π' with $\alpha' := \alpha^g$ and $\beta' := \beta^g$ then g must also map the refinement $\Pi \wedge \Delta$ to $\Pi' \wedge \Delta'$, where Δ' is the distance partition from α' in the orbital graph $\mathcal{X}(\alpha', \beta')$. If $\Pi \wedge \Delta$ and $\Pi' \wedge \Delta'$ have different sequences of cell sizes then we can conclude that there is no $g \in G(\mathcal{P})$ mapping Π to Π'.

Instead of Δ, we could have used a refinement of Δ as well, which can be obtained by partitioning the cells $\Delta[i]$ according to the number of neighbors in different cells (we described such refinements in detail at the beginning of Section 9.2, while sketching McKay's graph isomorphism program *nauty*). Of course, we have to consider whether the cost of computing this refinement is justified by the additional pruning of the search tree. As a compromise, in the GAP implementation the sets $\Delta[i]$ are partitioned according to the valencies of vertices, but further refinements are not computed.

We need an efficient method for computing Δ. For $0 \leq i \leq k$, let $\bar{\Delta}[i]$ be the set of vertices of $\mathcal{X}(\alpha, \beta)$ that can be reached by a walk of length i from α. If $\Delta[0], \ldots, \Delta[i-1]$ are already known then it is enough to compute $\bar{\Delta}[i]$, since $\Delta[i] = \bar{\Delta}[i] \setminus \bigcup_{j<i} \Delta[j]$. Also, since α and β are in the same H-orbit, we can fix $r \in H$ with $\alpha^r = \beta$.

Lemma 9.3.1. $\bar{\Delta}[i] = \alpha^{rH_\alpha \cdots rH_\alpha}$ *(i iterations of rH_α).*

Proof. We denote $rH_\alpha \cdots rH_\alpha$, with i iterations of rH_α, by $(rH_\alpha)^i$.

We prove the statement of the lemma by induction on i. For any $\gamma \in \Omega$, by definition $\gamma \in \bar{\Delta}[1]$ if and only if (α, γ) is an edge in $(\alpha, \beta)^H$. This happens if and only if there exists $h \in H$ with $\alpha^h = \alpha$ and $\beta^h = \gamma$. Since $\beta = \alpha^r$, the last statement is equivalent with the existence of $h \in H_\alpha$ such that $\alpha^{rh} = \gamma$, which is the definition of $\gamma \in \alpha^{rH_\alpha}$. Hence the lemma holds for $i = 1$.

Suppose that $\bar{\Delta}[i] = \alpha^{(rH_\alpha)^i}$ for some $i \geq 1$. For any $\gamma \in \Omega$, we have $\gamma \in \bar{\Delta}[i+1]$ if and only if there exists $\delta \in \bar{\Delta}[i]$ such that (δ, γ) is an edge in $\mathcal{X}(\alpha, \beta)$. By the inductive hypothesis, $\delta \in \bar{\Delta}[i]$ if and only if there exists $h \in (rH_\alpha)^i$ such that $\alpha^h = \delta$. Moreover, since h is an automorphism of $\mathcal{X}(\alpha, \beta)$, the condition (δ, γ) is an edge in $\mathcal{X}(\alpha, \beta)$ means that $\gamma = \beta_1^h$ for some neighbor β_1 of α. By the already proven case $i = 1$ of the lemma, β_1 is a neighbor of α if and only if $\beta_1 = \alpha^{rh_1}$ for some $h_1 \in H_\alpha$. In summary, $\gamma \in \bar{\Delta}[i+1]$ if and only if there exists $h \in (rH_\alpha)^i$ and $h_1 \in rH_\alpha$ such that $\gamma = \alpha^{h_1 h}$ (i.e., $\gamma \in (rH_\alpha)^{i+1}$). □

Lemma 9.3.1 implies that the cells $\Delta[i]$ are unions of H_α-orbits. This provides the opportunity to compute Δ working in a quotient graph Γ of $\mathcal{X}(\alpha, \beta)$, which may have significantly fewer vertices than $\mathcal{X}(\alpha, \beta)$. The vertices of Γ are the H_α-orbits on Ω; for $\gamma \in \Omega$, we denote by $[\gamma]$ the vertex γ^{H_α} of Γ. For $[\gamma], [\delta] \in V(\Gamma)$, the ordered pair $([\gamma], [\delta])$ is an edge of Γ if and only if there exists $\gamma' \in [\gamma]$ and $\delta' \in [\delta]$ such that $(\gamma', \delta') \in (\alpha, \beta)^H$.

Lemma 9.3.2. *Let $\gamma \in \Omega$ be arbitrary. Then there exists a walk of length i from α to γ in $\mathcal{X}(\alpha, \beta)$ if and only if there is a walk of length i in Γ from $[\alpha]$ to $[\gamma]$.*

Proof. One direction is clear. Namely, if $(\alpha = \gamma_0, \gamma_1, \ldots, \gamma_i = \gamma)$ is a walk in $\mathcal{X}(\alpha, \beta)$ from α to γ then by definition $([\gamma_0], \ldots, [\gamma_i])$ is a walk from $[\alpha]$ to $[\gamma]$ in Γ.

We prove the converse, that if there is a walk of length i from $[\alpha]$ to $[\gamma]$ in Γ then there is a walk of length i from α to γ in $\mathcal{X}(\alpha, \beta)$, by induction on i. The case $i = 0$ is trivial. Suppose that we know the statement for some i, and suppose that there is a walk of length $i+1$ from $[\alpha]$ to $[\gamma]$ for some $\gamma \in \Omega$. By definition, this means that there exist γ_j for $j \in [0, i]$ and γ'_j for $j \in [1, i+1]$ such that $(\gamma_j, \gamma'_{j+1}) \in (\alpha, \beta)^H$ for $j \in [0, i]$ and $\gamma'_j \in \gamma_j^{H_\alpha}$ for $j \in [1, i]$, with $\gamma_0 = \alpha$ and $\gamma'_{i+1} \in \gamma^{H_\alpha}$. Since there is a walk of length i from $[\alpha]$ to $[\gamma_i]$ in Γ, the inductive hypothesis implies that there is a walk of length i from α to γ_i in $\mathcal{X}(\alpha, \beta)$, and so there is a walk of length $i+1$ from α to γ'_{i+1} in $\mathcal{X}(\alpha, \beta)$. By Lemma 9.3.1, this means that $\gamma'_{i+1} = \alpha^h$ for some $h \in (rH_\alpha)^{i+1}$. Also, since γ and γ'_{i+1} are in the same H_α-orbit, $\gamma = (\gamma'_{i+1})^{h_1}$ for some $h_1 \in H_\alpha$. However,

$hh_1 \in (rH_\alpha)^{i+1}$ and $\alpha^{hh_1} = \gamma$, so Lemma 9.3.1 implies that there is a walk of length $i+1$ from α to γ in $\mathcal{X}(\alpha, \beta)$. □

It follows immediately from Lemma 9.3.2 that the sets $\bar{\Delta}[i]$ can be computed in Γ.

Another refinement method of [Theißen, 1997] is based on the fact that if α, β are contained in one-element cells of Π and $\Pi^g = \Pi'$, and $\alpha^g = \alpha'$, $\beta^g = \beta'$ for some $g \in G(\mathcal{P})$, then the smallest block of imprimitivity B of H containing α and β must be mapped by g to the smallest block of imprimitivity of H containing α' and β'. The block B can be computed as $B := \Omega\backslash\Delta[\infty]$ (see also Exercise 5.6).

9.4. Conjugacy Classes

Finding the conjugacy classes of a group is an important task, as it is an essential ingredient in the computation of the character table of the group. Listing conjugacy class representatives is also a less space-consuming alternative to listing all group elements. There are numerous methods for computing conjugacy classes, and the newest approaches use quite a bit of the machinery we have described in the previous chapters.

The output of the conjugacy class algorithms is a list of representatives of the classes. This list may be exponentially long compared to the input length, so the user is advised to do some structural exploration of the input group G before calling a conjugacy class algorithm in the major computer algebra systems. Obvious obstacles are when $Z(G)$ or G/G' are large and, in general, solvable groups may have numerous conjugacy classes. On the other end of the spectrum, nonabelian simple groups tend to have relatively few conjugacy classes.

Class Representatives by Random Sampling

The simplest algorithm for computing conjugacy classes, that of Butler and Cannon (cf. [Butler, 1979]), is to take random elements of the input group G until we find representatives for all classes. After a partial list $(g_1, \ldots, g_m)$ of class representatives is already constructed, we can test whether a new random element $g \in G$ is in the same class as one of the g_i by the algorithm of Example 4 in Section 9.1.2 or by the partition backtrack version of that algorithm. For each class representative g_i, we also compute $C_G(g_i)$ and $|g_i^G| = |G|/|C_G(g_i)|$; the algorithm terminates when $\sum_{i\le m} |g_i^G|$ reaches $|G|$.

The main drawback of the method described in the previous paragraph is that we have only a slim chance to find representatives for the small classes.

An obvious but very useful improvement is to test not only whether the new random element g is in a new conjugacy class but to test the powers of g as well (cf. a similar application of power maps in Section 5.3).

The ultimate word on finding class representatives by random sampling is in [Jerrum, 1995]. Instead of taking random elements of the input group G, Jerrum's method constructs a Markov chain M (cf. Section 2.2) with states $V := G$. For $g, h \in G$, the entry $p_{g,h}$ of the transition probability matrix P is defined as

$$p_{g,h} := \begin{cases} 1/|C_G(g)|, & h \in C_G(g) \\ 0, & \text{otherwise.} \end{cases}$$

Lemma 9.4.1. *The Markov chain M defined in the previous paragraph is irreducible and aperiodic.*

Proof. Given any two states $g, h \in G$, we can reach h from g in at most two steps since $p_{g,1} = 1/|C_G(g)| > 0$ and $p_{1,h} = 1/|G| > 0$. Hence M is irreducible. Also, $p_{1,1} = 1/|G| > 0$ and so M is aperiodic. □

The random walk on G corresponding to M is constructed by starting at a random element x_0 of G. After the sequence $(x_0, \ldots, x_n)$ is defined, we compute $C_G(x_n)$ and choose x_{n+1} as a uniformly distributed random element of $C_G(x_n)$.

Lemma 9.4.1 implies that M satisfies the conditions of Theorem 2.2.1, and so there is a stationary distribution vector $\mathbf{u} = (u[h] \mid h \in G)$ and a constant $\delta \in (0, 1)$ such that for all $m \geq 0$ and $h \in G$ we have $|\text{Prob}(x_m = h) - u[h]| \leq \delta^m$. Let k denote the number of conjugacy classes of G and define $u[h] := (k|h^G|)^{-1}$ for $h \in G$. Using that $|G| = |h^G||C_G(h)|$ for all $h \in G$, it is easy to see that $\mathbf{u} = \mathbf{u}P$ and $\sum_{h \in G} u[h] = 1$. Hence $\mathbf{u}$ is the stationary distribution vector of M. For each conjugacy class C of G, we have $\sum_{h \in C} u[h] = 1/k$. This means that for large enough m, x_m is close to being uniformly distributed among the conjugacy classes. Unfortunately, the usual caveat applies: For most groups, we do not have a good estimate for the parameter δ, so we do not know the speed of the convergence to the stationary distribution.

We note that [Jerrum, 1995] discusses a more general situation, when G permutes the elements of some domain Ω that is too big to list, and we would like to choose an orbit of G from the uniform distribution. Our description above corresponds to the case $\Omega = G$, with G acting by conjugation on itself. We leave the details of the more general argument as an exercise (cf. Exercise 9.4).

Representatives in Centralizers of p-Elements

A completely different approach for conjugacy class computations is described in [Butler, 1994]. The method is based on the following observation: Since every element $g \in G\backslash\{1\}$ has a power g^m of prime order and, of course, g^m commutes with g, we can find representatives for each conjugacy class of G in the centralizers of elements of prime order. Moreover, for a fixed prime p, all Sylow p-subgroups of G are conjugate; hence, for each prime p dividing the order of G, we can fix one Sylow p-subgroup S_p of G and find representatives of all classes of G in the centralizers of elements of these fixed S_p. Besides the difficulties with Sylow subgroup computations, a drawback of this approach is that there may be many more conjugacy classes of elements of order p in S_p then there are in G, since conjugation by the elements of G may fuse classes of S_p. Therefore, although S_p is solvable and so we may apply the method of [Mecky and Neubüser, 1989] from Section 7.3.2, we may have to compute a list of class representatives in S_p that is much longer than would be necessary to handle G.

Separation of the Solvable Radical

The latest approaches (cf. [Cannon and Souvignier, 1997; Hulpke, 2000]) combine the methods of separating computations in $G/O_\infty(G)$ and $O_\infty(G)$ (cf. Section 6.3) and extending the result through elementary abelian layers (Section 7.3.2). As noted at the end of Section 6.3, every group G has a series of characteristic subgroups $1 \le N_1 \le N_2 \le N_3 \le G$, where $N_1 = O_\infty(G)$, $N_2/N_1 = \mathrm{Soc}(G/N_1) \cong T_1 \times \cdots \times T_m$ is the direct product of nonabelian simple groups, $N_3/N_2 \lesssim \mathrm{Out}(T_1) \times \cdots \times \mathrm{Out}(T_m)$, and G/N_3 is a permutation group of degree m, corresponding to the conjugation action of G on $\{T_1, \ldots, T_m\}$. Given $G = \langle S \rangle \le \mathrm{Sym}(\Omega)$, both aforementioned references compute a faithful permutation representation of G/N_1 of degree at most $|\Omega|$, compute the conjugacy classes of G/N_1 in this new representation, and then obtain the conjugacy classes of G by the method of [Mecky and Neubüser, 1989], which we described in Section 7.3.2. The two papers also handle N_2/N_1 similarly, by computing the conjugacy classes of N_2/N_1 using the random sampling method described at the beginning of this section and then identifying the classes that are fused by the conjugation action of G/N_1 on N_2/N_1. The major difference is in the handling of classes that are subsets of $G\backslash N_2$. In [Cannon and Souvignier, 1997], random elements of G are taken until all conjugacy classes are hit. In [Hulpke, 2000], random sampling is used only in N_3; the classes in $G\backslash N_3$ are constructed by extending the ideas of [Mecky and Neubüser, 1989] into a nonabelian setting.

Exercises

9.1. Draw the search tree $\mathcal{T}$ for S_4, according to the base $B = (1, 2, 3)$. What vertices of $\mathcal{T}$ are visited during the computation of $C_{S_4}(\langle(1, 2)\rangle)$, if we use the restrictions in Section 9.1.1? What happens if we use instead the restrictions in Example 1 of Section 9.1.2? And if we combine all restrictions?

9.2. Repeat the computation of $C_{S_4}(\langle(1, 2)\rangle)$, using the partition method. Define a refinement process using the restrictions in Example 1 of Section 9.1.2.

9.3. Prove the following properties of ordered partitions:
(i) $(\Pi \wedge \Sigma) \wedge T = \Pi \wedge (\Sigma \wedge T)$.
(ii) $(\Pi \wedge \Sigma)^g = \Pi^g \wedge \Sigma^g$.
(iii) $\Pi \leq \Sigma \iff \Pi^g \leq \Sigma^g$.

9.4. [Jerrum, 1995] Let $G \leq \mathrm{Sym}(\Omega)$. We define a Markov chain M with states Ω, and with the following transition rule: Given a state $\alpha \in \Omega$, first compute a uniformly distributed random element g of G_α, and then compute a uniformly distributed random element β of the fixed point set $\{\gamma \in \Omega \mid \gamma^g = \gamma\}$ of g. Then we define β as the next state in M. Let k denote the number of orbits of G on Ω.

Prove that M is irreducible and aperiodic and that the stationary distribution vector $\mathbf{u} = (u[\alpha] \mid \alpha \in \Omega)$ satisfies $u[\alpha] = (k|\alpha^G|)^{-1}$ for all $\alpha \in \Omega$.

10

Large-Base Groups

The central theme of this book is the design and analysis of nearly linear-time algorithms. Given a permutation group $G = \langle S \rangle \leq \mathrm{Sym}(\Omega)$, with $|\Omega| = n$, such algorithms run in $O(|S| n \log^c |G|)$ time for some constant c. If, as happens most of the time in practice, $\log |G|$ is bounded from above by a polylogarithmic function of n then these algorithms run in time that is nearly linear as a function of the input length $|S|n$. In most cases, we did not make an effort to minimize the exponent c of $\log |G|$ in the running time since achieving the smallest possible exponent c was not the most pressing problem from either the theoretical or practical point of view. However, in families of groups where $\log |G|$ or, equivalently, the minimal base size of G is large, the nearly linear-time algorithms do not run as fast as their name indicates; in fact, for certain tasks, they may not be the asymptotically fastest algorithms at all. Another issue is the memory requirement of the algorithms, which again may be unnecessarily high. The purpose of this chapter is to describe algorithms that may be used in the basic handling of large-base groups. The practical limitation is their memory requirement, which is in most cases a quadratic function of n. The use of $\Theta(n^2)$ memory in storing an SGS for an arbitrary $G \leq \mathrm{Sym}(\Omega)$ seems to be unavoidable.

10.1. Labeled Branchings

A labeled branching is a data structure that uses a total of n permutations to code all transversals along a point stabilizer chain of some $G \leq \mathrm{Sym}(\Omega)$, with $|\Omega| = n$. This data structure supports membership testing in G with the same asymptotic efficiency as the storage of all transversal elements as permutations but is more space efficient: Storing all transversal elements may require $\Theta(n^3)$ memory. In this section we describe the definition and construction of labeled branchings based on [Jerrum, 1986].

Let $G \leq \mathrm{Sym}(\Omega)$, with $|\Omega| = n$, be an arbitrary permutation group, and let us fix an ordering $\omega_1 \prec \cdots \prec \omega_n$ of Ω. Let $G = G^{[1]} \geq G^{[2]} \geq \cdots \geq G^{[n-1]} \geq G^{[n]} = 1$ be the point stabilizer chain corresponding to this ordering; that is, $G^{[i]}$ is the subgroup comprised of the elements of G that fix $\{\omega_1, \ldots, \omega_{i-1}\}$ pointwise. A *labeled branching* $\mathcal{B}(\Omega, \vec{E})$ for G is a directed graph with vertex set Ω and edge set $\vec{E}$, and with a function $f : \vec{E} \to G$. For $\overrightarrow{(\omega_i, \omega_j)} \in \vec{E}$, we call the group element $f(\overrightarrow{(\omega_i, \omega_j)})$ the label of this edge. A labeled branching must satisfy the following properties:

(i) The underlying graph of $\mathcal{B}$ is a forest (i.e., disregarding the orientation of edges, each connected component of $\mathcal{B}$ is a tree).
(ii) Each connected component is a rooted tree, and all edges are directed away from the root.
(iii) For each $\overrightarrow{(\omega_i, \omega_j)} \in \vec{E}$, we have $i < j$.
(iv) Each $\overrightarrow{(\omega_i, \omega_j)} \in \vec{E}$ is labeled with some $g \in G^{[i]}$ satisfying $\omega_i^g = \omega_j$.

We denote by $f(\mathcal{B})$ the set of elements of G that are labels of some edge.

We shall use repeatedly the following simple characterization of branchings.

Lemma 10.1.1. *Let $\mathcal{B}(\Omega, \vec{E})$ be a directed graph such that for each $\overrightarrow{(\omega_i, \omega_j)} \in \vec{E}$, we have $i < j$. Then the underlying graph of $\mathcal{B}$ is a forest if and only if for each $\omega_k \in \Omega$ there exists at most one edge with endpoint ω_k.*

Proof. If the underlying graph of $\mathcal{B}$ is a forest then it is clear that each vertex is the endpoint of at most one edge. To prove the converse, suppose, on the contrary, that the underlying graph contains a cycle and let ω_k be the vertex with the largest index k in this cycle. Then both edges of the cycle incident with ω_k are directed toward ω_k, contradicting the assumption of the lemma. □

Let $\mathcal{T}$ be the *transitive closure* of $\mathcal{B}$. This means that $\mathcal{T}$ is a directed graph on Ω, with the property that $\overrightarrow{(\omega_i, \omega_j)}$ is an edge of $\mathcal{T}$ if and only if there is a directed path $(\omega_i = \omega_{i_1}, \ldots, \omega_{i_k} = \omega_j)$ in $\mathcal{B}$ connecting ω_i with ω_j. Note that since the connected components of $\mathcal{B}$ are trees, there is at most one such path. The labeling of $\mathcal{B}$ extends naturally to the edges of $\mathcal{T}$. Namely, if $(\omega_i = \omega_{i_1}, \ldots, \omega_{i_k} = \omega_j)$ is the unique path in $\mathcal{B}$ connecting ω_i with ω_j and $g_l = f(\overrightarrow{(\omega_{i_l}, \omega_{i_{l+1}})})$ for $l \in [1, k-1]$ are the labels of edges along this path, then we can define the label of $\overrightarrow{(\omega_i, \omega_j)}$ in $\mathcal{T}$ as the product $g_1 \cdots g_{k-1}$. Although $\mathcal{T}$ is not a forest, its edges and their labels satisfy properties (iii) and (iv) of the definition before the lemma. With a slight abuse of notation, we shall denote the extension of the function f that assigns labels to the edges of $\mathcal{T}$ by f as well, and we shall denote by $f(\mathcal{T})$ the set of elements of G that are labels of some edge in $\mathcal{T}$.

The reason for introducing labeled branchings is that we want to encode transversals $G^{[i]}$ mod $G^{[i+1]}$, for $1 \le i \le n-1$. We say that the labeled branching $\mathcal{B}$ *represents a transversal* $G^{[i]}$ *mod* $G^{[i+1]}$ if for all ω_j in the fundamental orbit $\omega_i^{G^{[i]}}$, either $i = j$ or $\overrightarrow{(\omega_i, \omega_j)}$ is an edge of $\mathcal{T}$. The transversal represented by $\mathcal{B}$ is $\{()\} \cup \{f(\overrightarrow{(\omega_i, \omega_j)}) \mid \omega_j \in \omega_i^{G^{[i]}} \backslash \{\omega_i\}\}$. We say that $\mathcal{B}$ *represents* G if it represents a transversal $G^{[i]}$ mod $G^{[i+1]}$ for all $i \in [1, n-1]$.

As an example, let us consider $G = \mathrm{Alt}([1, 4])$, with the ordering $1 \prec 2 \prec 3 \prec 4$ of the permutation domain. The directed graph with edge set $\{\overrightarrow{(1, 2)}, \overrightarrow{(2, 3)}, \overrightarrow{(2, 4)}\}$ and labels $f(\overrightarrow{(1, 2)}) = (1, 2)(3, 4)$, $f(\overrightarrow{(2, 3)}) = (2, 3, 4)$, $f(\overrightarrow{(2, 4)}) = (2, 4, 3)$ is a labeled branching $\mathcal{B}$ representing G, and its transitive closure $\mathcal{T}$ has the additional edge labels $f(\overrightarrow{(1, 3)}) = (1, 2)(3, 4) \cdot (2, 3, 4) = (1, 3, 2)$ and $f(\overrightarrow{(1, 4)}) = (1, 2)(3, 4) \cdot (2, 4, 3) = (1, 4, 2)$. The transversals represented by $\mathcal{B}$ are $\{(), (1, 2)(3, 4), (1, 3, 2), (1, 4, 2)\}$ for $G^{[1]}$ mod $G^{[2]}$, $\{(), (2, 3, 4), (2, 4, 3)\}$ for $G^{[2]}$ mod $G^{[3]}$, and $\{()\}$ for $G^{[3]}$ mod $G^{[4]}$.

Theorem 10.1.2. *Let $G \le \mathrm{Sym}(\Omega)$, and let $\omega_1 \prec \cdots \prec \omega_n$ be an arbitrary ordering of Ω. Then there exists a labeled branching representing G, with respect to the point stabilizer chain defined by this ordering.*

Proof. Let $G = G^{[1]} \ge G^{[2]} \ge \cdots \ge G^{[n-1]} \ge G^{[n]} = 1$ be the point stabilizer chain corresponding to the ordering $\omega_1 \prec \cdots \prec \omega_n$. We define a directed graph $\mathcal{T}$ with vertex set Ω by the following rule: For $1 \le i, j \le n$, $\overrightarrow{(\omega_i, \omega_j)}$ is an edge of $\mathcal{T}$ if and only if $i < j$ and $\omega_j \in \omega_i^{G^{[i]}}$. Then $\mathcal{T}$ is transitive, since if $\overrightarrow{(\omega_i, \omega_j)}$ and $\overrightarrow{(\omega_j, \omega_k)}$ are edges of $\mathcal{T}$ then $i < j < k$ and there exist $g_1 \in G^{[i]}$ with $\omega_i^{g_1} = \omega_j$ and $g_2 \in G^{[j]}$ with $\omega_j^{g_2} = \omega_k$. Hence $\omega_i^{g_1 g_2} = \omega_k$ and $g_1 g_2 \in G^{[i]}$, implying that $\overrightarrow{(\omega_i, \omega_k)}$ is an edge of $\mathcal{T}$.

Let $\vec{E}$ be a subset of the edge set of $\mathcal{T}$ such that the transitive closure of the graph $\mathcal{B}(\Omega, \vec{E})$ is $\mathcal{T}$ and $\vec{E}$ is a minimal set (subject to inclusion) with this property.

We claim that for each $\omega_k \in \Omega$, at most one edge in $\vec{E}$ has endpoint ω_k and so, by Lemma 10.1.1, the underlying graph of $\mathcal{B}$ is a forest. Suppose, on the contrary, that $\overrightarrow{(\omega_i, \omega_k)} \in \vec{E}$ and $\overrightarrow{(\omega_j, \omega_k)} \in \vec{E}$ for some $i < j$. Then there exist $g_1 \in G^{[i]}$ with $\omega_i^{g_1} = \omega_k$ and $g_2 \in G^{[j]}$ with $\omega_j^{g_2} = \omega_k$. We have $g_1 g_2^{-1} \in G^{[i]}$ and $\omega_i^{g_1 g_2^{-1}} = \omega_j$; hence, by definition, $\overrightarrow{(\omega_i, \omega_j)}$ is an edge of $\mathcal{T}$. Since the transitive closure of $\mathcal{B}$ is $\mathcal{T}$, there is a directed path $(\omega_i = \omega_{i_1}, \ldots, \omega_{i_l} = \omega_j)$ connecting ω_i with ω_j in $\mathcal{B}$, and $(\omega_i = \omega_{i_1}, \ldots, \omega_{i_l}, \omega_k)$ is a directed path connecting ω_i with ω_k. Therefore, if we delete $\overrightarrow{(\omega_i, \omega_k)}$ from $\vec{E}$, its transitive closure remains the same. However, this contradicts the minimality of $\vec{E}$.

So far, we have shown that $\mathcal{B}$ satisfies properties (i)–(iii) of being a labeled branching (for (ii), note that in each connected component the vertex with smallest index can be chosen as the root). By the definition of edges of $\mathcal{T}$, it is also clear that there are elements of G that can be chosen as the labels of the edges of $\mathcal{B}$ such that (iv) is satisfied. Moreover, since the transitive closure of $\mathcal{B}$ is $\mathcal{T}$, it is clear that $\mathcal{B}$ represents G. □

Now we describe a data structure for storing a labeled branching $\mathcal{B}$ that enables us to recover the labels of edges of the transitive closure $\mathcal{T}$ of $\mathcal{B}$ efficiently. First, we store an array B of length $n := |\Omega|$, containing integers from $[1, n]$. We define $B[k] := k$ if ω_k is the root of a tree in $\mathcal{B}$; otherwise, $B[k] := j$ for the integer j such that $\overrightarrow{(\omega_j, \omega_k)}$ is the unique edge in $\mathcal{B}$ with endpoint ω_k.

The most natural solution for storing the labels of edges of $\mathcal{B}$ is to define an array P of permutations of length n such that the kth entry $P[k]$ is the identity permutation if ω_k is the root of a tree in $\mathcal{B}$, and otherwise $P[k]$ is the label of the edge $\overrightarrow{(\omega_j, \omega_k)}$ for the parent ω_j of ω_k. However, if we need the label of an edge $\overrightarrow{(\omega_i, \omega_k)}$ of $\mathcal{T}$ then we have to construct the path from ω_i to ω_k in $\mathcal{B}$, and then we have to take the product of edge labels along this path. Although the construction of the path can be done in $O(n)$ time, starting at ω_k and using the array B to construct larger and larger finishing segments of the path, taking the product of edge labels may require $\Theta(n^2)$ time.

We can do better if we also define an array Q of n permutations as follows: Let $Q[k] := ()$ if ω_k is the root of a tree in $\mathcal{B}$, and otherwise let $Q[k]$ be the product of edge labels along the path from ω_t to ω_k, where ω_t is the root of the tree containing ω_k. Then the label of an edge $\overrightarrow{(\omega_i, \omega_k)}$ of $\mathcal{T}$ can be computed in $O(n)$ time, as the product $Q[i]^{-1}Q[k]$.

Given $i, j \in [1, n]$, we have two methods to decide whether $\overrightarrow{(\omega_i, \omega_j)}$ is an edge of $\mathcal{T}$. The first one is to use the array B and to check whether the unique path from ω_j to the root of the tree containing ω_j goes through ω_i. The second method computes $Q[i]^{-1}Q[j]$. It is clear that $\overrightarrow{(\omega_i, \omega_j)}$ is an edge of $\mathcal{T}$ if and only if $Q[i]^{-1}Q[j]$ fixes $\{\omega_1, \ldots, \omega_{i-1}\}$ pointwise. The second method also computes the label of the edge $\overrightarrow{(\omega_i, \omega_j)}$, in the case when this ordered pair is indeed an edge of $\mathcal{T}$.

If the labeled branching $\mathcal{B}$ represents G then the array Q can be used to test membership in G, in $O(n^2)$ time. Given $g \in \mathrm{Sym}(\Omega)$, we try to factor g as a product $g = r_{n-1} \cdots r_1$ of coset representatives, as in the original sifting procedure of Sims (cf. Section 4.1). If $r_1, \ldots, r_{i-1}$ and $g_i := g r_1^{-1} \cdots r_{i-1}^{-1}$ are already computed for some $i \in [1, n-1]$ then we compute $\omega_j := \omega_i^{g_i}$ and $Q[i]^{-1}Q[j]$. If $Q[i]^{-1}Q[j]$ fixes $\{\omega_1, \ldots, \omega_{i-1}\}$ pointwise then we define $r_i := Q[i]^{-1}Q[j]$ and $g_{i+1} := g_i r_i^{-1}$; otherwise we conclude that $g \notin G$.

10.1.1. Construction

In Theorem 10.1.2, we have shown that each permutation group can be represented by a labeled branching. However, the proof of that theorem did not provide an efficient algorithm for the construction of the labeled branching.

Theorem 10.1.3. *Given $G = \langle S \rangle \le \mathrm{Sym}(\Omega)$, with $|\Omega| = n$, and an ordering $\omega_1 \prec \cdots \prec \omega_n$ of Ω, a labeled branching representing G with respect to the point stabilizer chain defined by this ordering can be constructed in $O(|S|n^2 + n^5)$ time by a deterministic algorithm. The memory requirement is $O(|S|n + n^2)$.*

Proof. Let $G = G^{[1]} \ge G^{[2]} \ge \cdots \ge G^{[n-1]} \ge G^{[n]} = 1$, $G^{[i]} = G_{(\omega_1,\ldots,\omega_{i-1})}$, be the point stabilizer chain corresponding to the ordering $\omega_1 \prec \cdots \prec \omega_n$. The original Schreier–Sims algorithm, described in Section 4.2, constructs an SGS for G in a bottom-up manner. This means that, in intermediate steps, we have a correct strong generating set for a subgroup $H_i \le G^{[i]}$ for some index i, and then we try to construct an SGS for a subgroup $H_{i-1} \le G^{[i-1]}$ containing H_i.

In contrast, here we proceed in a top-down manner. Recursively for $i = 1, 2, \ldots, n$, we construct labeled branchings $\mathcal{B}_1, \ldots, \mathcal{B}_n$ such that $\mathcal{B}_i$ satisfies the following two properties:

(c1) $\mathcal{B}_i$ represents transversals $G^{[j]} \bmod G^{[j+1]}$, for all $j \in [1, i-1]$.
(c2) $f(\mathcal{B}_i) \cap G^{[i]}$ generates $G^{[i]}$.

At the end of the recursion, $\mathcal{B}_n$ represents G. During the construction, the branchings are stored using the arrays B, P, and Q, as described at the end of Section 10.1. When the construction is complete, we discard P.

Suppose that $\mathcal{B}_i$ is already constructed for some $i \in [1, n-1]$. Then we compute $\mathcal{B}_{i+1}$ by the following algorithm:

Step 1: Compute a generating set S_i for $G^{[i]}$ and a transversal T_i for $G^{[i]} \bmod G^{[i+1]}$.

Step 2: Modify $\mathcal{B}_i$ into a labeled branching $\mathcal{B}_i'$ such that $\mathcal{B}_i'$ represents T_i, maintaining the property that $\mathcal{B}_i'$ represents transversals $G^{[j]} \bmod G^{[j+1]}$, for all $j \in [1, i-1]$.

Step 3: For each Schreier generator g constructed from S_i and T_i (cf. Lemma 4.2.1), modify $\mathcal{B}_i'$ into a labeled branching $\mathcal{B}_i''$ such that $\langle f(\mathcal{B}_i)'' \cap G^{[i+1]} \rangle = \langle f(\mathcal{B}_i)' \cap G^{[i+1]}, g \rangle$, maintaining the property (c1) for all $j \in [1, i]$. Rename $\mathcal{B}_i''$ as $\mathcal{B}_i'$.

At the end of Step 3, $\mathcal{B}_{i+1}$ can be defined as the current version of $\mathcal{B}_i'$.

To start the recursion, we define the trivial branching $\mathcal{B}'_0$ by the arrays B, P, and Q with $B[j] := j$, $P[j] := ()$, and $Q[j] := ()$ for all $j \in [1, n]$, and we perform Step 3 using the elements g of the input generating set S.

Now we describe the three steps of the construction of $\mathcal{B}_{i+1}$ in more detail. In Steps 1 and 2, we work only with the arrays B and P.

In Step 1, we obtain S_i as $S_i := \{P[k] \mid i \leq B[k] < k\}$. Then we compute the orbit $\omega_i^{\langle S_i \rangle}$. The elements of the transversal T_i can be constructed during the orbit computation, as described at the end of Section 2.1.1. The time and memory requirements of Step 1 are $O(n^2)$.

In Step 2, we modify the arrays P and B. We process each $g \in T_i$. Let $\omega_k := \omega_i^g$. If $k = i$ then we do not do anything; otherwise, we define $P[k] := g$ and $B[k] := i$.

We have to show that this definition of $P[k]$ and $B[k]$ does not destroy the property that the labeled branching $\mathcal{B}'_i$, defined by B and P, represents transversals $G^{[j]} \bmod G^{[j+1]}$, for all $j \in [1, i-1]$. Let $j \leq i - 1$ be fixed, and let ω_l be an arbitrary element of the fundamental orbit $\omega_j^{G^{[j]}}$. If the unique path from ω_j to ω_l in $\mathcal{B}'_i$ (before the modification of $\mathcal{B}'_i$) does not go through ω_k then this path remains intact after the change of $P[k]$ and $B[k]$; therefore, it is enough to consider the case that the path from ω_j to ω_l contains ω_k. In this case, in particular, $\omega_k \in \omega_j^{G^{[j]}}$ and so there exists $g_j \in G^{[j]}$ with $\omega_j^{g_j} = \omega_k$. Then $\omega_j^{g_j g^{-1}} = \omega_i$ and $g_j g^{-1} \in G^{[j]}$, so there is a path $\omega_j = \omega_{i_1}, \ldots, \omega_{i_m} = \omega_i$ from ω_j to ω_i in $\mathcal{B}'_i$ (this path exists before and after the change of $P[k]$ and $B[k]$). Hence $\omega_j = \omega_{i_1}, \ldots, \omega_{i_m}, \omega_k$ is a path in $\mathcal{B}'_i$ after the change of $P[k]$ and $B[k]$, and so there is a path from ω_j to ω_l.

The time requirement of Step 2 is $O(n^2)$, and it is clear that after Step 2 the labeled branching represents the transversal T_i. After Step 2, we reconstruct the array Q. For $k = 1, 2, \ldots, n$, we define $Q[k] := ()$ if $B[k] = k$ and $Q[k] := Q[B[k]] \cdot P[k]$ if $B[k] < k$. The construction of Q requires $O(n^2)$ time. In Step 3, we shall use all three arrays B, P, and Q.

Before we describe Step 3, we introduce a quantity associated with labeled branchings. Let $\mathcal{B}$ be a labeled branching, stored in the arrays B, P, and Q as in the foregoing. The *length* $l(\mathcal{B})$ of $\mathcal{B}$ is defined as

$$l(\mathcal{B}) = \sum_{\{k \mid B[k] < k\}} (k - B[k]) + \sum_{\{k \mid B[k] = k\}} k.$$

In Step 3, we apply a modification of the sifting procedure described at the end of Section 10.1. Given a labeled branching $\mathcal{B}'_i$ that satisfies property (c1) for $i + 1$ (i.e., it represents transversals $G^{[j]} \bmod G^{[j+1]}$ for all $j \in [1, i]$), and given some $g \in G^{[i+1]}$, the sifting of g returns one of the following three outputs:

(i) a report that $g \in \langle f(\mathcal{B}'_i) \cap G^{[i+1]} \rangle$;
(ii) a labeled branching $\mathcal{B}''_i$ that satisfies property (c1) for $i+1$ and $l(\mathcal{B}'') < l(\mathcal{B}')$ and $\langle f(\mathcal{B}''_i) \cap G^{[i+1]} \rangle = \langle f(\mathcal{B}'_i) \cap G^{[i+1]}, g \rangle$; or
(iii) a labeled branching $\mathcal{B}''_i$ and $h \in G^{[i+1]}$ with the properties that $\mathcal{B}''_i$ satisfies (c1) for $i+1$ and $l(\mathcal{B}'') < l(\mathcal{B}')$, and $\langle f(\mathcal{B}''_i) \cap G^{[i+1]}, h \rangle = \langle f(\mathcal{B}'_i) \cap G^{[i+1]}, g \rangle$.

Some explanations are in order. In Section 4.2, in the original Schreier–Sims procedure, sifting a Schreier generator $g \in G^{[i+1]}$ in the already constructed Schreier tree data structure has two possible outcomes. One possibility is that g sifts through, which means that g can be discarded. This case corresponds to (i) above. The other possibility is that g has a nontrivial siftee h, which can be added to the strong generating set we construct. We try to apply the same strategy here and add the siftee h to $\mathcal{B}'_i$. However, a labeled branching is a more complicated data structure than a sequence of Schreier trees. Sometimes, we succeed in adding the siftee to $\mathcal{B}'_i$ and we are in case (ii) above; at other times, we do not succeed, and we are in case (iii). (In case (iii), the output h is not the siftee of g, just some other group element satisfying the property described there.) The progress we make in case (iii) is that the length of the labeled branching decreases, so, after attempting to sift h, sooner or later an output of type (i) or (ii) must occur.

Now we describe the sifting procedure. Given $g \in G^{[i+1]}$, we attempt the factorization of g as a product $g = r_{n-1} \cdots r_{i+1}$ of coset representatives, as described at the end of Section 10.1. If the factorization succeeds then we are in case (i) and we can terminate sifting. The other possibility is that sifting constructs a group element $g_l := g r_{i+1}^{-1} \cdots r_{l-1}^{-1}$ with the following properties:

$$g_l \text{ fixes } \{\omega_1, \ldots, \omega_{l-1}\} \text{ pointwise}, \; \omega_l^{g_l} = \omega_k \text{ for some } k > l, \\ \text{and there is no directed path from } \omega_l \text{ to } \omega_k \text{ in } \mathcal{B}'_i. \tag{10.1}$$

(Recall that we notice that there is no such path by the fact that $Q[l]^{-1}Q[k]$ does not fix $\{\omega_1, \ldots, \omega_{l-1}\}$ pointwise.) In some g_l satisfying (10.1) is constructed then sifting enters a second phase that tries to decrease the value of k. Namely, if there is an edge $\overrightarrow{(\omega_m, \omega_k)}$ in $\mathcal{B}'_i$ with $m > l$ then g_l is replaced by $g_l \cdot P[k]^{-1}$, and we redefine $k := m$. This new g_l also satisfies (10.1). The second phase terminates when $B[k] = k$ or $B[k] < l$.

If $B[k] = k$ then we add the edge $\overrightarrow{(\omega_l, \omega_k)}$ to $\mathcal{B}'_i$ with label g_l, and terminate sifting with an output in case (ii). Adding the edge to $\mathcal{B}'_i$ is done by defining $B[k] := l$ and $P[k] := g_l$.

If $m := B[k] < l$ then we delete the edge $\overrightarrow{(\omega_m, \omega_k)}$ and add the edge $\overrightarrow{(\omega_l, \omega_k)}$ with label g_l. Again, this modification is done by defining $B[k] := l$ and

$P[k] := g_l$. The same argument as in Step 2 shows that the new branching $\mathcal{B}_i''$ represents transversals $G^{[j]}$ mod $G^{[j+1]}$ for all $j \in [1, i]$. Moreover, if $m < i+1$ then $\langle f(\mathcal{B}_i'') \cap G^{[i+1]} \rangle = \langle f(\mathcal{B}_i') \cap G^{[i+1]}, g \rangle$ and we are in case (ii). If $i+1 < m < l$ then we also output the label of the deleted edge $\overrightarrow{(\omega_m, \omega_k)}$ as h, and we are in case (iii).

If sifting terminates with output in case (ii) or (iii) then after termination we recompute the array Q in the same way as was done after Step 2.

Finally, we estimate the time requirement of Step 3. One call of sifting and the possible recomputation of Q takes $O(n^2)$ time. For a fixed $i \geq 1$, the number of calls is less than $2n^2$, since we sift less than n^2 Schreier generators and less than n^2 permutations h obtained as an output in case (iii) of some previous call of sifting. The number of outputs in case (iii) is less than n^2 since at each output in case (ii) or (iii) the length of the labeled branching decreases, and the original length after Step 2 is at most $n(n-1)/2$. Similarly, for $i = 0$, the number of calls of sifting is less than $|S| + n^2$. Hence the total time requirement of Step 3, added up for all i, is $O(n^5 + |S|n^2)$. The memory requirement for the storing of S_i, T_i, B, P, and Q is $O(n^2)$. □

10.2. Alternating and Symmetric Groups

Before we invoke any form of the Schreier–Sims algorithm for an input group $G = \langle S \rangle \leq \mathrm{Sym}(\Omega)$, it is good to know whether $G = \mathrm{Alt}(\Omega)$ or $G = \mathrm{Sym}(\Omega)$. We shall call $\mathrm{Alt}(\Omega)$ or $\mathrm{Sym}(\Omega)$ the *giant* permutation groups on Ω. For giant inputs, the running times of the original Schreier–Sims algorithm and of the nearly linear-time version described in Section 4.5 are unnecessarily long, since there are faster, specialized algorithms that can handle these groups.

An algorithm for the fast recognition of giants is one of the first results in computational permutation group theory (cf. [Parker and Nikolai, 1958]). A variant of this algorithm is also described in [Cannon, 1984]. The method is based on the following result from [Jordan, 1873] (see also [Wielandt, 1964, Theorem 13.9]).

Theorem 10.2.1. *Let $G \leq \mathrm{Sym}(\Omega)$, with $|\Omega| = n$, be a primitive group, and let p be a prime satisfying $p < n-2$. If G contains a p-cycle then G is a giant.*

Corollary 10.2.2. *Let $G \leq \mathrm{Sym}(\Omega)$, with $|\Omega| = n$, be transitive, and let p be a prime satisfying $n/2 < p < n-2$. If G contains an element that has a cycle of length p then G is a giant.*

Proof. Let $g \in G$ have a cycle of length p. Since $p \nmid (n-p)!$, we have that $g^{(n-p)!}$ is a p-cycle. Moreover, G is primitive, since an imprimitive, transitive subgroup $H \le \mathrm{Sym}(\Omega)$ must satisfy $H \lesssim S_m \wr S_k$ for some $k, m \in [2, n/2]$, $km = n$. Hence H cannot contain an element of order divisible by p. □

To apply Corollary 10.2.2, we need an estimate for the proportion of elements of the giants having a cycle of length p for some prime p satisfying $n/2 < p < n-2$.

Lemma 10.2.3. *(a) Let q be an integer in the range $n/2 < q \le n$. The proportion of elements of S_n containing a cycle of length q is $1/q$. In A_n, this proportion is $1/q$ if $q \le n-2$, and it is 0 or $2/q$ if $q \in \{n-1, n\}$.*

(b) The proportion of elements of the giants that contain a cycle of length p for some prime p in the range $n/2 < p < n-2$ is asymptotically $\ln 2/\ln n$.

Proof. (a) There are $\binom{n}{q}(q-1)!(n-q)!$ elements of $\mathrm{Sym}([1,n])$ with a cycle of length q, since there are $\binom{n}{q}$ possibilities for the support Δ of this cycle, and on a fixed support, there are $(q-1)!$ cycles. The term $(n-q)!$ counts the number of possibilities for the restrictions of the permutations on $[1,n]\backslash\Delta$. We counted every permutation containing a cycle of length q exactly once, since $q > n/2$ implies that a permutation cannot contain two cycles of length q. Dividing by $n!$, we obtain the required proportion $1/q$. If $q \le n-2$ then a similar computation shows that there are $\binom{n}{q}(q-1)!(n-q)!/2$ elements of $\mathrm{Alt}([1,n])$ with a cycle of length q, and dividing by $n!/2$ gives the proportion $1/q$. If $q \in \{n-1, n\}$ then either none or all of the $\binom{n}{q}(q-1)!(n-q)!$ permutations of $\mathrm{Sym}([1,n])$ containing a cycle of length q are in $\mathrm{Alt}([1,n])$.

(b) No element of $\mathrm{Sym}([1,n])$ can have cycles of lengths p_1 and p_2 for two different primes greater than $n/2$. Therefore, the proportion of elements of the giants that contain a cycle of length p for some prime p in the range $n/2 < p < n-2$ is the sum of the proportions for each p. The sum of the reciprocals of primes less than x is asymptotically $\ln\ln x$ (see [Hardy and Wright, 1979, Theorem 427]); hence

$$\sum_{n/2<p<n-2} \frac{1}{p} \sim \ln\ln(n-2) - \ln\ln\frac{n}{2} = \ln\frac{\ln(n-2)}{\ln n - \ln 2} \sim \ln\left(1 + \frac{\ln 2}{\ln n}\right) \sim \frac{\ln 2}{\ln n}.$$

□

To decide whether the input $G = \langle S \rangle \le \mathrm{Sym}(\Omega)$ is a giant, first we compute whether G is transitive on Ω. If G is transitive then we take random elements, and we compute their cycle structure. If we find an element with a cycle of length p in the range $n/2 < p < n-2$ then we know with certainty that G is a giant. The symmetric and alternating groups are distinguished from each other by examining whether there is a generator of G that is an odd permutation. If none of $c \ln n / \ln 2$ random elements have a cycle of length p for any prime in the range $n/2 < p < n-2$ then, with probability about $1 - e^{-c}$, the input is not a giant. Since we have no SGS for G available, we have to use one of the methods of Section 2.2 for random element generation.

The algorithm described in the previous paragraph is a *nonconstructive recognition algorithm*. It gives an answer for the decision problem whether G is a giant but, in the case of an affirmative answer, it does not provide a way of expressing group elements in terms of the input generators. In a number of applications (for example, in Theorems 8.3.1 or 8.4.1, or when working with matrix groups) this is not enough: We need the *constructive recognition* of symmetric and alternating groups. We recall the definition of constructive recognition from Section 8.3, specialized for alternating and symmetric groups of known degree.

Let $G = \langle S \rangle$ be a black-box group, isomorphic to A_n or S_n for a given integer n. We say that G is *constructively recognizable* if there is a Las Vegas algorithm that decides whether $G \cong A_n$ or $G \cong S_n$ and finds a new set $S^* = \{s, t\}$ generating G and a homomorphism $\lambda : G \to \mathrm{Sym}([1, n])$, specified by the image of S^*, such that $\langle S^* \rangle$ satisfies the presentation

$$\left\langle s, t \mid s^n = t^2 = (st)^{(n-1)} = [t, s^j]^2 = 1 \text{ for } 2 \le j \le n/2 \right\rangle \qquad (10.2)$$

in the case $G \cong S_n$,

$$\langle s, t \mid s^{n-2} = t^3 = (st)^n = (ts^{-k}ts^k)^2 = 1 \text{ for } 1 \le k \le (n-3)/2 \rangle \qquad (10.3)$$

in the case $G \cong A_n$ with n odd, and

$$\langle s, t \mid s^{n-2} = t^3 = (st)^{n-1} = [t, s]^2 = 1 \rangle \qquad (10.4)$$

in the case $G \cong A_n$ with n even. Moreover, there are deterministic algorithms for the following:

(i) Given $g \in G$, find $\lambda(g)$ and a straight-line program of length $O(n \log n)$ from S^* to g.
(ii) Given $h \in \mathrm{Sym}([1, n])$, decide whether or not $h \in \lambda(G)$; and, if it is, find $\lambda^{-1}(h)$ and a straight-line program of length $O(n \log n)$ from $\lambda(S^*)$ to h.

We note that the presentation (10.2) is from the book [Coxeter and Moser, 1957], and the presentations (10.3) and (10.4) are from [Carmichael, 1923]. The permutations $s = (1, 2, \ldots, n)$ and $t = (1, 2)$ satisfy (10.2), whereas $s = (3, 4, \ldots, n)$ and $t = (1, 2, 3)$ satisfy (10.3), and $s = (1, 2)(3, 4, \ldots, n)$ and $t = (1, 2, 3)$ satisfy (10.4).

The rest of this section is devoted to the description of the constructive recognition of alternating and symmetric groups from [Beals et al., 2002]. The material in Sections 10.2.1–10.2.3 is reproduced from this paper, © 2002 American Mathematical Society. Reprinted by Permission. Although the algorithm works in the black-box group setting, so it can be used to construct an isomorphism between G and a standard copy of A_n or S_n for any permutation group or matrix group representation G of these groups, there is a much simpler version if G is a giant (i.e., G is a permutation group acting on a domain of size n). We shall comment on this simpler version in Section 10.2.4.

Let ξ be an upper bound on the time required per element to construct independent, nearly uniformly distributed random elements of G and let μ be an upper bound on the time required for each group operation in G.

Theorem 10.2.4. (*a*) *Given an integer* $n \geq 7$, *a black-box group* $G = \langle S \rangle$ *isomorphic to* A_n *or* S_n, *and an upper bound* $\varepsilon > 0$ *for the probability of failure of the algorithm,* G *can be constructively recognized in* $O(\log \varepsilon^{-1}(\xi n + \mu |S| n \log n))$ *time, with probability at least* $1 - \varepsilon$.

The time requirement of (i) *is* $O(\mu n \log n)$, *whereas constructing the straight-line program in* (ii) *is in* $O(n \log n)$ *time and the computation of the inverse image is in* $O(\mu n \log n)$ *time. The data structures underlying* (i) *and* (ii) *require the storage of* $O(\log n)$ *elements of* G.

(*b*) *Given an arbitrary black-box group* $G = \langle S \rangle$, *the algorithm described in part* (*a*) *can be used as a Monte Carlo algorithm to decide whether* G *is alternating or symmetric of a given degree* n. *If the algorithm returns the answer "yes" then the answer is correct. The time requirement is* $O(\log \varepsilon^{-1}(\xi n + \mu |S| n \log n))$.

We shall prove Theorem 10.2.4 in Sections 10.2.2 and 10.2.3 (cf. the summary at the end of Section 10.2.3).

10.2.1. Number Theoretic and Probabilistic Estimates

In this section, we collect some estimates that are needed in the proof of Theorem 10.2.4. We shall use the following notation: If the cycle decomposition of an element $h \in S_n$ consists of m_i cycles of length i, where $0 \leq m_i \leq n$ for $1 \leq i \leq n$, then we say that h has *cycle type* $1^{m_1} \cdots n^{m_n}$. If $m_i = 0$ then we omit i^{m_i} from this expression. Note that if $n \neq 6$ then $\mathrm{Aut}(A_n) \cong S_n$, so the cycle type of $\lambda(g)$ is the same at any isomorphism $\lambda : G \to S_n$ or A_n. Hence we can talk about the *cycle type of elements of* G.

The first four lemmas are of a number theoretic nature. In most cases, there are better asymptotic results but, for the algorithmic applications, we need estimates that are valid for all values of n.

Lemma 10.2.5. *For all n, the number of divisors of n is at most $48\, n^{1/3}/2520^{1/3}$, and the sum of the divisors of n is at most $n(3+\ln n)/2$.*

Proof. For each $\delta > 0$, the number of divisors of n is at most $C_\delta n^\delta$ for a constant C_δ. An algorithm for computing the value of C_δ is described in [Niven et al., 1991, pp. 395–396]; $C_{1/3} = 48/2520^{1/3} \approx 3.527$.

We use the trivial estimate $\sum_{i=1}^{\lceil\sqrt{n}\rceil-1} i < n/2$ for the sum of the divisors less than $\sqrt{n}$. The sum Σ of divisors greater than or equal to $\sqrt{n}$ satisfies

$$\Sigma \leq n + \frac{n}{2} + \frac{n}{3} + \cdots + \frac{n}{\lfloor\sqrt{n}\rfloor} < n\left(1 + \int_1^{\sqrt{n}} \frac{1}{t}dt\right) = n\left(1 + \frac{\ln n}{2}\right).$$

Adding these two estimates, we obtain the second assertion of the lemma. □

Lemma 10.2.6. *Let $x = np$, where p is prime, with $n > p^2$ and all prime divisors of n greater than p. Let D denote the set of divisors of x that are not greater than n. Then $|D| \leq 96\, n^{1/3}/2520^{1/3} < 8n^{1/3}$ and*

$$\sum_{d\in D} d \leq n\frac{p+3-2(p+1)\ln p + (p+1)\ln n}{2}.$$

Proof. Because $p \nmid n$, the number of divisors of x is twice the number of divisors of n. Thus Lemma 10.2.5 yields $|D| \leq 96\, n^{1/3}/2520^{1/3}$.

Also, because of the restriction on the prime factorization of n, all divisors of x except np are at most n, and $\sum_{d\in D} d = (p+1)\sum_{d|n} d - pn$. Hence it is enough to estimate the sum of the divisors of n. We use a refinement of the argument in Lemma 10.2.5. For the sum of divisors less than $\sqrt{n}$, we still use the trivial estimate $n/2$. However, for the larger divisors, we note that since the largest proper divisor of n is at most $n/(p+1)$, the sum Σ of divisors greater or equal than $\sqrt{n}$ satisfies

$$\begin{aligned}\Sigma &\leq n + \frac{n}{p+1} + \frac{n}{p+2} + \cdots + \frac{n}{\lfloor\sqrt{n}\rfloor}\\ &< n\left(1 + \int_p^{\sqrt{n}} \frac{1}{t}dt\right) = n\left(1 + \frac{\ln n}{2} - \ln p\right).\end{aligned}$$

Combining these estimates, we obtain the assertion of the lemma. □

Lemma 10.2.7. *Let $x = ny$, and let D denote the set of divisors of x that are not greater than n. Let $k > 1$. Then*

$$\sum_{d \in D} d^k \le n^k \left(1 + \frac{y}{k-1}\right).$$

Proof. All divisors $d \in D$ are of the form $d = x/s$ for some $s \ge y$. Therefore,

$$\sum_{d \in D} d^k \le \sum_{s=y}^{x} \left(\frac{x}{s}\right)^k < \left(\frac{x}{y}\right)^k + x^k \int_y^\infty \frac{1}{t^k} dt$$

$$= \left(\frac{x}{y}\right)^k + x^k \cdot \frac{1}{y^{k-1}(k-1)} = \left(\frac{x}{y}\right)^k \left(1 + \frac{y}{k-1}\right). \qquad \square$$

Lemma 10.2.8. *Let $x = m!\,(n-m)$. If $n > m + m!\,m$ then the largest divisor of x not exceeding n is $n - m$.*

Proof. Suppose that there is a divisor k of x with $n - m < k \le n$. Then $(k, n-m) \le m$, and so the least common multiple of k and $n - m$ is at least $k(n-m)/m$. This implies $x \ge k(n-m)/m > k(m!\,m)/m > m!\,(n-m)$, which is a contradiction. $\square$

The proof of the next lemma is quite simple and, since it is similar to the proof of Lemma 10.2.3(a), it is omitted.

Lemma 10.2.9. *For $n \ge 10$, each column of the following table lists a cycle type in the first row and the proportion of elements of that cycle type in S_n in the second row.*

cycle type	n^1	$1^1(n-1)^1$	$3^1(n-3)^1$
proportion	n^{-1}	$(n-1)^{-1}$	$\frac{1}{3}(n-3)^{-1}$
cycle type	$1^1 3^1 (n-4)^1$	$1^2 3^1 (n-5)^1$	$1^3 3^1 (n-6)^1$
proportion	$\frac{1}{3}(n-4)^{-1}$	$\frac{1}{6}(n-5)^{-1}$	$\frac{1}{18}(n-6)^{-1}$

In A_n, the proportions are 0 or twice the proportions in S_n, depending on whether the permutations with the desired cycle type are odd or even.

Let x be the product of the cycle lengths in a cycle type occurring in Lemma 10.2.9. The rest of this section is devoted to estimating the conditional probability that a randomly chosen element $h \in A_n$ or S_n satisfying $h^x = 1$ has the cycle type that defined x. This is the key step in the analysis of the algorithm proving Theorem 10.2.4.

For integers n and x, we define $T_n(x) := \{h \in S_n \mid h^x = 1\}$ and $N_n(x) := |T_n(x)|$.

Theorem 10.2.10. *Let m be a fixed nonnegative integer. For any $\varepsilon > 0$ there exists a bound $n(\varepsilon)$ such that if h is a uniformly distributed random permutation from S_n for some $n > n(\varepsilon)$ then the probability that $h^{m!\,(n-m)} = 1$ is less than $(1+\varepsilon)/n$.*

Proof. If h is a uniformly distributed random permutation from S_n then $\text{Prob}(h^{m!\,(n-m)} = 1) = N_n(m!\,(n-m))/n!$. Therefore, we have to prove the upper bound $(1+\varepsilon)n!/n$ for $N_n(m!\,(n-m))$. We can suppose that $n > 8+8m!\,m$. We denote the set of divisors of $m!(n-m)$ by D.

The basic strategy of the proof is as follows. For $h \in S_n$, $h^{m!(n-m)} = 1$ if and only if the length of each cycle of h is a divisor of $m!(n-m)$. This gives us too many conditions to handle, so, instead, we fix a number k and consider the set of those h that satisfy the property that the lengths of the cycles intersecting the first k points of the permutation domain are divisors of $m!(n-m)$. We shall compute an upper estimate for the number of such permutations in terms of n, m, and k (cf. (10.7)), and we are lucky enough that k can be chosen to be a constant (depending on m and ε, but independent of n) so that for large enough n, this upper estimate is less than $(1+\varepsilon)n!/n$.

Let $k \geq 3$ be fixed, let $\Pi = \{P_1, \ldots, P_l\}$ be a fixed partition of $\{1, \ldots, k\}$ into l nonempty parts for some $l \leq k$, and let $d_1, d_2, \ldots, d_l$ be a sequence of elements from D such that $\sum_{i=1}^{l} d_i \leq n$. First, we estimate from above the number $N(k, \Pi, d_1, \ldots, d_l)$ of permutations $h \in T_n(m!\,(n-m))$ for which h has cycles $C_1, \ldots, C_l$ of lengths $d_1, \ldots, d_l$, respectively, such that $C_i \cap \{1, 2, \ldots, k\} = P_i$ for $1 \leq i \leq l$.

We can choose the support set of C_1 in $\binom{n-k}{d_1-|P_1|}$ ways, and then the cycle C_1 itself in $(d_1-1)!$ ways. Recursively, if $C_1, \ldots, C_{i-1}$ are already defined, the support set of C_i can be chosen in

$$\binom{n-k-\sum_{j=1}^{i-1}(d_j-|P_j|)}{d_i-|P_i|}$$

ways, and then the cycle C_i can be chosen in $(d_i-1)!$ ways. Finally, after $C_1, \ldots, C_l$ are defined, the rest of the permutation can be chosen in at most $(n-\sum_{i=1}^{l} d_i)!$ ways. Multiplying these numbers, we obtain

$$N(k, \Pi, d_1, \ldots, d_l) \leq \frac{(n-k)!\prod_{i=1}^{l}(d_i-1)!}{\prod_{i=1}^{l}(d_i-|P_i|)!} \leq (n-k)!\prod_{i=1}^{l} d_i^{|P_i|-1}. \qquad (10.5)$$

For a fixed partition Π, let $N(k, \Pi)$ denote the sum of all $N(k, \Pi, d_1, \dots, d_l)$ obtained above for all choices $d_1, \dots, d_l$ of divisors of $m!\,(n-m)$. If $l = 1$ then we obtain $N(k, \Pi) \le (n-k)! \sum_{d \in D} d^{k-1}$, which, by Lemmas 10.2.8 and 10.2.7, is at most

$$(n-k)!(n-m)^{k-1}\left(1 + \frac{m!}{k-2}\right). \tag{10.6}$$

For $l \ge 2$, we use the estimates that, by Lemma 10.2.5, a sequence $d_1, \dots, d_l$ from D can be chosen in at most $(8(m!\,(n-m))^{1/3})^l$ ways and for each sequence trivially $\prod_{i=1}^{l} d_i^{|P_i|-1} \le n^{\sum_i (|P_i|-1)} = n^{k-l}$. Hence, since $n \ge 8 + 8m!\,m$,

$$\begin{aligned} N(k, \Pi) &\le (n-k)!\big(8(m!\,(n-m))^{1/3}\big)^l n^{k-l} \\ &\le (n-k)!\big(8(m!(n-m))^{1/3}\big)^2 n^{k-2} < 64m!\,(n-k)!\,n^{k-4/3}. \end{aligned}$$

We estimate $(n-k)!$ by

$$(n-k)! = \frac{n!}{n^k}\frac{n}{n-1}\frac{n}{n-2}\cdots\frac{n}{n-k+1} < \frac{n!}{n^k}\left(1 + \frac{k}{n-k}\right)^k < \frac{n!}{n^k}e^{\frac{k}{n-k}}$$

and the number of partitions of $\{1, 2, \dots, k\}$ by k^k (note that each partition can be obtained as the sets where some function $f : \{1, 2, \dots, k\} \to \{1, 2, \dots, k\}$ takes constant values). Combining these estimates with the observation that each element of $T_n(m!\,(n-m))$ is counted exactly once in $\sum_\Pi N(k, \Pi)$, we obtain that

$$N_n(m!\,(n-m)) \le n!\left[\left(1 + \frac{m!}{k-2}\right)e^{\frac{k}{n-k}}\frac{1}{n} + 64m!k^k e^{\frac{k}{n-k}}\frac{1}{n^{4/3}}\right]. \tag{10.7}$$

Given $\varepsilon > 0$, we choose k such that

$$\left(1 + \frac{m!}{k-2}\right)e^{\frac{k}{k^2-k}} < 1 + \frac{\varepsilon}{2}.$$

After that, we choose $n_0 > k^2$ such that

$$64m!\,k^k e^{\frac{k}{n_0-k}} < \frac{\varepsilon}{2}n_0^{1/3}.$$

Then, for $n > \max\{8 + 8m!\,m, n_0\}$, we have $N_n(m!\,(n-m)) < (1+\varepsilon)n!/n$.

□

Corollary 10.2.11. *Let $\varepsilon > 0$ and $n > n(\varepsilon)$.*

(a) *Let h be a uniformly distributed random element of $T_n(m!\,(n-m))$. Then, with probability greater than $1-\varepsilon$, h contains a cycle of length $n-m$.*

(*b*) *Let* $2 \le s \le m$, *and let* h *be a uniformly distributed random element of* $T_n((n-m)s)$. *Then, with probability greater than* $(1-\varepsilon)/m!$, *the cycle structure of* h *is* $1^{m-s}s^1(n-m)^1$.

Proof. (a) This is immediate from Lemma 10.2.3(a) and Theorem 10.2.10.

(b) $T_n((n-m)s) \subseteq T_n(m!\,(n-m))$, so it has at most $(1+\varepsilon)n!/n$ elements. Out of these, there are $n!/(s(n-m)(m-s)!) > n!/(m!\,n)$ with cycle structure $1^{m-s}s^1(n-m)^1$, so the proportion of elements in $T_n((n-m)s)$ with the required cycle structure is greater than $(1-\varepsilon)/m!$. □

Corollary 10.2.11 covers all cases occurring in Lemma 10.2.9, since if $m \in \{0, 1\}$ then part (a) of the corollary can also be interpreted as a conditional probability of permutations with the required cycle structure. Note that if x is odd then $T_n(x) \subseteq A_n$, so if $(n-m)s$ is odd then the conditional probability that an element $h \in A_n$ has cycle structure $1^{m-s}s^1(n-m)^1$ given that $h^{(n-m)s} = 1$ is the same as the corresponding conditional probability in S_n. However, for our algorithmic application, we need a lower bound for the conditional probability that is valid for all values of n.

Theorem 10.2.12. *Let* $n \ge 5$ *and let* h *be a randomly selected permutation from* S_n. *Let* $x = n$ *or* $x = (n-m)s$ *for one of the cycle types* $1^{m-s}s^1(n-m)^1$ *described in Lemma* 10.2.9. *Then, given that* $h^x = 1$, *the conditional probability that* h *is an* n*-cycle, or* h *has cycle structure* $1^{m-s}s^1(n-m)^1$, *is at least* $1/180$.

Proof. Using the notation of the proof of Theorem 10.2.10, we derive a tighter upper bound for $N_n(x)$ by evaluating (10.5) more carefully in the case $k = 3$. First, we suppose that $n > 50$.

There is one partition Π_3 of $\{1, 2, 3\}$ with three parts, and (10.5) gives $N(3, \Pi_3, d_1, d_2, d_3) \le (n-3)!\,d_1^0 d_2^0 d_3^0 = (n-3)!$. By Lemmas 10.2.5 and 10.2.6,

$$N(3, \Pi_3) \le \frac{96^3}{2520}(n-m)\,(n-3)! \le \frac{96^3}{2520}\,n\,(n-3)!.$$

There are three partitions Π_2 of $\{1, 2, 3\}$ with two parts, and (10.5) gives $N(3, \Pi_2, d_1, d_2) \le (n-3)!\,d_1 d_2^0$. Hence, using both statements of Lemmas 10.2.5 and 10.2.6,

$$N(3, \Pi_2) \le (n-3)!\frac{96}{2520^{1/3}}n^{4/3}\frac{6 - 8\ln 3 + 4\ln n}{2}.$$

In this estimate, we used Lemma 10.2.6 with $p = 3$, since for $n > 50$ this value gives the largest upper bound.

Finally, there is one partition Π_1 of $\{1, 2, 3\}$ with one part, $N(3, \Pi_1, d_1) \leq (n-3)!\, d_1^2$ and, by Lemmas 10.2.7 and 10.2.8, $N(3, \Pi_1) \leq 4n^2\,(n-3)!$. (Again, the case $y = 3$ yields the largest upper estimate.) Adding these estimates, we obtain

$$\begin{aligned} N_n(x) &\leq (n-3)! \left(4n^2 + \frac{3 \cdot 96}{2520^{1/3}} n^{4/3} (3 - 4 \ln 3 + 2 \ln n) + \frac{96^3}{2520} n \right) \\ &= f(n)(n-1)!. \end{aligned}$$

The function $f(n)$ is monotone decreasing for $n \geq 50$.

We claim that $N_n(x) \leq 10(n-1)!$ for all $n \geq 5$. We have $f(301) < 10$, so it is enough to check that $N_n(x) \leq 10(n-1)!$ for $5 \leq n \leq 300$. This can be done by a GAP program, using the following recursion to compute $N_n(x)$: For $1 \leq k \leq n$, let $r_x(k)$ denote the number of permutations $h \in S_k$ such that $h^x = 1$, and initialize $r_x(0) = 0$. Then

$$r_x(k) = \sum_{d|x} \binom{k-1}{d-1} (d-1)!\, r_n(k-d),$$

which can be seen by partitioning the permutations according to the length of the cycle containing the first point of the permutation domain. We have $N_n(x) = r_x(n)$.

By Lemma 10.2.9 and by checking the cases $n \leq 9$, which are not covered by the lemma, the number of permutations with the required cycle structure is at least $(n-1)!/18$. Hence the proportion of these elements in $T_n(x)$ is at least $1/180$. □

Although Theorem 10.2.12 gives a positive constant lower estimate for the conditional probability that can be used in the design of an algorithm with asymptotic running time described in Theorem 10.2.4, the constant $1/180$ is too small for an efficient implementation. In fact, we cannot expect a good constant from an argument covering all cases together, since $N_n(n) \geq (n-1)!$ and the number of permutations with cycle type $1^3 3^1 (n-6)^1$ is only about $(n-1)!/18$. Therefore, in the range $n \leq 300$, we have computed the value of $N_n(x)$ explicitly, and this can be used for obtaining better estimates for the number of iterations in an implementation. We summarize these computations in the next lemma. Without doubt, the constant $1/7$ bounds the conditional probabilities for larger values of n as well, for all required cycle types. We also note that [Warlimont, 1978] gives an asymptotic estimate with a very precise error term for $N_n(n)$.

Lemma 10.2.13. *Let* $8 \le n \le 300$.

(*a*) *If* $n \notin \{8, 12, 24\}$ *then the conditional probability is greater than* $1/2$ *for the event that an element of* S_n *of order dividing* n *is an* n*-cycle. For all* n*, the conditional probability is greater than* $1/4$.

(*b*) *If* n *is even then the conditional probability is greater than* $1/2$ *for the event that an element of* A_n *of order dividing* $n-1$ *has cycle type* $1^1(n-1)^1$.

(*c*) *If* $n \notin \{31, 61, 91, 121, 151, 181, 211, 241, 271\}$ *then the conditional probability is greater than* $1/3$ *for the event that an element of* A_n *of order dividing* $3(n-m)$ *for the* m *values described in Lemma* 10.2.9 *has cycle type* $1^{m-3}3^1(n-m)^1$*. For all* n*, the conditional probability is greater than* $1/7$.

10.2.2. Constructive Recognition: Finding the New Generators

Given a black-box group $G = \langle S \rangle$ isomorphic to A_n or S_n, in this section we describe an algorithm that constructs $s, t \in G$ such that the subgroup $\langle s, t \rangle$ satisfies the presentation (10.3) or (10.4), if n is odd or even, respectively.

We construct random elements of G (the number of which are needed will be computed in the proof of Theorem 10.2.17) to find $a \in G$ satisfying $a^{n-k} = 1$, where $k = 0$ if n is odd, and $k = 1$ if n is even. Also, we construct random elements $c \in G$ to find one satisfying $c^{3(n-m)} = 1$, where m is given in the following table:

$n \mod 6$	0	1	2	3	4	5
m	5	6	3	4	3	4

(10.8)

Then, by Theorem 10.2.12, the cycle type of a is $1^k(n-k)^1$ with probability at least $1/180$, and the cycle type of $b := c^{n-m}$ is $1^{n-3}3^1$ with probability at least $1/180$.

The following two lemmas describe algorithms that, given $a, b \in G$ as in the previous paragraph, construct $s, t \in G$ as required in (10.3) or (10.4). These algorithms are of Las Vegas type, which means that if the input group elements a, b have cycle type $1^k(n-k)^1$ and $1^{n-3}3^1$, respectively, then the output group elements s, t behave as required, or the algorithm reports failure. However, if the input elements do not have the prescribed cycle types then the algorithms may return an incorrect answer. If the output group elements s, t are incorrect then this fact is noticed when we check whether $\langle s, t \rangle$ satisfies the presentation (10.3) or (10.4).

Lemma 10.2.14. *Let* n *be odd and* $n \ge 7$*. Suppose that* $a \in G$ *has cycle type* n^1 *and* $b \in G$ *has cycle type* $1^{n-3}3^1$*. Then in* $O(\xi n + \mu n)$ *time, it is possible to* *construct* $s, t \in G$ *such that* $\langle s, t\rangle$ *satisfies the presentation* (10.3) *for* A_n. *This algorithm is of Las Vegas type and succeeds with probability at least* 3/4.

Proof. Let us fix a homomorphism $\lambda : G \to \mathrm{Sym}([1, n])$ such that $\lambda(a) = (1, 2, \ldots, n)$. Our first goal is to show that, with probability at least 3/4, among $1 + n/3$ random conjugates of b we find one, c, satisfying $\lambda(c) = (i, i+1, k)$ or $\lambda(c) = (i+1, i, k)$ with $1 \le i, k \le n$ and $k \notin \{i, i+1\}$, where the numbers are taken modulo n.

Suppose c is a conjugate of b satisfying $[c, c^a] \neq 1$. We claim that $\lambda(c)$ has support set $\{i, i+1, k\}$ as desired. Indeed, as c is a conjugate of b, it satisfies $\lambda(c) = (i, j, k)$ for some triple $\{i, j, k\}$ and $\lambda(c^a) = (i+1, j+1, k+1)$. Now c and c^a do not commute if and only if the sets $\{i, j, k\}$ and $\{i+1, j+1, k+1\}$ intersect. Hence it follows that two of i, j, k are consecutive numbers modulo n.

Next we show that the probability that we cannot find such an element c among $1 + n/3$ random conjugates of b is less than 1/4. There are $\binom{n}{3}$ possible support sets for a 3-cycle, and out of these, $n(n-3)$ contain two consecutive numbers modulo n. Hence one random conjugate succeeds with probability $n(n-3)/\binom{n}{3} = 6(n-3)/((n-1)(n-2))$, and the probability that $1 + n/3 > 2(n-1)(n-2)/(6(n-3))$ random conjugates succeed is greater than

$$1 - \left(1 - \frac{6(n-3)}{(n-1)(n-2)}\right)^{\frac{2(n-1)(n-2)}{6(n-3)}} > 1 - \frac{1}{e^2} > \frac{3}{4}.$$

The rest of the algorithm is deterministic and runs in $O(\mu)$ time. Without loss of generality, we can suppose that $\mathrm{supp}(\lambda(c)) = \{1, 2, k\}$ for some k with $3 \le k \le n-1$. The next goal is to construct $t \in G$ such that $\lambda(t) = (1, 2, 3)$.

If $k = 3$ then cc^a is an involution whereas if $4 \le k \le n-1$ then cc^a has order 5. Hence these two cases can be distinguished in $O(\mu)$ time.

Suppose first that $k = 3$. We can distinguish the cases $\lambda(c) = (1, 2, 3)$ and $\lambda(c) = (1, 3, 2)$ by computing $x := c^{a^2}$ and $y := c^x$. In the first case $\lambda(y) = (1, 2, 4)$ and in the second case $\lambda(y) = (1, 5, 2)$, which can be distinguished by checking whether $[y, y^{a^2}] = 1$. After that, t is defined as $t := c$ or $t := c^2$, respectively.

Suppose next that $4 \le k \le n-1$. The case $k \in \{4, n-1\}$ can be distinguished from the case $5 \le k \le n-2$ by checking whether $[c, c^{a^2}] = 1$. If $5 \le k \le n-2$ then $\lambda(c) = (1, 2, k)$ can be distinguished from $\lambda(c) = (1, k, 2)$ by computing $x := c^a$ and $y := c^x$. In the first case $\lambda(y) = (1, 3, k)$ and in the second case $\lambda(y) = (1, k, k+1)$, which can be distinguished by checking whether $[y, y^a] = 1$. If it turns out that $\lambda(c) = (1, 2, k)$ then define $t := [c^2, x]$. If $\lambda(c) = (1, k, 2)$ then define $t := [c, x^2]$.

If $k \in \{4, n-1\}$ then the cases $\lambda(c) = (1, 2, 4)$, $\lambda(c) = (1, 4, 2)$, $\lambda(c) = (1, 2, n-1)$, and $\lambda(c) = (2, 1, n-1)$ are also distinguished by computing $x := c^a$ and $y := c^x$. In the four cases, $\lambda(y) = (1, 3, 4)$, $\lambda(y) = (1, 4, 5)$, $\lambda(y) = (n-1, 1, 3)$, and $\lambda(y) = (n-1, n, 1)$, respectively. The third of these is distinguished from the others as the only one with $[y, y^a] = 1$. The second one is distinguished among the remaining three as the only one with $[y, y^{a^2}] = 1$. Finally, the first and fourth are distinguished by the order of yy^a. If $\lambda(c) = (1, 2, 4)$ or $\lambda(c) = (1, 2, n-1)$ then define $t := [c^2, x]$. If $\lambda(c) = (1, 4, 2)$ or $\lambda(c) = (2, 1, n-1)$ then define $t := [c, x^2]$.

Finally, output $s := at^2$, which satisfies $\lambda(s) = (3, 4, \ldots, n)$, and t. □

Lemma 10.2.15. *Let n be even and $n \geq 10$. Suppose that $a \in G$ has cycle type $1^1(n-1)^1$ and $b \in G$ has cycle type $1^{n-3}3^1$. Then, in $O(\xi n + \mu n)$ time, it is possible to construct $s, t \in G$ such that $\langle s, t\rangle$ satisfies the presentation* (10.4) *for A_n. This algorithm is of Las Vegas type and succeeds with probability at least* 3/4.

Proof. Let us fix a homomorphism $\lambda : G \to \mathrm{Sym}([1, n])$ such that $\lambda(a) = (2, 3, \ldots, n)$. Our first goal is to show that, with probability at least 3/4, among $2n/3$ random conjugates of b we find one, c, satisfying $\lambda(c) = (1, i, j)$ with $2 \leq i, j \leq n$.

Suppose c is a conjugate of b satisfying $[c, c^a] \neq 1$, $[c, c^{a^2}] \neq 1$, and $[c, c^{a^4}] \neq 1$. We claim that $\lambda(c)$ has support set $\{1, i, j\}$ as desired. Indeed, as c is a conjugate of b, if $1 \in \mathrm{supp}(\lambda(c))$ then c does not commute with c^{a^m} for any m. However, if $\mathrm{supp}(\lambda(c)) = \{i, j, k\} \subseteq \{2, 3, \ldots, n\}$ then c commutes with c^a if no two of i, j, k are consecutive in the $(n-1)$-cycle $\lambda(a)$, and in the other cases, it is easy to check that c commutes with c^{a^2} or c^{a^4}.

Next we show that the probability that we cannot find such an element c among $2n/3$ random conjugates of b is less than 1/4. There are $\binom{n}{3}$ possible support sets for a 3-cycle, and out of these, $\binom{n-1}{2}$ contain 1. One random conjugate succeeds with probability $\binom{n-1}{2}/\binom{n}{3} = 3/n$. Hence the probability that $2n/3$ random conjugates succeed is greater than $1 - (1 - 3/n)^{2n/3} > 1 - 1/e^2 > 3/4$.

Define $t := [c^a, c]$. Then $\lambda(t) = (1, i, i+1)$ and, without loss of generality, we may suppose $\lambda(t) = (1, 2, 3)$. Then $s := at$ satisfies $\lambda(s) = (1, 2)(3, 4, \ldots, n)$. Output s and t. □

Lemma 10.2.16. *Given $s, t \in G$, it can be checked in $O(\mu n)$ time whether $\langle s, t\rangle$ satisfies the presentation for A_n given in* (10.3) *and* (10.4).

Proof. The case when n is even, as well as the evaluation of the relators s^{n-2}, t^3, and $(st)^n$ in the odd case, is clear. In the case when n is odd, we evaluate the relators $(ts^{-k}ts^k)^2$ for $k = 1, \ldots, \lfloor (n-3)/2 \rfloor$ in $\lfloor (n-3)/2 \rfloor$ rounds: In the kth round, we use the input s^{k-1} and we output s^k and $(ts^{-k}ts^k)^2$. One round requires only a constant number of group operations. □

Theorem 10.2.17. *Given a black-box group $G = \langle S \rangle$ isomorphic to A_n or S_n and an error probability $\varepsilon > 0$, group elements $s, t \in G$ such that $\langle s, t \rangle$ satisfies the presentation for A_n given in* (10.3) *or* (10.4) *can be constructed by a Las Vegas algorithm, with probability at least* $1 - \varepsilon$, *in* $O(\log \varepsilon^{-1}(\xi n + \mu n \log n))$ *time.*

Proof. By Lemma 10.2.9, among $2n$ uniformly distributed random elements $a \in G$ we can find one satisfying $a^{n-k} = 1$, with the appropriate $k \in \{0, 1\}$, with probability at least $1 - (1 - 1/n)^{2n} > 1 - 1/e^2 > 3/4$. Similarly, among $36n$ uniformly distributed random elements $c \in G$ we can find one satisfying $c^{3(n-m)} = 1$ for the value m described in (10.8), with probability greater than $3/4$. Constructing a, c and taking the appropriate powers can be done in $O(\xi n + \mu n \log n)$ time. By Theorem 10.2.12, a has cycle type $1^k(n-k)^1$ and c^{n-m} has cycle type $1^{n-3}3^1$ with probability at least $1/180^2$. By applying the appropriate one of Lemmas 10.2.14 and 10.2.15, and then Lemma 10.2.16, if a and c have the correct cycle type then s and t are constructed in $O(\xi n + \mu n)$ time, with probability at least $3/4$. Hence the entire procedure takes $O(\xi n + \mu n \log n)$ time and succeeds with probability greater than $(3/4)^3/180^2$.

Repeating this procedure up to $\lceil (4/3)^3 180^2 \ln(\varepsilon^{-1}) \rceil$ times, we construct s and t with probability at least $1 - \varepsilon$, in $O(\log \varepsilon^{-1}(\xi n + \mu n \log n))$ time. □

Remark 10.2.18. We note that, in view of Lemma 10.2.13, for "practical" values of n the number $\lceil (4/3)^3 180^2 \ln(\varepsilon^{-1}) \rceil$ can be replaced by $\lceil (4/3)^3 7^2 \ln(\varepsilon^{-1}) \rceil$ in the expression for the number of iterations in the proof of Theorem 10.2.17. This reduction does not really matter if the input group is indeed isomorphic to A_n or S_n, since the expected number of iterations depends on the true conditional probabilities that an element of order dividing $(n-m)s$ has cycle type $1^{n-m-s}s^1(n-m)^1$ for the appropriate values of s and m, and it does not depend on how badly or well we estimate these conditional probabilities. However, if the algorithm is used as a Monte Carlo algorithm to test whether an unknown input group G is isomorphic to A_n or S_n then the better constants ensure earlier termination in the case of a negative answer.

10.2.3. Constructive Recognition: The Homomorphism λ

Given a black-box group $G = \langle S \rangle$ isomorphic to A_n or S_n, the algorithm described in the proof of Theorem 10.2.17 constructs $s, t \in G$ such that $\langle s, t \rangle \cong A_n$ and it satisfies the presentation in (10.3) or (10.4). In this section we construct a homomorphism $\lambda : G \to \mathrm{Sym}([1, n])$ by specifying the images of s and t and by giving a procedure that constructs the image of any $z \in G$. The algorithm will detect if $G \cong S_n$, and in this case it replaces s and t by two new elements s_1, t_1 such that $\langle s_1, t_1 \rangle$ satisfies (10.2). We shall also describe a procedure that, given an arbitrary element $z \in G$, computes a straight-line program reaching z from s and t (or from s_1 and t_1 in the case $G \cong S_n$).

Recall that for $n > 6$ we have $\mathrm{Aut}(A_n) \cong S_n$, and so, for any $g \in G$, the cycle structure of $\lambda(g)$ is the same for any faithful homomorphism $\lambda : G \to \mathrm{Sym}([1, n])$. Therefore, without loss of generality, we can assume that $\lambda(s) = (3, 4, \ldots, n)$ or $\lambda(s) = (1, 2)(3, 4, \ldots, n)$, depending on the parity of n, and $\lambda(t) = (1, 2, 3)$.

We start with the easier inverse problem: Finding straight-line programs reaching any $p \in A_n$ from $g := \lambda(s)$ and $h := \lambda(t)$.

Lemma 10.2.19. *Given $p \in A_n$, a straight-line program of length $O(n \log n)$ reaching p from g and h can be constructed in $O(n \log n)$ time by a deterministic algorithm.*

Proof. Let $g_1 := g$ and $h_1 := h$. Recursively for $i = 1, \ldots, n-3$, define

$$h_{i+1} = \begin{cases} h_i^{-1} g_i^{-1} h_i g_i h_i & \text{if } n-i \text{ is even} \\ h_i^{-1} g_i^{-1} h_i^2 g_i h_i & \text{if } n-i \text{ is odd,} \end{cases}$$

$$g_{i+1} = \begin{cases} g_i h_{i+1} & \text{if } n-i \text{ is even} \\ g_i h_i^2 h_{i+1}^{-1} & \text{if } n-i \text{ is odd.} \end{cases} \tag{10.9}$$

Then $h_i = (i, i+1, i+2)$ for all $i \in [1, n-2]$, and $g_i = (i, i+1)(i+2, i+3, \ldots, n)$ or $g_i = (i+2, i+3, \ldots, n)$, depending on the parity of $n - i$. It is clear from the recursion that all g_i, h_i can be obtained by a single straight-line program of length $O(n)$ from g and h. Hence it is enough to write a straight-line program of length $O(n \log n)$ from $T := \{g_i, h_i \mid 1 \le i \le n-2\}$ to p.

Write p as the product of transpositions by decomposing each cycle $(c_1, \ldots, c_l)$ of p as $(c_1, \ldots, c_l) = (c_1, c_2)(c_1, c_3) \cdots (c_1, c_l)$. Since $p \in A_n$, we have an even number of transpositions in this product. By inserting $(n-1, n)(n-1, n)$ between the $(2k-1)$st and $(2k)$th transposition for all k, the permutation p is written as the product of less than n permutations of the form $(i, j)(n-1, n)$ or $(n-1, n)(i, j)$ and it is enough to show that any such permutation can be obtained by a straight-line program of length $O(\log n)$ from T.

For any $k \in [i+2, n]$, we have $g_i^{-(k-i-2)} h_i g_i^{k-i-2} = (i, i+1, k)$ or $g_i^{-(k-i-2)} h_i g_i^{k-i-2} = (i+1, i, k)$, and these 3-cycles can be reached by straight-line programs of length $O(\log n)$ from T. Hence it is enough to observe that

$$(i, i+1)(n-1, n) = (i, n, i+1)(i, i+1, n-1)(i, n, i+1)$$

and, for $j \in [i+2, n-2]$, $(i, j)(n-1, n) = (i, i+1, j) \cdot (i, i+1)(n-1, n) \cdot (i, j, i+1)$. Finally, if $i < j$ and $j \in \{n-1, n\}$ then $(i, j)(n-1, n)$, $(n-1, n)(i, j) \in \langle (i, n-1, n) \rangle$ and $(i, n-1, n) = (i, n-1, i+1)(i, i+1, n)$. □

Corollary 10.2.20. *Given $p \in A_n$, the inverse image $\lambda^{-1}(p) \in G$ can be computed in $O(\mu n \log n)$ time by a deterministic algorithm. At any moment during the execution of this algorithm, we have to store only a constant number of elements of G.*

Proof. We simply evaluate in G the straight-line program reaching p from g and h, starting with s and t. The storage requirement can be satisfied by ordering the cycles of p according to the smallest element contained in them and decomposing each cycle as $(c_1, \ldots, c_l) = (c_1, c_2)(c_1, c_3) \cdots (c_1, c_l)$ using its smallest element c_1. Then transpositions (i, j) with $i < j$ for some fixed i occur consecutively, and all later transpositions (i_1, j_1) have $i_1, j_1 > i$. Therefore, evaluating the straight-line programs reaching the preimages of the elements $(i, j)(n-1, n)$ and $(n-1, n)(i, j)$ successively as required in the proof of Lemma 10.2.19, we have to store $\lambda^{-1}(g_i)$ and $\lambda^{-1}(h_i)$ only for a fixed i at any given time. Note that an inverse image $\lambda^{-1}(g_i^{k-i-2})$ can be computed by repeated squaring, storing only a constant number of elements of G at any time. After processing all (i, j) with this fixed i, we compute $\lambda^{-1}(g_{i+1})$ and $\lambda^{-1}(h_{i+1})$ by the formulas in (10.9), and we discard $\lambda^{-1}(g_i)$ and $\lambda^{-1}(h_i)$. □

The main idea of the construction of $\lambda(z)$ for an arbitrary $z \in G$ is the following. We need to define an n-element set Ω on which G acts, and then we need to identify Ω with $\{1, 2, \ldots, n\}$. We define Ω as a set of unordered pairs $\{a, b\} \subseteq G$ such that both a and b have cycle type $1^{n-3}3^1$, satisfying the requirement that the supports of $\lambda(a)$ and $\lambda(b)$ intersect in exactly one point $i \in [1, n]$. In this way, $\{a, b\}$ can be identified with i, and Ω can be identified with $[1, n]$.

Representing Ω in this way creates two problems. First, we do not want to store $2n$ elements of G, so the set Ω is not computed explicitly. Elements of Ω, when needed, will be reached by straight-line programs from s and t. Second, for an arbitrary $z \in G$ the image $\{a^z, b^z\}$ of a point $\{a, b\} \in \Omega$ is not necessarily an element of Ω. Thus we need to be able to identify the intersection of the supports of $\lambda(a^z)$ and $\lambda(b^z)$ with $\text{supp}(\lambda(c)) \cap \text{supp}(\lambda(d))$ for some $\{c, d\} \in \Omega$ in a way different from simply checking whether $\{a^z, b^z\} = \{c, d\}$. We shall narrow down the possibilities for $\text{supp}(\lambda(a^z)) \cap \text{supp}(\lambda(b^z))$ to a constant number by taking commutators of a^z and b^z with certain elements $x_1, \ldots, x_m$ of G of order five, utilizing the fact that an element of order five and a three-cycle in S_n commute if and only if their supports are disjoint.

Now we describe the construction of the elements $x_1, \ldots, x_m$. Let $n = 5k + r$ where $0 \le r \le 4$, let $m' := \lceil \log_2(k+1) \rceil$, and put $m := 2m'$. For $1 \le j \le m'$, we define partitions $p_j = \{P_{j,0}, P_{j,1}\}$ of the set $\{1, \ldots, k\}$ into two parts such that the common refinement of these partitions is the trivial one. Namely, for each $i \in [1, k]$, we compute the binary expansion $b_1^{(i)} b_2^{(i)} \ldots b_{m'}^{(i)}$ of i. For $i \in [1, k]$ and $j \in [1, m]$, let

$$\varepsilon(j, i) = \begin{cases} b_j^{(i)} & \text{if } j \le m' \\ 1 - b_{j-m'}^{(i)} & \text{if } j \ge m' + 1. \end{cases} \tag{10.10}$$

Then, for $1 \le j \le m'$, we define $P_{j,0} := \{i \mid \varepsilon(j, i) = 0\}$ and $P_{j,1} := \{i \mid \varepsilon(j, i) = 1\}$.

We define elements $a_1, \ldots, a_k$ of G with cycle type $1^{n-5}5^1$ as follows: Working in the group $A := \langle \{s^{-e} t s^e \mid 0 \le e \le 7\} \rangle \cong A_{10}$, in $O(\mu)$ time we construct $a_1, a_2 \in \langle s, t \rangle$ with $\lambda(a_1) = (1, 2, 3, 4, 5)$ and $\lambda(a_2) = (6, 7, 8, 9, 10)$. We also compute $c := s^5$ and *define, but do not compute*, $a_i := c^{-(i-2)} a_2 c^{i-2}$ for $3 \le i \le k$. Note that $\lambda(a_i) = (5i - 4, 5i - 3, 5i - 2, 5i - 1, 5i)$.

Finally we define $x_j := \prod_{i=1}^{k} a_i^{\varepsilon(j,i)}$ for $1 \le j \le m$. Thus each x_j has order five. For $j \le m'$, we have $\{i \mid \text{supp}(\lambda(a_i)) \subseteq \text{supp}(\lambda(x_j))\} = P_{j,1}$. Similarly, for $j > m'$, we have $\{i \mid \text{supp}(\lambda(a_i)) \subseteq \text{supp}(\lambda(x_j))\} = P_{j-m',0}$ and so the supports of $\lambda(x_j)$ and $\lambda(x_{j+m'})$ are disjoint for all $j \in [1, m']$.

Note that, for any $J \subseteq \{1, 2, \ldots, m'\}$,

$$\left| \bigcap_{j \in J} \text{supp}(\lambda(x_j)) \cap \bigcap_{j \in \{1,2,\ldots,m'\} \setminus J} \text{supp}(\lambda(x_{j+m'})) \right| \in \{0, 5\}. \tag{10.11}$$

If this intersection is nonempty, then it is the support of $\lambda(a_i)$ for the unique i whose binary expansion contains 1s exactly in the positions given in J.

Lemma 10.2.21. *Given a_1, a_2, and c as defined in the preceding paragraphs, $x_1, \ldots, x_m$ can be computed in $O(\mu n \log n)$ time, storing only $O(\log n)$ elements of G at any moment of the computation.*

Proof. Initialize $x_j := 1$ for $1 \le j \le m$. Iteratively for $i = 1, \ldots, k$, compute $x_j := x_j a_i^{\varepsilon(j,i)}$ for all j, compute $a_{i+1} = a_i^c$, and discard a_i. □

Lemma 10.2.22. *Given $z \in G$ and $\{x_1, \ldots, x_m\}$ as constructed in the foregoing, for any $l \in \{1, \ldots, n\}$ the image $l^{\lambda(z)}$ can be determined in $O(\mu \log n)$ time by a deterministic algorithm.*

Proof. We choose some i with $1 \leq i \leq n-4$ such that $l \in \{i, i+1, i+2, i+3, i+4\}$, and we determine $i^{\lambda(z)}, (i+1)^{\lambda(z)}, (i+2)^{\lambda(z)}, (i+3)^{\lambda(z)}$, and $(i+4)^{\lambda(z)}$ simultaneously. First, we construct the ten elements $t_{\{i_1,i_2,i_3\}}$ of G with the property that $\mathrm{supp}(\lambda(t_{\{i_1,i_2,i_3\}})) = \{i_1, i_2, i_3\}$, for all three-element subsets $\{i_1, i_2, i_3\} \subseteq \{i, i+1, i+2, i+3, i+4\}$. This can be done by constructing these ten elements $t_{\{i_1,i_2,i_3\}}$ for $6 \leq i_1, i_2, i_3 \leq 10$ in the group $A := \langle\{s^{-e}ts^e \mid 0 \leq e \leq 7\}\rangle \cong A_{10}$ in $O(\mu)$ time and conjugating the result by s^{i-6}.

Let $j \in [1, m']$ be fixed. We construct the twenty commutators $[x_j, (t_{\{i_1,i_2,i_3\}})^z]$, $[x_{j+m'}, (t_{\{i_1,i_2,i_3\}})^z]$. At least one of these twenty commutators is trivial, since at least three elements of $\{i, i+1, i+2, i+3, i+4\}^{\lambda(z)}$ are outside $\mathrm{supp}(\lambda(x_j))$ or $\mathrm{supp}(\lambda(x_{j+m'}))$, and two permutations with disjoint support commute.

Suppose, for example, that $[x_j, (t_{\{i,i+1,i+2\}})^z] = 1$; in the other nineteen cases, we can use an analogous argument. That $[x_j, (t_{\{i,i+1,i+2\}})^z] = 1$ means that

$$i^{\lambda(z)}, (i+1)^{\lambda(z)}, (i+2)^{\lambda(z)} \in \mathrm{supp}(\lambda(x_{j+m'})) \cup \{5k+1, \ldots, 5k+r\}.$$

We also want to compute which of the sets $\mathrm{supp}(\lambda(x_j)) \cup \{5k+1, \ldots, 5k+r\}$ and $\mathrm{supp}(\lambda(x_{j+m'})) \cup \{5k+1, \ldots, 5k+r\}$ contain the other two images $(i+3)^{\lambda(z)}$ and $(i+4)^{\lambda(z)}$. If $[x_j, (t_{\{i,i+1,i+3\}})^z] \neq 1$ then $(i+3)^{\lambda(z)} \in \mathrm{supp}(\lambda(x_j)) \cup \{5k+1, \ldots, 5k+r\}$, since $\{i, i+1, i+3\}^{\lambda(z)}$ can intersect $\mathrm{supp}(\lambda(x_j))$ only in the point $(i+3)^{\lambda(z)}$. If $[x_j, (t_{\{i,i+1,i+3\}})^z] = 1$ then $(i+3)^{\lambda(z)} \in \mathrm{supp}(\lambda(x_{j+m'})) \cup \{5k+1, \ldots, 5k+r\}$. We can decide similarly which of these two sets contains $(i+4)^{\lambda(z)}$.

Having performed the commutator calculations of the previous two paragraphs for all $j \in [1, m']$, by (10.11) we have at most $5+r \leq 9$ possibilities for $l^{\lambda(z)}$. To finish the algorithm, we need a procedure that decides whether $l^{\lambda(z)} = \bar{\imath}$ for a fixed $\bar{\imath} \in [1, n]$, in $O(\mu \log n)$ time.

We construct $s_1, s_2, s_3, s_4, s_5 \in \langle s, t\rangle$ of cycle type $1^{n-3}3^1$, such that $\mathrm{supp}(\lambda(s_{d_1})) \cap \mathrm{supp}(\lambda(s_{d_2})) = \{\bar{\imath}\}$ for any two distinct $d_1, d_2 \in [1, 5]$. Again, these s_d can be obtained as conjugates of appropriate elements of A. We also pick $t_{\{i_1,i_2,i_3\}}$ and $t_{\{i'_1,i'_2,i'_3\}}$ from our collection of ten group elements such that $\{i_1, i_2, i_3\} \cap \{i'_1, i'_2, i'_3\} = \{l\}$. We claim that $l^{\lambda(z)} = \bar{\imath}$ if and only if $[s_d, (t_{\{i_1,i_2,i_3\}})^z] \neq 1$ for at least four $d \in [1, 5]$ and $[s_d, (t_{\{i'_1,i'_2,i'_3\}})^z] \neq 1$ for at least four $d \in [1, 5]$. Indeed, if $l^{\lambda(z)} = \bar{\imath}$ then $\mathrm{supp}(\lambda((t_{\{i_1,i_2,i_3\}})^z))$ intersects but is not equal to at least four of the sets $\mathrm{supp}(\lambda(s_d))$ and so $(t_{\{i'_1,i'_2,i'_3\}})^z$ and s_d do not commute. The same argument works for $(t_{\{i'_1,i'_2,i'_3\}})^z$ as well. Conversely, if $[s_d, (t_{\{i_1,i_2,i_3\}})^z] \neq 1$ for at least four $d \in [1, 5]$ then $\mathrm{supp}(\lambda((t_{\{i_1,i_2,i_3\}})^z))$ intersects four three-element sets with pairwise intersection $\bar{\imath}$. This can happen only if $\bar{\imath} \in \{i_1, i_2, i_3\}^{\lambda(z)}$, and similarly $\bar{\imath} \in \{i'_1, i'_2, i'_3\}^{\lambda(z)}$. □

Lemma 10.2.23. *(a) Given s, t and $x_1, \ldots, x_m$, it can be decided in $O(\mu|S|n \log n)$ time whether $G \cong A_n$ or $G \cong S_n$.*

(b) If it turns out that $G \cong S_n$ then generators s_1, t_1 for G satisfying (10.2) *can be computed in $O(\mu n \log n)$ time.*

(c) If $G \cong S_n$ then, for any permutation $p \in S_n$, a straight-line program of length $O(n \log n)$ reaching p from $\lambda(s_1)$ and $\lambda(t_1)$ can be written by a deterministic algorithm, in $O(n \log n)$ time. The inverse image $\lambda^{-1}(p)$ can be computed in $O(\mu n \log n)$ time, storing only a constant number of elements of G at any moment during the computation.

Proof. (a) By Lemma 10.2.22, the images $\lambda(z)$ of all input generators $z \in S$ can be computed in $O(\mu|S|n \log n)$ time. All $\lambda(z)$ are even permutations if and only if $G \cong A_n$.

(b) Suppose that we have found $z_0 \in S$ such that $q := \lambda(z_0)$ is an odd permutation. By Corollary 10.2.20, $z_1 := \lambda^{-1}(q \cdot (1, 2))$ can be computed in $O(\mu n \log n)$ time, and then $t_1 := z_0^{-1} z_1$ satisfies $\lambda(t_1) = (1, 2)$. Depending on the parity of n, we compute $z_2 := s$ or $z_2 := t_1 s$ such that $\lambda(z_2) = (3, 4, \ldots, n)$. Finally, we compute $s_1 := z_2 t$, which satisfies $\lambda(s_1) = (1, 2, \ldots, n)$.

(c) Depending on the parity of p, we compute a straight-line program reaching p or $p \cdot (1, 2)$ from $\lambda(s)$ and $\lambda(t)$, as described in Lemma 10.2.19. Then, it is enough to observe that $\lambda(s)$ and $\lambda(t)$ can be reached from $\lambda(s_1)$ and $\lambda(t_1)$ by a straight-line program of constant length, since $t = [s_1, t_1]$ and $s = s_1 t^2$ if n is odd, and $s = t_1 s_1 t^2$ if n is even.

The evaluation of this straight-line program in G is done as described in Corollary 10.2.20. □

Lemma 10.2.24. *Given any $z \in G$, a straight-line program of length $O(n \log n)$ and reaching z from s_1, t_1 in the case $G \cong S_n$ and from s, t in the case $G \cong A_n$ can be computed in $O(\mu n \log n)$ time.*

Proof. Using the algorithm described in the proof of Lemma 10.2.22, we compute $\lambda(z)$. By Lemmas 10.2.19 and 10.2.23(c), we can write a straight-line program reaching $\lambda(z)$ from $\lambda(s), \lambda(t)$ or $\lambda(s_1), \lambda(t_1)$, respectively. The same straight-line program reaches z from s, t or s_1, t_1. □

Summary of the Proof of Theorem 10.2.4(a)

The decision procedure for whether $G \cong A_n$ or $G \cong S_n$ is described in Lemma 10.2.23(a), and the new generators for G are constructed in the proof of Theorem 10.2.17 for $G \cong A_n$ and in the proof of Lemma 10.2.23(b) for $G \cong S_n$. Given $g \in G$, the construction of $\lambda(g)$ is described in Lemma 10.2.22, and the straight-line program reaching g is constructed in Lemma 10.2.24. Finally, the inverse image of a permutation is constructed in Lemma 10.2.19 and Corollary 10.2.20 in the case $G \cong A_n$ and in Lemma 10.2.23(c) in the case $G \cong S_n$.

Proof of Theorem 10.2.4(b)

Given $G = \langle S \rangle$ of unknown isomorphism type, we attempt to construct $s, t, x_1, \ldots, x_m$ (and s_1, t_1, if necessary). If the construction fails then, with high probability, G is not isomorphic to A_n or S_n. If the construction succeeds and s_1, t_1 have been computed then we check that $\langle s_1, t_1 \rangle$ satisfies (10.2) (recall that it has been checked during the construction that $\langle s, t \rangle$ satisfies either (10.3) or (10.4) as appropriate).

Finally, we write straight-line programs from s, t or s_1, t_1 to each generator $z \in S$, as described in the proof of Lemma 10.2.24, and evaluate the straight-line programs. If the construction of the straight-line program succeeds and the evaluated value is equal to z for all $z \in S$ then we know with certainty that G is A_n or S_n. If not, then G is not isomorphic to A_n or S_n.

10.2.4. Constructive Recognition: The Case of Giants

If the input black-box group in Theorem 10.2.4 acts as a giant on a set Δ (i.e., $G^\Delta \cong A_n$ or $G^\Delta \cong S_n$, where $n := |\Delta|$) then the constructive recognition algorithm for G^Δ becomes much simpler than the general procedure described in the previous sections.

From Section 10.2.1, we need only the trivial Lemma 10.2.9, since most of the work was needed to recognize elements that may have certain desired cycle types. Here, since the input action is the natural one on Δ, it is trivial to decide whether some randomly selected $z \in G$ acts as a permutation with cycle type n^1, $1^1(n-1)^1$, or $1^{n-m-3}3^1(n-m)^1$ on Δ, as required in the proof of Theorem 10.2.17. The rest of the algorithm in the proof of Theorem 10.2.17 is also simplified, although to a lesser extent: In Lemmas 10.2.14 and 10.2.15, we can recognize more easily whether elements of cycle type $1^{n-3}3^1$ move consecutive points in the ordering of Δ defined by the n-cycle or $(n-1)$-cycle we have constructed. Checking the presentations (10.3) and (10.4) can be omitted. The time requirement of computing elements $s, t \in G$ such that $s|_\Delta, t|_\Delta$ satisfy (10.3) or (10.4) is $O(\xi n + \mu n)$.

Most of the algorithms in Section 10.2.3, including the computation and storage of $x_1, \ldots, x_m$, are also unnecessary, since it is trivial to determine the image $\lambda(z) = z|_\Delta$ of any $z \in G$. All we have to keep are the routines for computing and evaluating straight-line programs.

In the next section, we shall apply the constructive recognition of giants as a subroutine in a strong generating set construction algorithm, in a situation where a subgroup H acts as a giant on a subset of the permutation domain. In this setting, we would like to express the parameter ξ as a function of the size of the input permutation domain. However, nearly uniformly distributed random elements of H are not available, and appealing to Theorem 2.2.4 would increase the running time too much. With this application in mind, we describe a constructive recognition algorithm from [Babai et al., 1988]. This algorithm uses randomization only by applying the random subproduct method (cf. Section 2.3).

Theorem 10.2.25. *Suppose that $H = \langle S \rangle \leq \mathrm{Sym}(\Omega)$ and a subset $\Delta \subseteq \Omega$ are given, such that Δ is an orbit of H and $H|_\Delta = \mathrm{Alt}(\Delta)$. Let $n := |\Omega|$ and $k := |\Delta|$, and let $\Delta = \{\delta_1, \ldots, \delta_k\}$. Moreover, let $s = (\delta_1, \delta_2)(\delta_3, \delta_4, \ldots, \delta_k)$ or $s = (\delta_3, \delta_4, \ldots, \delta_k)$ if k is even or odd, respectively, and let $t = (\delta_1, \delta_2, \delta_3)$. Then, given an arbitrary $d > 0$, there is a Las Vegas algorithm that, with probability greater than $1 - k^d$, computes $g, h \in H$ such that $g|_\Delta = s$ and $h|_\Delta = t$. The time requirement of the algorithm is $O(d(nk^2 \log^2 k + |S| n \log k))$ and the storage requirement is $O(kn + |S|n)$.*

Proof. Let $c \geq 45$ be a sufficiently large but fixed constant, depending on d. For $i = 1, 2, \ldots, k$, we construct recursively a subset $S_i \subseteq H$ and permutations $g_i, h_i \in H$ such that

(i) $|S_i| = c \log k$, $\langle S_i \rangle|_\Delta$ stabilizes $\{\delta_1, \ldots, \delta_{i-1}\}$ pointwise, and with high probability $\langle S_i \rangle|_{\{\delta_i, \ldots, \delta_k\}} = \mathrm{Alt}(\{\delta_i, \ldots, \delta_k\})$.
(ii) For all $j \in [1, i-1]$, we have $\delta_j^{g_i} = \delta_j^s$ and $\delta_j^{h_i} = \delta_j^t$.

If this recursion succeeds than we output $g := g_k$ and $h := h_k$.

We initialize $g_1 := ()$, $h_1 := ()$, and S_1 as the set of $c \log k$ random subproducts made from S, where the constant c is determined such that, with probability at least $1 - k^{d+1}$, the group $\langle S_1 \rangle$ acts transitively on the set of 6-tuples made from distinct elements of Δ. By Lemma 2.3.7, such a c exists, and by the inequality (2.9) in the proof of that lemma, c can be chosen as a linear function of d. One of the most celebrated consequences of the classification of finite simple groups is that the only 6-transitive permutation groups are the giants. Therefore, if $\langle S_1 \rangle$

acts transitively on the 6-tuples then $\langle S_1\rangle|_\Delta = \mathrm{Alt}(\Delta)$. The time requirement of this initialization step is $O(d|S|n\log k)$.

Suppose that S_i, g_i, and h_i are already constructed for some $i \in [1, k-6]$. We compute and store a transversal T_i for $\langle S_i\rangle \bmod \langle S_i\rangle_{\delta_i}$, and we choose transversal elements a, b such that $g_{i+1} := ag_i$ and $h_{i+1} := bh_i$ satisfy $\delta_i^{g_{i+1}} = \delta_i^s$ and $\delta_i^{h_{i+1}} = \delta_i^t$. We construct the elements of the set S_{i+1} as random subproducts, made from the Schreier generators defined by S_i and T_i (cf. Lemma 4.2.1). Again, we construct $c\log k$ random subproducts. If $\langle S_i\rangle|_{\{\delta_i,\ldots,\delta_k\}} = \mathrm{Alt}(\{\delta_i,\ldots,\delta_k\})$ then the same 6-transitivity argument that we used in the case of S_1 shows that, with probability at least $1-k^{d+1}$, $\langle S_{i+1}\rangle|_{\{\delta_{i+1},\ldots,\delta_k\}} = \mathrm{Alt}(\{\delta_{i+1},\ldots,\delta_k\})$.

If $i \in [k-5, k-1]$ then we compute T_i, g_{i+1}, and h_{i+1} as in the previous paragraph and let S_{i+1} consist of all Schreier generators defined by S_i and T_i.

For a fixed i, the construction of T_i can be done by an orbit algorithm (cf. Section 2.1.1) in $O(dnk\log k)$ time. To stay within the memory usage asserted in the statement of the theorem, we do not compute and store the Schreier generators explicitly. Whenever a particular Schreier generator is needed in a random subproduct, we compute it from S_i and T_i. Even with this restriction, the construction of one random subproduct can be performed in $O(dnk\log k)$ time, and the construction of S_{i+1} is accomplished in $O(dnk\log^2 k)$ time. Therefore, performing all k steps of the recursion can be done in $O(d(nk^2\log^2 k + |S|n\log k))$ time and the probability of failure is at most $k/k^{d+1} = 1/k^d$. □

10.3. A Randomized Strong Generator Construction

For arbitrary input groups $G = \langle S\rangle \le \mathrm{Sym}(\Omega)$, with $|\Omega| = n$, the worst-case time complexity of any deterministic version of the Schreier–Sims SGS construction presented in this book is at least $O(n^5 + |S|n^2)$ (cf. Theorems 4.2.4, 5.2.3, and 10.1.3). The term n^5 is necessary, as [Knuth, 1991] contains examples where the running time is indeed $\Theta(n^5)$.

Although the n^5 bound achieved some notoriety, it can be broken: [Babai et al., 1997b] contains a deterministic SGS construction with running time $O^\sim(n^4)$ (recall that the notation $O^\sim$ hides a factor $\log^c n$ for some constant c). In this section, we present a Monte Carlo SGS construction from [Babai et al., 1995], which can be considered as an elementary version of the algorithm of [Babai et al., 1997b]. Some paragraphs in this section are reproduced from [Babai et al., 1995], copyright © 1995 by Academic Press; reprinted by permission of the publisher. Randomization makes the algorithm simpler and it also lowers the running time one more order of magnitude, to $O^\sim(n^3)$.

The algorithm constructs an SGS in two phases. Given $G = \langle S\rangle \le \mathrm{Sym}(\Omega)$, with $|\Omega| = n$, in the first phase it constructs a subgroup chain $G = G_1 \ge$

$G_2 \geq \cdots \geq G_m = 1$, which, in general, is *not* the point stabilizer subgroup chain relative to any ordering of Ω. Each G_i has its own permutation domain Ω_i. We have $\Omega_1 := \Omega$. For $i \in [2, m]$, either $\Omega_i = \Omega_{i-1}$ or $\Omega_i = \Omega_{i-1} \cup \Sigma_{i-1}$, where Σ_{i-1} is a minimal block system on some orbit $\Delta_{i-1} \subseteq \Omega_{i-1}$ of G_{i-1}. The group G_i is either the stabilizer of a point $\alpha_{i-1} \in \Omega_i$ or the pointwise stabilizer of an orbit $\Delta_{i-1} \subseteq \Omega_{i-1}$ of G_{i-1}, on which G_{i-1} acts as a giant.

Roughly speaking, the algorithm always tries to compute generators for the stabilizer of a point in the smallest orbit of the group at hand. Moreover, it tries to work in primitive groups; hence, if the action on the smallest orbit is imprimitive then it adds blocks of imprimitivity to the permutation domain and first computes the pointwise stabilizer of these blocks. The last consideration is that if the primitive group at hand is a giant then, counting transitivity, the algorithm recognizes it and switches to a routine designed especially to handle the giants. The version we present here depends on the classification of finite simple groups, since it utilizes the fact that the only 6-transitive permutation groups are the giants. If we used instead of that the weaker, but elementary result from [Wielandt, 1934], that the only $3 \log n$-transitive subgroups of S_n are A_n and S_n, then the running time would increase by a factor $\log n$.

The algorithm constructs sets $U_i \subseteq G_i$. If G_i is the stabilizer of a point $\alpha_{i-1} \in \Omega_i$ then U_{i-1} is a transversal G_{i-1} mod G_i, whereas if G_i is a pointwise stabilizer of an orbit Δ_{i-1} on which G_{i-1} acts as a giant then $U_{i-1} = \{s, t\}$ for two permutations $s, t \in G_{i-1}$, such that $s|_{\Delta_{i-1}}, t|_{\Delta_{i-1}}$ satisfy the appropriate one of the presentations (10.2)–(10.4). For each $i \in [1, m]$, the set $\bigcup_{j \geq i} U_j|_{\Omega_i}$ generates G_i. This finishes the outline of the first phase of the algorithm.

In the second phase, the algorithm constructs a point stabilizer chain relative to some ordering of Ω. Using the sets U_i, we can choose uniformly distributed random cosets of G_i in G_{i-1}. This is clear if U_{i-1} is a transversal G_{i-1} mod G_i. If G_i is the pointwise stabilizer of an orbit Δ_{i-1} then we construct a random element p of $\mathrm{Alt}(\Delta_{i-1})$ or $\mathrm{Sym}(\Delta_{i-1})$, depending on the action of G_{i-1} on Δ_{i-1}, by applying the algorithm of Exercise 2.1 or 2.2. Then, by the algorithms described in the proofs of Corollary 10.2.20 and Lemma 10.2.23(c), we can construct $g \in \langle U_{i-1} \rangle$ with the property $g|_{\Delta_{i-1}} = p$.

Since uniformly distributed random cosets of G_i in G_{i-1} are available for all i, we can construct uniformly distributed random elements of G, as the product of the restrictions of random coset representatives to Ω. Therefore, we can apply the algorithm described in the proof of Lemma 5.4.1 to construct an SGS.

We describe the first phase of the algorithm in more detail. Throughout the algorithm, the points of the permutation domain are marked by integers, denoting the priority of which ones we want to stabilize first. Initially, each point has mark 0; if G is not transitive then the points of the smallest orbit get mark 1; if the action on this orbit is not primitive, a system of blocks of imprimitivity is added with mark 2; etc. Higher mark means higher priority.

The main routine is a recursive procedure **SGS**$[H, \Psi]$, where H is a group acting on the set Ψ. In the previous notation, H is G_{i-1}, and Ψ is the corresponding domain Ω_{i-1} with the fixed points deleted. The sets U_i are collected in an array U. The purpose of the following pseudocode is to describe how to choose the next point to be fixed and how to detect that the group acts as a giant on an orbit. The subroutines that compute generators for the next group in the subgroup chain will be described later.

Initially, we set the global variable n to $|\Omega|$, U to the empty list, and `transitivity` to 0. We set the mark $m(\alpha) = 0$ for all $\alpha \in \Omega$. Then **SGS**$[G, \Omega]$ is called. All internal calculations on permutations are carried out on all n points and possibly additional points (such as points corresponding to the action on blocks).

SGS$[H, \Psi]$

Input: generators for $H \leq \mathrm{Sym}(\Psi)$; global variables n, `transitivity`, U

Output: U

```
Ψ := Ψ\{α ∈ Ψ | α^H = {α}}
if Ψ = ∅ then stop
mark := max{m(α) | α ∈ Ψ}
O := {α ∈ Ψ | m(α) = mark}
Case:
    H^O is intransitive:
        transitivity := 0
        let O_1 := smallest orbit of H^O
        for α ∈ O_1, m(α) := mark + 1
        call SGS[H, Ψ]
    H^O is imprimitive:
        compute a minimal block system B on O
        Ψ := B ∪ Ψ
        for α ∈ B, m(α) := mark + 1
        transitivity := 0
        call SGS[H, Ψ]
    H^O is primitive:
        transitivity := transitivity + 1
        if transitivity ≥ 6 then
            (* H^O is Sym(O) or Alt(O) *)
            compute s, t ∈ H with s|_O, t|_O satisfying one of
```

(10.2)–(10.4)
add $\{s, t\}$ to U
$H :=$ pointwise stabilizer of $\mathcal{O}$ in H
call SGS$[H, \Psi]$
else
$\alpha :=$ first element of $\mathcal{O}$
add a transversal H mod H_α to U
compute generators for H_α
call SGS$[H_\alpha, \Psi]$

Next, we describe the auxiliary routines that compute generators for the groups G_i and estimate the time requirement of these routines. First, we establish that the permutation domains Ω_i cannot grow too large.

Lemma 10.3.1. *For each i, $|\Omega_i| \leq 2n$.*

Proof. The procedure **SGS**$[G, \Omega]$ defines naturally a forest $\mathcal{F}$, whose vertices are the elements of $\bigcup_i \Omega_i$. The leaves of $\mathcal{F}$ are the elements of Ω. Any other vertex v of $\mathcal{F}$ is a block of imprimitivity for a subgroup of G, acting on some previously defined vertices of $\mathcal{F}$. The children of v are the elements of the block v. Since every nonleaf vertex has at least two children, the total number of vertices is less than $2n$. □

Lemma 10.3.1 implies that the cost of permutation multiplications is $O(n)$ in all groups G_i. As a second preliminary step, we observe that the length of any subgroup chain in S_n is less than $3n/2$ (cf. [Cameron et al., 1989]), and so $m < 3n/2$.

Lemma 10.3.2. *Suppose that $G_i = \langle S_i \rangle$ is already computed, and suppose that $G_{i+1} = (G_i)_{\alpha_i}$ for some point $\alpha_i \in \Omega_{i+1}$. Let $k := |\alpha_i^{G_i}|$. Then there is a Monte Carlo algorithm that computes a transversal G_i mod G_{i+1} and a generating set S_{i+1} of size $O(n)$ for G_{i+1} in $O(nk|S_i|\log n)$ time. The memory requirement is $O(n \min\{n, k|S_i|\})$.*

Proof. A transversal U_i for G_i mod G_{i+1} can be computed by an orbit algorithm in $O(nk + k|S_i|)$ time. If $k|S_i| \leq n$ then we construct S_{i+1} as the set of Schreier generators defined by S_i and U_i. If $k|S_i| > n$ then we construct the elements of the set S_{i+1} as random subproducts, made from these Schreier generators (cf. the proof of Theorem 10.2.25 for a similar argument). In the latter case, we do not compute and store the Schreier generators explicitly. Whenever a particular Schreier generator is needed in a random subproduct, we compute it from S_i

and U_i. Since the length of any subgroup chain of G_i is less than $3n/2$, the number of Schreier generators is $k|S_i|$, and one permutation multiplication can be performed in $O(n)$ time, Theorem 2.3.6 implies that the construction of S_{i+1} takes $O(kn|S_i|\log n)$ time. □

Lemma 10.3.3. *Suppose that $G_i = \langle S_i \rangle$ is already computed, and suppose that G_{i+1} is the pointwise stabilizer of an orbit $\Delta_i \subseteq \Omega_i$ on which G_i acts as a giant. Let $k := |\Delta_i|$. Then there is a Las Vegas algorithm that computes $s, t \in G_i$ such that $s|_{\Delta_i}$ and $t|_{\Delta_i}$ satisfy the appropriate one of the presentations* (10.2)–(10.4). *The time requirement is $O(n^2 \log^4 n + nk^2 \log^2 n + |S_i| n \log n)$ and the memory requirement is $O(n^2 \log^2 n + |S_i| n)$.*

Proof. First, we determine whether G_i acts on Δ_i as $\mathrm{Alt}(\Delta_i)$ or $\mathrm{Sym}(\Delta_i)$. This can be done in $O(k|S_i|)$ time, by examining whether there is some $g \in S_i$ such that $g|_{\Delta_i}$ is an odd permutation. If $G_i|_{\Delta_i} = \mathrm{Sym}(\Delta_i)$ then we compute $H := G'$ by the Monte Carlo algorithm described in the proof of Theorem 2.4.8 in $O(n^2 \log^4 n + |S_i| n \log n)$ time. Note that the memory requirement is $O(n^2 \log^2 n)$, since the normal closure part of the algorithm first contructs $O(n \log^2 n)$ generators. If $G_i|_{\Delta_i} = \mathrm{Alt}(\Delta_i)$ then we define $H := G$.

In the next step, we compute $g, h \in H$ such that $g|_{\Delta_i}$ and $h|_{\Delta_i}$ satisfy (10.3) or (10.4), as described in the proof of Theorem 10.2.25. The time requirement for this step is $O(n^2 \log n + nk^2 \log^2 n)$ if $G_i|_{\Delta_i} = \mathrm{Sym}(\Delta_i)$ (and so we constructed $O(n)$ generators for $H = G'$) and $O(|S_i| n \log n + nk^2 \log^2 n)$ in the case $G_i|_{\Delta_i} = \mathrm{Alt}(\Delta_i)$.

Finally, if $G_i|_{\Delta_i} = \mathrm{Alt}(\Delta_i)$ then we output $s := g$ and $t := h$. If $G_i|_{\Delta_i} = \mathrm{Sym}(\Delta_i)$ then we compute $s, t \in G_i$ satisfying (10.2), as described in the proof of Lemma 10.2.23(b). This computation takes $O(kn \log n)$ time. □

Suppose that we are in the situation described in Lemma 10.3.3: $G_i = \langle S_i \rangle$ is already computed, G_{i+1} is the pointwise stabilizer of an orbit $\Delta_i \subseteq \Omega_i$ on which G_i acts as a giant, and $s, t \in G_i$ satisfy the appropriate one of the presentations (10.2)–(10.4). For $g \in S_i$, let $\mathrm{slp}(g)$ be a straight-line program reaching $g|_{\Delta_i}$ from $s|_{\Delta_i}$ and $t|_{\Delta_i}$, as described in the proof of Lemmas 10.2.19 and 10.2.23(c). Let $\bar{g}$ denote the evaluation of $\mathrm{slp}(g)$, starting from s and t, and let $\tilde{S}_i := \{g\bar{g}^{-1} \mid g \in S_i\}$. Finally, let $E(s, t)$ be the set of permutations obtained by evaluating the relators of the appropriate one of (10.2)–(10.4) (for example, if $G_i|_{\Delta_i} = \mathrm{Alt}(\Delta_i)$ and $k = |\Delta_i|$ is even then $E(s, t) = \{s^{k-2}, t^3, (st)^{k-1}, [t, s]^2\}$).

Lemma 10.3.4. *With the notation introduced in the previous paragraph, we have $G_{i+1} = \langle (\tilde{S}_i \cup E(s, t))^{G_i} \rangle$.*

Proof. Let $K := \langle (\tilde{S}_i \cup E(s,t))^{G_i} \rangle$. By definition, $K \lhd G_i$, and clearly $K \leq G_{i+1}$. We also have $G_i = \langle K, s, t \rangle$, since any $g \in S_i$ can be written as $= (g\bar{g}^{-1})\bar{g}$, with $g\bar{g}^{-1} \in K$ and $\bar{g} \in \langle s, t \rangle$.

Let $\bar{G}_i := G_i/K$. On one hand, $|\bar{G}_i| \geq |G_i/G_{i+1}|$, since $K \leq G_{i+1}$; on the other hand, $|\bar{G}_i| \leq |G_i/G_{i+1}|$, since $E(s,t) \subseteq K$, and so $\bar{G}_i$ is a homomorphic image of G_i/G_{i+1}. Therefore, $K = G_{i+1}$. □

Lemma 10.3.5. *Suppose that $G_i = \langle S_i \rangle$ is already computed, and suppose that G_{i+1} is the pointwise stabilizer of an orbit $\Delta_i \subseteq \Omega_i$ on which G_i acts as a giant. Let $k := |\Delta_i|$. Then there is a Monte Carlo algorithm that computes $O(n)$ generators for G_{i+1} in $O(n^2 \log^4 n + kn^2 \log^2 n + |S_i| n \log n)$ time. The memory requirement is $O(n^2 \log^2 n + |S_i| n)$.*

Proof. If $|S_i| > 15n$ then, using the algorithm described in the proof of Theorem 2.3.6, we construct a generating set of size $O(n)$ for G_i in $O(|S_i| n \log n)$ time. Hence, from now on, we suppose that $|S_i| \in O(n)$.

First, we compute $s, t \in G_i$ such that $s|_{\Delta_i}$ and $t|_{\Delta_i}$ satisfy the appropriate one of the presentations (10.2)–(10.4). By Lemma 10.3.3, this can be done in $O(n^2 \log^4 n + k^2 n \log^2 n)$ time.

Our next goal is to construct $\tilde{S}_i$ and $E(s,t)$. For each $g \in S_i$, we compute slp(g) and $\bar{g}$. By Corollary 10.2.20 and Lemma 10.2.23(c), the time requirement for that is $O(kn \log n)$. Hence, $\tilde{S}_i$ can be obtained in $O(kn^2 \log n)$ time. The set $E(s,t)$ can be constructed in $O(kn)$ time (cf. the proof of Lemma 10.2.16).

Finally, we compute the normal closure of $\langle \tilde{S}_i \cup E(s,t) \rangle$ in G_i. By Corollary 2.4.6, the time requirement of this step is $O(n^2 \log^4 n)$. □

For the overall running time estimate of the algorithm **SGS**$[G, \Omega]$, we need a lemma that may be interesting on its own.

Lemma 10.3.6. *Let $G \leq \mathrm{Sym}(\Omega)$ be a primitive permutation group. Let $\alpha \in \Omega$, and suppose that G_α has an orbit of length $d \geq 2$. Then every nontrivial subgroup of G_α has a nontrivial orbit of length at most d.*

Proof. Let $\Delta \subseteq \Omega$ denote an orbit of G_α of length d. Consider the orbital graph $\mathcal{X}$ corresponding to Δ. Recall that $\mathcal{X}$ is a directed graph with vertex set Ω and edge set $(\alpha, \beta)^G$ for some (and so for all) $\beta \in \Delta$. Note that, in $\mathcal{X}$, every vertex has out-degree d. Clearly G acts as automorphisms of $\mathcal{X}$. Since G is primitive, $\mathcal{X}$ must be strongly connected (cf. Exercise 5.6).

If $H \le G_\alpha$ is nontrivial then there exists a point γ not fixed by H. Let us consider a directed path $\alpha = \alpha_0, \alpha_1, \ldots, \alpha_l = \gamma$ in $\mathcal{X}$, connecting α to such a point γ. Let α_i be the first point on this path not fixed by H (necessarily $i \ge 1$). Then all points in the orbit α_i^H are out-neighbors of α_{i-1} in $\mathcal{X}$ (since $\alpha_{i-1}^H = \{\alpha_{i-1}\}$), and so $|\alpha_i^H| \le d$. □

Theorem 10.3.7. *Given* $G = \langle S \rangle \le \mathrm{Sym}(\Omega)$, *with* $|\Omega| = n$, *the running time of the algorithm* **SGS**$[G, \Omega]$ *is* $O(n^3 \log^4 n + |S| n \log n)$.

Proof. By Theorem 2.3.6, in $O(|S| n \log n)$ time we can construct $O(n)$ generators for G. Hence, because all subroutines construct $O(n)$ generators for subgroups, we may suppose that, at each recursive call, the input of **SGS**$[H, \Psi]$ contains $O(n)$ generators.

During the run of **SGS**$[G, \Omega]$, the number of recursive calls of **SGS**$[H, \Psi]$ is $O(n)$. At each call, we have to find the orbits of the input group and and test primitivity on a domain of size $O(n)$. By Theorem 2.1.1 and Corollary 5.5.9, one computation of the orbit structure can be done in $O(n^2)$ time and one test of primitivity in $O(n^2 \log n)$ time. Therefore, the total time spent in orbit and primitivity computations is $O(n^3 \log n)$.

Next, we estimate the time spent in computing G_{i+1} for the values of i when G_i acts as a giant on some orbit $\Delta_i \subseteq \Omega_i$, and G_{i+1} is the pointwise stabilizer of Δ_i. For different i, these orbits Δ_i are disjoint, so Lemma 10.3.1 implies $\sum_i |\Delta_i| \le 2n$ and Lemma 10.3.5 implies that the total time spent for these i values is $O(n^3 \log^4 n)$.

Finally, we estimate the time requirement of computing G_{i+1} for the values of i when $G_{i+1} = (G_i)_{\alpha_i}$ for some $\alpha_i \in \Omega_{i+1}$. Let $I \subseteq [1, m]$ denote the set of indices i in this category, and let k_i be the length of the orbit $\alpha_i^{G_i}$ in Ω_{i+1}. We claim that

$$\sum_{i \in I} k_i \le 10 n \log n. \tag{10.12}$$

Let $\alpha \in \Omega_m$ be arbitrary, and let us consider the subsequence $i_1 < \cdots < i_l$ of indices, $i_j \in I$ for $j \in [1, l]$, such that α is in the orbit $\Delta_{i_j} := \alpha_{i_j}^{G_{i_j}}$. We have $\Delta_{i_1} \supsetneqq \Delta_{i_2} \supsetneqq \cdots \supsetneqq \Delta_{i_l}$. We claim that $l \le 5 \log n$; that is, α occurs in at most $5 \log n$ orbits of points α_i with $i \in I$. Since there are at most $2n$ possibilities for α, this new claim implies (10.12) by counting the incidences of orbits $\alpha_i^{G_i}$, $i \in [1, m]$, and the points in them in two ways.

We claim that, for each $j \in [1, l-1]$, either $k_{i_{j+1}} = k_{i_j} - 1$ or $k_{i_{j+1}} \le k_{i_j}/2$ and that the former cannot occur more than five times in a row. Clearly, this implies the claimed bound $5 \log n$.

For each $j \in [1, l]$, $G_{i_j}|_{\Delta_{i_j}}$ is primitive. There are three possibilities.

Case 1: $G_{i_j}|_{\Delta_{i_j}}$ is doubly transitive and the point stabilizer $(G_{i_j})_{\alpha_{i_j}}$ acts primitively on $\Delta_{i_j}\backslash\{\alpha_{i_j}\}$. In this case, $k_{i_{j+1}} = k_{i_j} - 1$. If this continues more than five times in a row then the algorithm recognizes that $(G_{i_j})_{\alpha_{i_j}}$ acts as a giant on $\Delta_{i_j}\backslash\{\alpha_{i_j}\}$ and in the next recursive call of **SGS**$[H, \Psi]$ all points of $\Delta_{i_j}\backslash\{\alpha_{i_j}\}$ are stabilized.

Case 2: $G_{i_j}|_{\Delta_{i_j}}$ is doubly transitive and the point stabilizer $(G_{i_j})_{\alpha_{i_j}}$ acts imprimitively on $\Delta_{i_j}\backslash\{\alpha_{i_j}\}$. Then in the next recursive call of **SGS**$[H, \Psi]$ a minimal block system Σ_{i_j} is added to Ω_{i_j}, and $G_{i_{j+1}}$ is a subgroup of the pointwise stabilizer of the action of $(G_{i_j})_{\alpha_{i_j}}$ on Σ_{i_j}. Hence $\Delta_{i_{j+1}}$ is a subset of a block in Σ_{i_j} and so $k_{i_{j+1}} \leq (k_{i_j} - 1)/2$.

Case 3: $G_{i_j}|_{\Delta_{i_j}}$ is primitive but not doubly transitive. Then $(G_{i_j})_{\alpha_{i_j}}$ acts intransitively on $\Delta_{i_j}\backslash\{\alpha_{i_j}\}$, and its smallest orbit is of size at most $(k_{i_j} - 1)/2$. The point α may not be in this smallest orbit but, by Lemma 10.3.6, $k_{i_{j+1}} \leq (k_{i_j} - 1)/2$ still holds. This finishes the proof of (10.12).

Lemma 10.3.2 and (10.12) imply that the total time requirement of computing G_{i+1} for $i \in I$ is $O(n^3 \log^2 n)$. We note that (10.12) also implies that the output of **SGS**$[G, \Omega]$ consists of $O(n \log n)$ permutations, which require $O(n^2 \log n)$ storage (but the memory requirement of the entire algorithm is slightly bigger, $O(n^2 \log^2 n)$). □

Our last task is to give the time requirement of the second phase of the algorithm, the construction of a strong generating set relative to some ordering of the original permutation domain Ω. Recall that the output U of **SGS**$[G, \Omega]$ is used to generate uniformly distributed random elements of G.

If $G_{i+1} = (G_i)_{\alpha_i}$ for some $\alpha_i \in \Omega_{i+1}$ then we have a transversal G_i mod G_{i+1} stored in U, so a uniformly distributed random coset representative can be written down in $O(n)$ time. If G_{i+1} is the pointwise stabilizer of an orbit $\Delta_i \subseteq \Omega_i$, with $|\Delta_i| = k_i$, on which G_i acts as a giant then Corollary 10.2.20 and Lemma 10.2.23(c) imply that a coset representative G_i mod G_{i+1} can be constructed in $O(k_i n \log n)$ time, as described at the beginning of this section. Therefore, a uniformly distributed random element of G is constructed in $O(n^2 \log n)$ time, and Lemma 5.4.1 implies that an SGS for G can be constructed in $O(n^3 \log^4 n)$ time.

Bibliography

[Acciaro and Atkinson, 1992] Acciaro, V., and Atkinson, M. D. (1992). A new algorithm for testing the regularity of a permutation group. *Congr. Numer.*, 90:151–160.

[Atkinson, 1975] Atkinson, M. (1975). An algorithm for finding the blocks of a permutation group. *Math. Comp.*, 29:911–913.

[Atkinson and Neumann, 1990] Atkinson, M. D., and Neumann, P. M. (1990). Computing Sylow subgroups of permutation groups. *Congr. Numer.*, 79:55–60.

[Atkinson et al., 1984] Atkinson, M., Hassan, R., and Thorne, M. (1984). Group theory on a micro-computer. In Atkinson, M., editor, *Computational Group Theory*, pages 275–280, Academic Press, London.

[Babai, 1979] Babai, L. (1979). Monte Carlo algorithms for graph isomorphism testing. Technical report DMS 79–10, Univ. de Montréal.

[Babai, 1991] Babai, L. (1991). Local expansion of vertex-transitive graphs and random generation in finite groups. In *Proc. 23rd ACM* Symposium on Theory of Computing, pages 164–174, ACM Press, New York.

[Babai, 1992] Babai, L. (1992). Bounded round interactive proofs in finite groups. *SIAM J. Discrete Math.*, 5:88–111.

[Babai, 1997] Babai, L. (1997). Randomization in group algorithms: Conceptual questions. In *Groups and Computation II*, volume 28 of *Amer. Math. Soc. DIMACS Series*, pages 1–17.

[Babai and Beals, 1999] Babai, L., and Beals, R. (1999). A polynomial-time theory of black box groups I. In Campbell, C. M., Robertson, E. F., Ruskuc, N., and Smith, G. C., editors, *Groups St. Andrews 1997 in Bath, I*, volume 260 of *London Math. Soc. Lecture Note Ser.*, pages 30–64, Cambridge University Press, Cambridge.

[Babai and Moran, 1988] Babai, L., and Moran, S. (1988). Arthur–Merlin games: A randomized proof system and a hierarchy of complexity classes. *J. Comp. Syst. Sci.*, 36:254–276.

[Babai and Szemerédi, 1984] Babai, L., and Szemerédi, E. (1984). On the complexity of matrix group problems, I. In *Proc. 25th IEEE Symposium on Foundations of Computer Science*, pages 229–240, IEEE Press, Washington.

[Babai et al., 1983] Babai, L., Kantor, W. M., and Luks, E. M. (1983). Computational complexity and the classification of finite simple groups. In *Proc. 24th IEEE Symposium on Foundations of Computer Science*, pages 162–171, IEEE Press, Washington.

[Babai et al., 1987] Babai, L., Luks, E. M., and Seress, Á. (1987). Permutation groups in NC. In *Proc. 19th ACM Symposium on Theory of Computing*, pages 409–420, ACM Press, New York.

[Babai et al., 1988] Babai, L., Luks, E. M., and Seress, Á. (1988). Fast management of permutation groups. In *Proc. 29th IEEE Symposium on Foundations of Computer Science*, pages 272–282, IEEE Press, Washington.

[Babai et al., 1991] Babai, L., Cooperman, G., Finkelstein, L., and Seress, Á. (1991). Nearly linear time algorithms for permutation groups with a small base. In *Proc. of International Symposium on Symbolic and Algebraic Computation* ISSAC '91, pages 200–209, ACM Press, New York.

[Babai et al., 1993] Babai, L., Luks, E. M., and Seress, Á. (1993). Computing composition series in primitive groups. In *Groups and Computation*, volume 11 of *Amer. Math. Soc. DIMACS Series*, pages 1–16.

[Babai et al., 1995] Babai, L., Cooperman, G., Finkelstein, L., Luks, E. M., and Seress, Á. (1995). Fast Monte Carlo algorithms for permutation groups. *J. Comp. Syst. Sci.*, 50:296–308.

[Babai et al., 1997a] Babai, L., Goodman, A. J., Kantor, W. M., Luks, E. M., and Pálfy, P. P. (1997a). Short presentations for finite groups. *J. Algebra*, 194:97–112.

[Babai et al., 1997b] Babai, L., Luks, E. M., and Seress, Á. (1997b). Fast management of permutation groups I. *SIAM J. Computing*, 26:1310–1342.

[Beals, 1993a] Beals, R. (1993a). Computing blocks of imprimitivity for small-base groups in nearly linear time. In *Groups and Computation*, volume 11 of *Amer. Math. Soc. DIMACS Series*, pages 17–26.

[Beals, 1993b] Beals, R. M. (1993b). An elementary algorithm for computing the composition factors of a permutation group. In *Proc. of International Symposium on Symbolic and Algebraic Computation* ISSAC '93, pages 127–134, ACM Press, New York.

[Beals and Babai, 1993] Beals, R., and Babai, L. (1993). Las Vegas algorithms for matrix groups. In *Proc. 34rd IEEE Symposium on Foundations of Computer Science*, pages 427–436, IEEE Press, Washington.

[Beals and Seress, 1992] Beals, R. M., and Seress, Á. (1992). Structure forest and composition factors for small base groups in nearly linear time. In *Proc. 24th ACM Symposium on Theory of Computing*, pages 116–125, ACM Press, New York.

[Beals et al., 2002] Beals, R., Leedham-Green, C., Niemeyer, A., Praeger, C., and Seress, Á. (2002). A black-box group algorithm for recognizing finite symmetric and alternating groups, I. Transactions Amer. Math. Soc. To appear.

[Bosma et al., 1997] Bosma, W., Cannon, J., and Playoust, C. (1997). The Magma algebra system I: The user language. *J. Symbolic Comput.*, 24:235–265.

[Brassard and Bratley, 1988] Brassard, G., and Bratley, P. (1988). *Algorithmics Theory and Practice*, Prentice–Hall, Englewood Cliffs, NJ.

[Bratus, 1999] Bratus, S. (1999). *Recognition of Finite Black Box Groups*. PhD thesis, Northeastern University.

[Bratus and Pak, 2000] Bratus, S., and Pak, I. (2000). Fast constructive recognition of a black box group isomorphic to S_n or A_n using Goldbach's Conjecture. *J. Symbolic Comput.*, 29:33–57.

[Brown et al., 1989] Brown, C. A., Finkelstein, L., and Purdom, P. W. (1989). A new base change algorithm for permutation groups. *SIAM J. Computing*, 18:1037–1047.

[Butler, 1979] Butler, G. (1979). *Computational Approaches to Certain Problems in the Theory of Finite Groups*. PhD thesis, University of Sydney.

[Butler, 1982] Butler, G. (1982). Computing in permutation and matrix groups II: Backtrack algorithm. *Math. Comp.*, 39:671–680.

[Butler, 1983] Butler, G. (1983). Computing normalizers in permutation groups. *J. Algorithms*, 4:163–175.

[Butler, 1985] Butler, G. (1985). Effective computation with group homomorphisms. *J. Symbolic Comput.*, 1:143–157.

[Butler, 1991] Butler, G. (1991). *Fundamental Algorithms for Permutation Groups*, volume 559 of *Lecture Notes in Computer Science*, Springer-Verlag, Berlin.

[Butler, 1994] Butler, G. (1994). An inductive schema for computing conjugacy classes in permutation groups. *Math. Comp.*, 62:363–383.

[Butler and Cannon, 1989] Butler, G., and Cannon, J. J. (1989). Computing in permutation and matrix groups III: Sylow subgroups. *J. Symbolic Comput.*, 8:241–252.

[Butler and Cannon, 1991] Butler, G., and Cannon, J. J. (1991). Computing Sylow subgroups using homomorphic images of centralizers. *J. Symbolic Comput.*, 12:443–458.

[Cameron, 1999] Cameron, P. J. (1999). *Permutation Groups*, Cambridge University Press, Cambridge.

[Cameron et al., 1984] Cameron, P. J., Neumann, P. M., and Saxl, J. (1984). On groups with no regular orbits on the set of subsets. *Archive Math.*, 43:295–296.

[Cameron et al., 1989] Cameron, P. J., Solomon, R., and Turull, A. (1989). Chains of subgroups in symmetric groups. *J. Algebra*, 127:340–352.

[Cannon, 1984] Cannon, J. (1984). A computational toolkit for finite permutation groups. In *Proc. Rutgers Group Theory Year,* 1983–1984, pages 1–18, Cambridge University Press, Cambridge.

[Cannon and Souvignier, 1997] Cannon, J., and Souvignier, B. (1997). On the computation of conjugacy classes in permutation groups. In *Proc. of International Symposium on Symbolic and Algebraic Computation* ISSAC '97, pages 392–399, ACM Press, New York.

[Cannon, 1971] Cannon, J. J. (1971). Computing local structure of large finite groups. In Birkhoff, G., and Marshall Hall, J., editors, *Computers in Algebra and Number Theory*, volume 4 of *Proc. Amer. Math. Soc.*, pages 161–176, Amer. Math. Soc., Providence, RI.

[Cannon, 1973] Cannon, J. J. (1973). Construction of defining relators for finite groups. *Discrete Math.*, 5:105–129.

[Cannon and Holt, 1997] Cannon, J. J., and Holt, D. F. (1997). Computing chief series, composition series and socles in large permutation groups. *J. Symbolic Comput.*, 24:285–301.

[Cannon and Holt, 2002] Cannon, J. J., and Holt, D. F. (2002). Computing maximal subgroups of finite groups. Submitted.

[Cannon et al., 1997] Cannon, J. J., Cox, B. C., and Holt, D. F. (1997). Computing Sylow subgroups in permutation groups. *J. Symbolic Comput.*, 24:303–316.

[Carmichael, 1923] Carmichael, R. (1923). Abstract definitions of the symmetric and alternating groups and certain other permutation groups. *Quart. J. Math.*, 49:226–270.

[Celler et al., 1995] Celler, F., Leedham-Green, C. R., Murray, S. H., Niemeyer, A. C., and O'Brien, E. (1995). Generating random elements of a finite group. *Comm. Algebra*, 23:4931–4948.

[Celler et al., 1990] Celler, F., Neubüser, J., and Wright, C. R. B. (1990). Some remarks on the computation of complements and normalizers in soluble groups. *Acta Appl. Math.*, 21:57–76.

[Chernoff, 1952] Chernoff, H. (1952). A measure of asymptotic efficiency for tests of a hypothesis based on the sum of observations. *Ann. Math. Statistics*, 23:493–507.

[Conway et al., 1985] Conway, J., Curtis, R., Norton, S., Parker, R., and Wilson, R. (1985). *Atlas of Finite Groups*, Clarendon Press, Oxford.

[Cooperman and Finkelstein, 1992] Cooperman, G., and Finkelstein, L. (1992). A fast cyclic base change for permutation groups. In *Proc. of International Symposium on Symbolic and Algebraic Computation* ISSAC '92, pages 224–232, ACM Press, New York.

[Cooperman and Finkelstein, 1993] Cooperman, G., and Finkelstein, L. (1993). Combinatorial tools for computational group theory. In *Groups and Computation*, volume 11 of *Amer. Math. Soc. DIMACS Series*, pages 53–86.

[Cooperman et al., 1989] Cooperman, G., Finkelstein, L., and Luks, E. M. (1989). Reduction of group constructions to point stabilizers. In *Proc. of International Symposium on Symbolic and Algebraic Computation* ISSAC '89, pages 351–356, ACM Press, New York.

[Cooperman et al., 1990] Cooperman, G., Finkelstein, L., and Sarawagi, N. (1990). A random base change algorithm for permutation groups. In *Proc. of International Symposium on Symbolic and Algebraic Computation* ISSAC '90, pages 161–168, ACM Press, New York.

[Cooperman et al., 1997] Cooperman, G., Finkelstein, L., and Linton, S. A. (1997). Recognizing $GL_n(2)$ in non-standard representation. In *Groups and Computation II*, volume 28 of *Amer. Math. Soc. DIMACS Series*, pages 85–100.

[Cooperstein, 1978] Cooperstein, B. N. (1978). Minimal degree for a permutation representation of a classical group. *Israel J. Math.*, 30:213–235.

[Coxeter and Moser, 1957] Coxeter, H., and Moser, W. (1957). *Generators and Relations for Discrete Groups*, 4th edition, volume 14 of *Ergeb. Math. Grenzgeb.* Springer-Verlag, Berlin.

[Dixon, 1968] Dixon, J. D. (1968). The solvable length of a solvable linear group. *Math. Z.*, 107:151–158.

[Dixon and Mortimer, 1996] Dixon, J. D., and Mortimer, B. (1996). *Permutation Groups*. Graduate Texts in Math., Springer-Verlag, Berlin.

[Easdown and Praeger, 1988] Easdown, D., and Praeger, C. E. (1988). On minimal faithful permutation representations of finite groups. *Bull. Aust. Math. Soc.*, 38:207–220.

[Eick, 1997] Eick, B. (1997). Special presentations for finite soluble groups and computing (pre-)Frattini subgroups. In *Groups and Computation II*, volume 28 of *Amer. Math. Soc. DIMACS Series*, pages 101–112.

[Eick and Hulpke, 2001] Eick, B., and Hulpke, A. (2001). Computing the maximal subgroups of a permutation group I. In Kantor, W. M., and Seress, Á., editors, *Groups and Computation III*, volume 8 of *OSU Mathematical Research Institute Publications*, pages 155–168, de Gruyter, Berlin.

[Erdős and Rényi, 1965] Erdős, P., and Rényi, A. (1965). Probabilistic methods in group theory. *J. d'Analyse Math.*, 14:127–138.

[Feller, 1968] Feller, W. (1968). *An Introduction to Probability Theory and Its Applications*, volume 1, 3rd edition, Wiley, New York.

[Furst et al., 1980] Furst, M., Hopcroft, J., and Luks, E. M. (1980). Polynomial-time algorithms for permutation groups. In *Proc. 21st IEEE Symposium on Foundations of Computer Science*, pages 36–41, IEEE Press, Washington.

[GAP, 2000] GAP (2000). *GAP – Groups, Algorithms, and Programming, Version 4.2.* The GAP Group, Aachen, St Andrews (`http://www-gap.dcs.st-and.ac.uk/~gap`).

[Garey and Johnson, 1979] Garey, M. R., and Johnson, D. S. (1979). *Computers and Intractability, A Guide to the Theory of NP-completeness*, Freeman, New York.

[Gebhardt, 2000] Gebhardt, V. (2000). Constructing a short defining set of relations for a finite group. *J. Algebra*, 233:526–542.

[Guralnick, 1983] Guralnick, R. M. (1983). Subgroups of prime power index in a simple group. *J. Algebra*, 81:304–311.

[Guralnick and Kantor, 2000] Guralnick, R. M., and Kantor, W. M. (2000). The probability of generating a simple group. *J. Algebra*, 234:743–792.

[Hardy and Wright, 1979] Hardy, G. H., and Wright, E. M. (1979). *An Introduction the Theory of Numbers*, 5th edition, Clarendon Press, Oxford.

[Hoffmann, 1982] Hoffmann, C. M. (1982). *Group-theoretic Algorithms and Graph Isomorphism*, volume 136 of *Lecture Notes in Computer Science*, Springer-Verlag, Berlin.

[Holt, 1991] Holt, D. F. (1991). The computation of normalizers in permutation groups. *J. Symbolic Comput.*, 12:499–516.

[Holt, 1997] Holt, D. F. (1997). Representing quotients of permutation groups. *Quart. J. Math. Oxford Ser. (2)*, 48:347–350.

[Holt, 2001] Holt, D. F. (2001). Computing automorphism groups of finite groups. In Kantor, W. M., and Seress, Á., editors, *Groups and Computation III*, volume 8 of *OSU Mathematical Research Institute Publications*, pages 201–208, de Gruyter, Berlin.

[Holt and Rees, 1994] Holt, D. F., and Rees, S. (1994). Testing modules for irreducibility. *J. Austral. Math. Soc. Ser. A*, 57:1–16.

[Hulpke, 1993] Hulpke, A. (1993). *Zur Berechnung von Charaktertafeln*. Diplomarbeit, RWTH Aachen.

[Hulpke, 1996] Hulpke, A. (1996). *Konstruktion transitiver Permutationsgruppen*. PhD thesis, RWTH Aachen.

[Hulpke, 2000] Hulpke, A. (2000). Conjugacy classes in finite permutation groups via homomorphic images. *Math. Comp.*, 69:1633–1651.

[Hulpke and Seress, 2001] Hulpke, A., and Seress, Á. (2001). Short presentations for three-dimensional unitary groups. *J. Algebra*, 245:719–729.

[Huppert, 1967] Huppert, B. (1967). *Endliche Gruppen I*, volume 134 of *Grundlehren Math. Wiss.*, Springer-Verlag, Berlin.

[Ivanyos and Lux, 2000] Ivanyos, G., and Lux, K. (2000). Treating the exceptional cases of the Meataxe. *Exp. Math.*, 9:373–381.

[Jerrum, 1986] Jerrum, M. (1986). A compact representation for permutation groups. *J. Algorithms*, 7:60–78.

[Jerrum, 1995] Jerrum, M. (1995). Computational Pólya theory. In *Surveys in Combinatorics, 1995*, pages 103–118, Cambridge University Press, Cambridge.

[Jordan, 1873] Jordan, C. (1873). Sur la limite de transitivité des groupes non alternés. *Bull. Soc. Math. France*, 1:40–71.

[Kantor, 1985a] Kantor, W. M. (1985a). Some consequences of the classification of finite simple groups. In McKay, J., editor, *Finite Groups – Coming of Age*, volume 45 of *Contemporary Mathematics*, pages 159–173, Amer. Math. Soc., Providence, RI.

[Kantor, 1985b] Kantor, W. M. (1985b). Sylow's theorem in polynomial time. *J. Comp. Syst. Sci.*, 30:359–394.

[Kantor, 1990] Kantor, W. M. (1990). Finding Sylow normalizers in polynomial time. *J. Algorithms*, 11:523–563.

[Kantor, 1991] Kantor, W. M. (1991). Finding composition factors of permutation groups of degree $n \leq 10^6$. *J. Symbolic Comput.*, 12:517–526.

[Kantor and Luks, 1990] Kantor, W. M., and Luks, E. M. (1990). Computing in quotient groups. In *Proc. 22nd ACM Symposium on Theory of Computing*, pages 524–534, ACM Press, New York.

[Kantor and Magaard, 2002] Kantor, W. M., and Magaard, K. (2002). Black-box exceptional groups of Lie type. In preparation.

[Kantor and Penttila, 1999] Kantor, W. M., and Penttila, T. (1999). Reconstructing simple group actions. In Cossey, J., Miller, C. F., Neumann, W. D., and Shapiro, M., editors, *Geometric Group Theory Down Under*, pages 147–180, de Gruyter, Berlin.

[Kantor and Seress, 1999] Kantor, W. M., and Seress, Á. (1999). Permutation group algorithms via black box recognition algorithms. In Campbell, C. M., Robertson, E. F., Ruskuc, N., and Smith, G. C., editors, *Groups St. Andrews 1997 in Bath, II*, volume 261 of *London Math. Soc. Lecture Note Ser.*, pages 436–446, Cambridge University Press, Cambridge.

[Kantor and Seress, 2002] Kantor, W. M., and Seress, Á. (2002). Computing with matrix groups. Submitted.

[Kantor and Seress, 2001] Kantor, W. M., and Seress, Á. (2001). Black box classical groups. *Memoirs Amer. Math. Soc.*, 149(708).

[Kantor and Taylor, 1988] Kantor, W. M., and Taylor, D. E. (1988). Polynomial-time versions of Sylow's theorem. *J. Algorithms*, 9:1–17.

[Kantor et al., 1999] Kantor, W. M., Luks, E. M., and Mark, P. D. (1999). Sylow subgroups in parallel. *J. Algorithms*, 31:132–195.

[Kimmerle et al., 1990] Kimmerle, W., Lyons, R., Sandling, R., and Teague, D. N. (1990). Composition factors from the group ring and Artin's theorem on orders of simple groups. *Proc. London Math. Soc.* (3), 60:89–122.

[Knuth, 1969] Knuth, D. E. (1969). *The Art of Computer Programming. Volume 2: Seminumerical Algorithms*, Addison-Wesley, Reading, MA.

[Knuth, 1973] Knuth, D. E. (1973). *The Art of Computer Programming. Volume 3: Sorting and Searching*, Addison-Wesley, Reading, MA.

[Knuth, 1991] Knuth, D. E. (1991). Notes on efficient representation of perm groups. *Combinatorica*, 11:57–68.

[Laue et al., 1984] Laue, R., Neubüser, J., and Schoenwaelder, U. (1984). Algorithms for finite soluble groups and the SOGOS system. In Atkinson, M., editor, *Computational Group Theory*, pages 105–135, Academic Press, London.

[Leedham-Green and Soicher, 1990] Leedham-Green, C., and Soicher, L. (1990). Collection from the left and other strategies. *J. Symbolic Comput.*, 9:665–675.

[Leedham-Green, 2001] Leedham-Green, C. R. (2001). The computational matrix group project. In Kantor, W. M., and Seress, Á., editors, *Groups and Computation III*, volume 8 of *OSU Mathematical Research Institute Publications*, pages 229–247, de Gruyter, Berlin.

[Leon, 1980a] Leon, J. S. (1980a). Finding the order of a permutation group. In Cooperstein, B., and Mason, G., editors, *Finite Groups*, volume 37 of *Proc. Sympos. Pure Math.*, pages 511–517, Amer. Math. Soc., Providence, RI.

[Leon, 1980b] Leon, J. S. (1980b). On an algorithm for finding a base and strong generating set for a group given by generating permutations. *Math. Comp.*, 35:941–974.

[Leon, 1991] Leon, J. S. (1991). Permutation group algorithms based on partitions, I: Theory and algorithms. *J. Symbolic Comput.*, 12:533–583.

[Leon, 1997] Leon, J. S. (1997). Partitions, refinements, and permutation group computation. In *Groups and Computation II*, volume 28 of *Amer. Math. Soc. DIMACS Series*, pages 123–158.

[Luks, 1982] Luks, E. M. (1982). Isomorphism of graphs of bounded valence can be tested in polynomial time. *J. Comp. Syst. Sci.*, 25:42–65.

[Luks, 1987] Luks, E. M. (1987). Computing the composition factors of a permutation group in polynomial time. *Combinatorica*, 7:87–99.

[Luks, 1990] Luks, E. M. (1990). Lectures on Polynomial-Time Computation in Groups. Lecture notes, Northeastern University.

[Luks, 1993] Luks, E. M. (1993). Permutation groups and polynomial-time computation. In *Groups and Computation*, volume 11 of *Amer. Math. Soc. DIMACS Series*, pages 139–175.

[Luks and Seress, 1997] Luks, E. M., and Seress, Á. (1997). Computing the Fitting subgroup and solvable radical of small-base permutation groups in nearly linear time. In *Groups and Computation II*, volume 28 of *Amer. Math. Soc. DIMACS Series*, pages 169–181.

[Mark, 1993] Mark, P. D. (1993). Parallel computation of Sylow subgroups in solvable groups. In *Groups and Computation*, volume 11 of *Amer. Math. Soc. DIMACS Series*, pages 177–187.

[McIver and Neumann, 1987] McIver, A., and Neumann, P. M. (1987). Enumerating finite groups. *Quart. J. Math. Oxford*, 38:473–488.

[McKay, 1981] McKay, B. D. (1981). Practical graph isomorphism. *Congr. Numer.*, 30:45–87.

[Mecky and Neubüser, 1989] Mecky, M., and Neubüser, J. (1989). Some remarks on the computation of conjugacy classes of soluble groups. *Bull. Aust. Math. Soc.*, 40:281–292.

[Morje, 1995] Morje, P. (1995). *A Nearly Linear Algorithm for Sylow Subgroups of Small-Base Permutation Groups*. PhD thesis, The Ohio State University.

[Morje, 1997] Morje, P. (1997). On nearly linear time algorithms for Sylow subgroups of small-base permutation groups. In *Groups and Computation II*, volume 28 of *Amer. Math. Soc. DIMACS Series*, pages 257–272.

[Neubüser, 1982] Neubüser, J. (1982). An elementary introduction to coset table methods in computational group theory. In Campbell, C. M., and Robertson, E. F., editors, *Groups – St Andrews 1981*, volume 71 of *London Math. Soc. Lecture Note Ser.*, pages 1–45, Cambridge University Press, Cambridge.

[Neumann, 1979] Neumann, P. M. (1979). A lemma that is not Burnside's. *Math. Scientist*, 4:133–141.

[Neumann, 1986] Neumann, P. M. (1986). Some algorithms for computing with finite permutation groups. In Robertson, E., and Campbell, C., editors, *Proc. of Groups – St Andrews 1985*, volume 121 of *London Math. Soc. Lecture Note Ser.*, pages 59–92, Cambridge University Press, Cambridge.

[Niven et al., 1991] Niven, I., Zuckerman, H. S., and Montgomery, H. L. (1991). *An Introduction to the Theory of Numbers*, 5th edition, Wiley, New York.

[Pak, 2001] Pak, I. (2001). What do we know about the product replacement algorithm? In Kantor, W. M., and Seress, Á., editors, *Groups and Computation III*, volume 8 of *OSU Mathematical Research Institute Publications*, pages 301–347, de Gruyter, Berlin.

[Parker and Nikolai, 1958] Parker, E. T., and Nikolai, P. J. (1958). A search for analogues of the Mathieu groups. *Math. Tables Aids Comput.*, 12:38–43.

[Parker, 1984] Parker, R. (1984). The computer calculation of modular characters (the Meat-Axe). In Atkinson, M., editor, *Computational Group Theory*, pages 267–274, Academic Press, London.

[Passman, 1968] Passman, D. (1968). *Permutation Groups*. Benjamin, New York.

[Praeger and Saxl, 1980] Praeger, C. E., and Saxl, J. (1980). On the orders of primitive permutation groups. *Bull. London Math. Soc.*, 12:303–307.

[Pyber, 1993] Pyber, L. (1993). Asymptotic results for permutation groups. In *Groups and Computation*, volume 11 of *Amer. Math. Soc. DIMACS Series*, pages 197–219.

[Rákóczi, 1995] Rákóczi, F. (1995). Fast recognition of the nilpotency of permutation groups. In *Proc. of International Symposium on Symbolic and Algebraic Computation* ISSAC '95, pages 265–269, ACM Press, New York.

[Rákóczi, 1997] Rákóczi, F. (1997). *Data Structures and Algorithms for Computing in Nilpotent and Solvable Permutation Groups*. PhD thesis, University of Oregon.

[Rónyai, 1990] Rónyai, L. (1990). Computing the structure of finite algebras. *J. Symbolic Comput.*, 9:355–373.

[Rose, 1965] Rose, J. (1965). Abnormal depth and hypereccentric length in finite soluble groups. *Math. Z.*, 90:29–40.

[Rotman, 1995] Rotman, J. J. (1995). *An Introduction to the Theory of Groups*, 4th edition, Springer-Verlag, Berlin.

[Schönert and Seress, 1994] Schönert, M., and Seress, Á. (1994). Finding blocks of imprimitivity in small-base groups in nearly linear time. In *Proc. of International Symposium on Symbolic and Algebraic Computation* ISSAC '94, pages 154–157, ACM Press, New York.

[Scott, 1980] Scott, L. L. (1980). Representations in characteristic p. In *The Santa Cruz Conference on Finite Groups*, volume 37 of *Proc. Symposium Pure Math.*, pages 319–331, Amer. Math. Soc., Providence, RI.

[Sedgewick, 1988] Sedgewick, R. (1988). *Algorithms*, 2nd edition, Addison–Wesley, Reading, MA.

[Seress, 1997] Seress, Á. (1997). An introduction to Computational Group Theory. *Notices Amer. Math. Soc.*, 44:671–679.

[Seress, 1998] Seress, Á. (1998). Nearly linear time algorithms for permutation groups: an interplay between theory and practice. *Acta Appl. Math.*, 52:183–207.

[Seress and Weisz, 1993] Seress, Á., and Weisz, I. (1993). PERM: A program computing strong generating sets. In *Groups and Computation I*, volume 11 of *Amer. Math. Soc. DIMACS Series*, pages 269–276.

[Sims, 1970] Sims, C. C. (1970). Computational methods in the study of permutation groups. In *Computational Problems in Abstract Algebra*, pages 169–183, Pergamon Press, Oxford.

[Sims, 1971a] Sims, C. C. (1971a). Computation with permutation groups. In *Proc. Second Symposium on Symbolic and Algebraic Manipulation*, pages 23–28, ACM Press, New York.

[Sims, 1971b] Sims, C. C. (1971b). Determining the conjugacy classes of permutation groups. In Birkhoff, G., and Marshall Hall, J., editors, *Computers in Algebra and Number Theory*, volume 4 of *Proc. Amer. Math. Soc.*, pages 191–195, Amer. Math. Soc., Providence, RI.

[Sims, 1978] Sims, C. C. (1978). Some group theoretic algorithms. In Dold, A., and Eckmann, B., editors, *Topics in Algebra*, volume 697 of *Lecture Notes in Math.*, pages 108–124, Springer-Verlag, Berlin.

[Sims, 1990] Sims, C. C. (1990). Computing the order of a solvable permutation group. *J. Symbolic Comput.*, 9:699–705.

[Sims, 1994] Sims, C. C. (1994). *Computation with Finitely Presented Groups*, Cambridge University Press, Cambridge.

[Suzuki, 1962] Suzuki, M. (1962). On a class of doubly transitive groups. *Ann. Math.*, 75:105–145.

[Tarjan, 1975] Tarjan, R. E. (1975). Efficiency of a good but not linear set union algorithm. *J. Assoc. Comput. Mach.*, 22:215–225.

[Taylor, 1992] Taylor, D. E. (1992). *The Geometry of the Classical Groups*, volume 9 of *Sigma Series in Pure Mathematics*, Heldermann-Verlag, Berlin.

[Theißen, 1997] Theißen, H. (1997). *Eine Methode zur Normalisatorberechnung in Permutationsgruppen mit Anwendungen in der Konstruktion primitiver Gruppen*. PhD thesis, RWTH Aachen.

[Vaughan-Lee, 1990] Vaughan-Lee, M. (1990). Collection from the left. *J. Symbolic Comput.*, 9:725–733.

[Warlimont, 1978] Warlimont, R. (1978). Über die Anzahl der Lösungen von $x^n = 1$ in der symmetrischen Gruppe S_n. *Archive Math.*, 30:591–594.

[Wielandt, 1934] Wielandt, H. (1934). Abschätzungen für den Grad einer Permutationsgruppe von vorgeschriebenem Transitivitätsgrad. *Schrift. Math. Sem. Inst. Angew. Math. Univ. Berlin*, 2:151–174.

[Wielandt, 1964] Wielandt, H. (1964). *Finite Permutation Groups*, Academic Press, New York.

Index

Note: Numerals in boldface type indicate where a notion or notation is defined.